AAPG REPRINT SERIES NO. 11

Abnormal Subsurface Pressure

Selected Papers Reprinted from

AAPG BULLETIN

Published by The American Association of Petroleum Geologists
Tulsa, Oklahoma, U.S.A., 1974

Contents

 Published, July 1974. Third printing, May 1982.
Library of Congress Catalog Card No. 74-82871. ISBN: 0-89181-536-8

Abnormal Subsurface Pressure: Preface

Fluids move in response to pressure differentials. Fundamental to the better understanding of the occurrence of oil and gas is a knowledge of the fluid-pressure history of sedimentary basins. The time and place of hydrocarbon accumulation will become more predictable as our ability to determine the time and direction of underground fluid movement increases.

Abnormal pressure is that pressure measured in reservoir rocks that is greater than (or less than) the equivalent weight of a column of formation water extending to the surface. Abnormal-pressure studies are relatively new to the field of hydrodynamics; the first definitive paper included in this volume is dated 1953. Regardless of the cause of abnormal pressure, its preservation is a function of rock permeability. The top of a zone of abnormal pressure is related to lithology and can have a dramatic effect on local structure in a basin. The occurrence of abnormal pressure is worldwide. If the depth at which it will be reached is not predicted accurately, the cost of drilling can be increased significantly.

The origin, detection, measurement, and occurrence of abnormal pressure in sedimentary rocks are the subject of this volume. The subject encompasses such a broad spectrum of physical and chemical principles that the published literature has been quite diverse. Many of the landmark papers in this field were first published by the AAPG.

All the papers were originally printed in the AAPG *Bulletin* during the period 1953–1973, and were selected by me. The papers were assembled by E. M. Tidwell of the AAPG Editorial Staff.

George B. Vockroth
Chevron Oil Company
New Orleans, Louisiana
April 24, 1974

Reprinted from:
BULLETIN OF THE AMERICAN ASSOCIATION OF PETROLEUM GEOLOGISTS
VOL. 37, NO. 2 (FEBRUARY, 1953), PP. 410–432, 16 FIGS.

GEOLOGICAL ASPECTS OF ABNORMAL RESERVOIR PRESSURES IN GULF COAST LOUISIANA[1]

GEORGE DICKINSON[2]
Houston, Texas

ABSTRACT

High-pressure zones commonly make drilling of wells most difficult in a belt about 50 miles wide along the coastal plain northwest of the Gulf of Mexico from the Rio Grande to the Mississippi Delta. This study is an attempt to link geological factors with occurrences of abnormal pressure in order to provide a better understanding of their origin.

Abnormal pressure has been defined as any pressure which exceeds the hydrostatic pressure of a column of water containing 80,000 parts per million total solids.

Dangerously abnormal pressures occur commonly in isolated porous reservoir beds in thick shale sections developed below the main sand series. Their locations are controlled by the regional facies change in the Gulf Coast Tertiary province, and they appear to be independent of depth and geological age of the formation.

The high pressures are caused by compaction of the shales under the weight of the overburden which is equivalent to approximately one pound per square inch per foot depth. Difference in density between gas and water causes abnormal pressure where hydrocarbon accumulations occur above water, irrespective of whether the water is at normal or abnormal pressures. The magnitude of this pressure depends on the structural elevation above the source of pressure in the water and may cause very high pressure gradients in isolated sand bodies. However, the trend of pressures in the Gulf Coast region indicates that maximum pressures probably do not exceed 90 per cent of the overburden pressure.

The abrupt increase in pressure above normal hydrostatic pressure commonly occurs over a very short vertical interval which makes control difficult. Successful drilling through abnormal pressures involves cementing casing below the main sand series and above the high pressure zones so that heavy mud may be used without loss of circulation.

INTRODUCTION

Drilling operations in the coastal plain northwest of the Gulf of Mexico commonly encounter high-pressure zones which are most difficult to control. These zones of excessive pressure are widely distributed in a belt 35–75 miles wide along the coast from the Rio Grande in the southwest to the Mississippi Delta in the east, a distance of approximately 800 miles. This belt coincides approximately with the area of Pleistocene and Recent formations (Fig. 1).

There has been only limited success in drilling through high-pressure zones to prospective reservoir rocks thought to be favorably located for the accumulation of oil and gas. An adequate understanding of the origin of pressure in reservoir formations becomes, therefore, increasingly important as shallow objectives become fewer and as attainable drilling depths increase.

The present study of the geological aspects of the problem attempts to link geological factors with occurrences of abnormal pressure.

[1] Read before the Third World Petroleum Congress, The Hague, May, 1951, and published in the proceedings. Manuscript received, June 14, 1952. Published by permission of the Shell Oil Company.

[2] Chief production geologist, technical services department, Shell Oil Company. The writer expresses appreciation to the management of the Shell Oil Company for permission to publish this paper. Thanks are due also to members of the staffs of the exploration and production departments in both the regional office, Houston, and the several offices in the New Orleans area who contributed suggestions and assistance in assembling the information and in preparation of the enclosures.

The Gulf Coast region of Louisiana (Fig. 1) was chosen for this purpose since it is part of a relatively simple geological province favorable for analysis.

STRATIGRAPHY

The general stratigraphic column of the Tertiary (Fig. 5) is overlain by sediments of Recent and Pleistocene age which in some places exceed 3,000 feet in thickness. In the inland part of the area a few wells penetrated the Eocene, for example in the Bear and Bannister districts (Fig. 7).

A continental-shelf environment similar to that now prevailing probably continued throughout the Tertiary. The distribution of the various geological units follows consistent trends which nearly parallel the present coast line except in the area of the Mississippi Delta where the coast has been built out into the Gulf of Mexico. Sedimentation was almost continuous and all the major stratigraphic units in the subsurface thicken and become progressively more marine in character from the outcrop toward the Gulf of Mexico. For example, the Frio thickens from about 1,700 feet in the Bannister wells to more than 4,200 feet at Iowa about 30 miles downdip. In general the change from mainly sandy sediments to marine shales occurs at progressively higher stratigraphic levels from the lower Frio inland to high in the Miocene in the coastal zone (Figs. 7 and 8). However, the detailed facies studies of Lowman (27)[3] have shown that while the general change of each zone is from fresh-water facies farthest shoreward through brackish-water facies and shallow-marine facies to progressively deeper-marine facies, the successive change is affected by rhythmic cycles caused by transgressions and regressions of the sea.

STRUCTURE

The regional structure of the Gulf Coast consists of a homocline dipping gently gulfward. Surface dips are very slight but increase with depth because of the increasing thickness of the sediments toward the Gulf. Regional faulting is typically downthrown toward the coast and is possibly connected with the depositional environment and the increasing amount of compaction of the more argillaceous sediments in that direction. The area is typified by numerous salt domes in which the salt may be anywhere from the surface in piercement-type domes to below the depth reached by drilling at present. Some of these salt domes appear to be connected with the regional faulting but in some places other faults dip inland and appear to connect between domes. The salt domes have characteristic fault patterns which are caused by local uplifting of the formations, but there is little evidence of any other tectonic forces acting in the area under review so that the effect of compaction of the sediments is easily recognizable.

NORMAL PRESSURE GRADIENT

Throughout the Gulf Coast region most wells encounter subsurface pressures which, when measured at the oil-water or gas-water interface, approximate

[3] References given at end of paper.

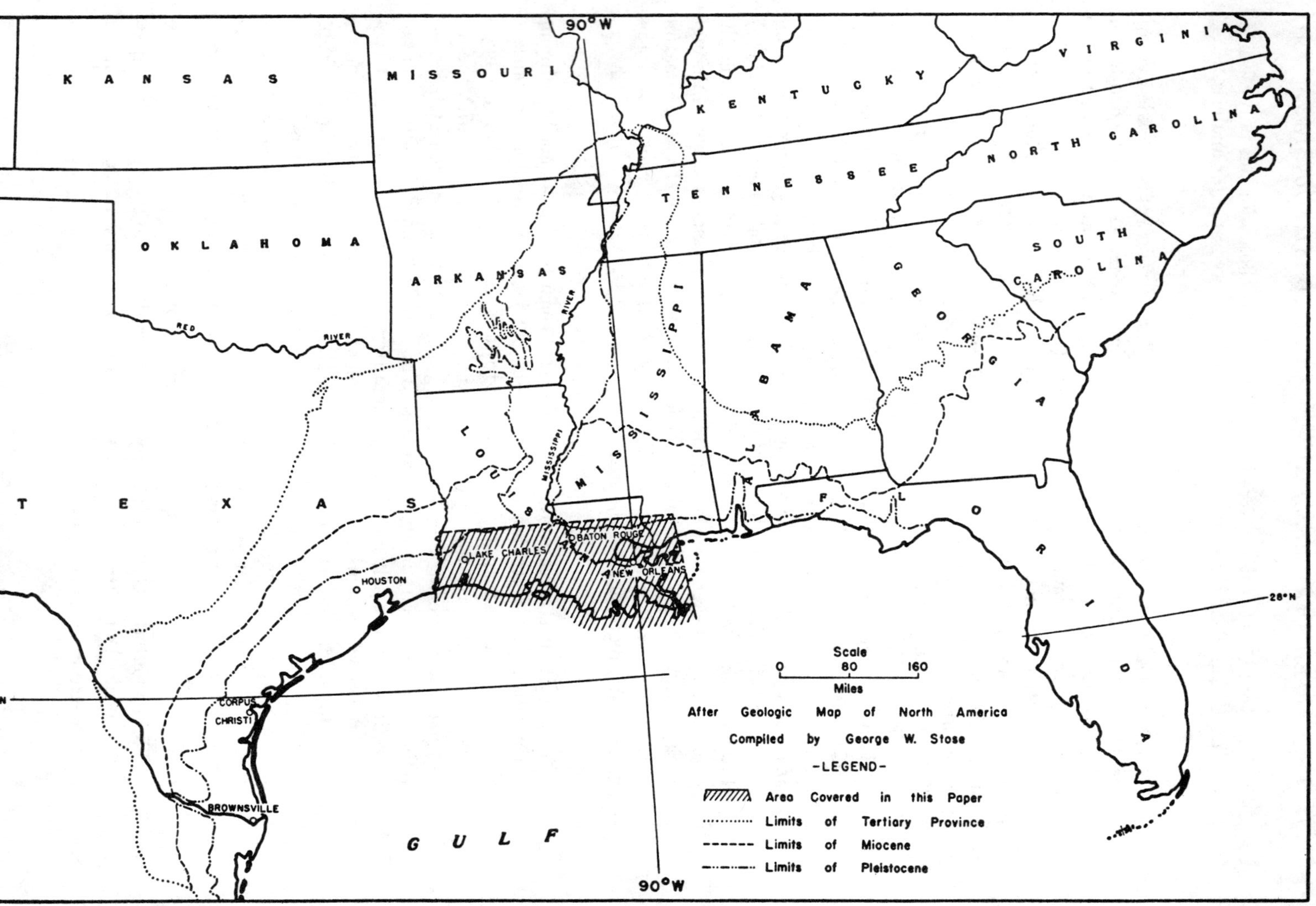

FIG. 1.—Index map.

very closely the hydrostatic pressure of a column of water containing 80,000 parts per million total solids, or a pressure gradient of 0.465 pound per square inch per foot depth .This gradient has been established in a range from the surface to about 16,000 feet in Queen Bess Island (Fig. 2).

OCCURRENCES OF ABNORMAL PRESSURE

The available abnormal pressure measurements were plotted in a chart of which Figure 2 is a simplified version and were numbered to correspond with their locations as shown in Figure 3. Actual measurements of abnormal pressures encountered in a well are rare so that it is usually necessary to estimate the bottom-hole pressure from testing and production data or from the mud weight in the hole at the time the abnormal pressure was encountered, compared with that required to control the pressure. Where actual bottom-hole pressure measurements are available for comparison, it appears that the former method is reasonably reliable, although somewhat low pressures result; whereas the latter method appears to give pressures which are about 10 per cent too high (Fig. 4). All pressures estimated from mud weights have, therefore, been reduced by this amount. However, this correction factor is based on very few data, and it is possible that it may vary with hole size since the swabbing action induced when pulling drill pipe necessitates an increasing pressure differential as the hole size is decreased.[4]

Many of the abnormal pressure occurrences, which were reviewed, flowed salt water with no oil, but it is probable that solution gas was present (17) although gas was not always reported. In the case of most of the high-pressure gas and oil accumulations the depth of the oil-water interface is not known, so that, depth for depth, the abnormal pressures may be higher than if the zone contained salt water only. However, a study of the pressure gradients given in Figure 2 shows that the highest pressures known have a pressure gradient of about 0.87 pound per square inch per foot depth, or about 1.87 normal hydrostatic pressure, irrespective of whether the reservoir contains salt water or gas and oil.

Abnormal pressures are encountered in formations ranging in age from the upper Miocene in the Mississippi Delta area, to the base of the Oligocene in a strip extending from Baton Rouge to the Lake Charles area. Figure 3 shows the locations of all abnormal pressure occurrences for which data were available, and the geological zone in which the first abnormal pressure was recorded. It is apparent from this map that these geological zones follow trends which agree closely with the "bay line" of Lowman (27) and with the established producing trends of the region (Lowman, Fig. 6). When plotted on a stratigraphic correlation chart (Fig. 5), the grouping of the occurrences of abnormal pressure is even more striking, so that some geological control seems to be indicated.

[4] According to J. M. Bugbee, 800–1,200 pounds per square inch overbalancing mud pressure is required in a 6-inch hole compared with only 200–500 pounds per square inch in a 8½-inch hole.—Oral communication.

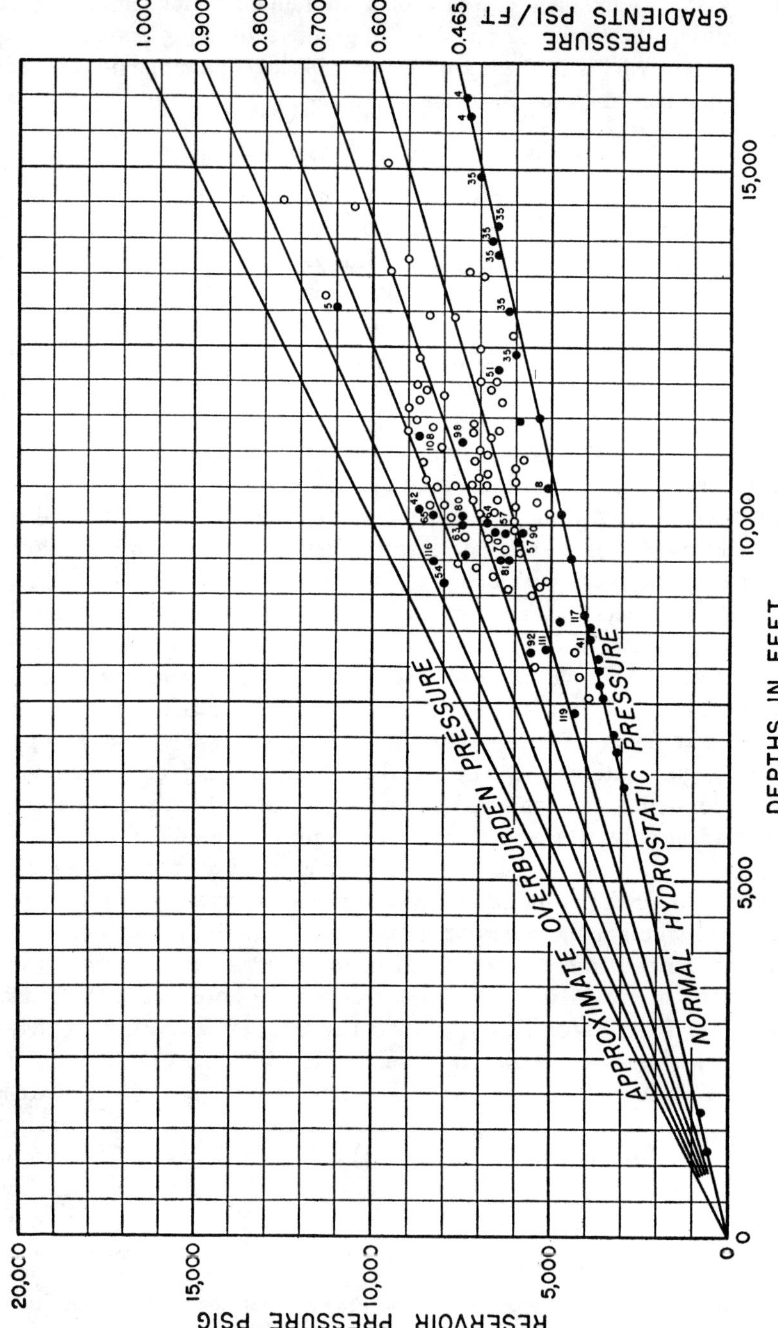

FIG. 2.—Reservoir pressure *versus* depth for Louisiana Gulf Coast wells. Solid circles: measured pressures. Open circles: estimated pressures.

Cannon and Craze (12) and Cannon and Sullins (13) of the Humble Oil and Refining Company, after reviewing a large number of abnormal pressure occurrences, concluded that depth alone seemed to be the governing factor regardless of the age of the formation. However, in the latter paper adjacent normal and abnormal pressures in the same formation were attributed to depositional and faulting characteristics, but this geological aspect was not further pursued.

The change from normal hydrostatic to abnormal pressure for some wells is shown in Figure 6. A study of the available data appears to lead to the conclusion that once the zone containing abnormal pressure has been reached, the pressure will increase suddenly, as in Iowa and Manilla Village, or somewhat less rapidly, as in Chalkley, South Roanoke, and La Pice. No reliable examples of gradual pressure increase over an appreciable depth range were found. However, the use of progressively increasing mud weight in many wells probably indicates that a gradual pressure increase does occur.

In order to investigate possible geological control of abnormal pressures, a detailed study was made of the logs of all the wells known to have encountered abnormal pressure and of many near-by wells with normal pressure. The results of this study are illustrated by three diagrammatic stratigraphic sections (Figs. 7, 8, and 9), drawn through a series of typical wells across the west, east, and delta areas of the region. These sections show that abnormal pressure commonly occurs only below the base of the main sand development in or below a major shaly series. Even though most of the abnormal pressure occurrences reviewed conform with the conditions shown in the cross sections, high pressure may also be found in the main sand series where conditions are favorable for isolation of sand bodies by faulting or lensing-out of the sand, for example, in Darrow, Lirette, and Venice.

The change in facies from mainly sand to mainly shale occurs at the base of the Frio in the northern part of the area under review but gradually climbs the stratigraphic section until it reaches high in the Miocene in the Mississippi Delta area. These changes of facies and of the accompanying fauna have been described by Lowman (27). As a result of the present study, it appears that a knowledge of the depth at which the main facies change takes place is an important factor in forecasting the depth at which abnormal pressures may be encountered in exploration wells.

Undoubtedly abnormal pressures have been encountered in wells located between the occurrences shown in Figure 3, but the data are not readily available. However, there are also other deep wells which penetrated the same formations without encountering high-pressure reservoirs. It is apparent, therefore, that other factors must be present in addition to the shaly facies with lenticular sands.

Regardless of the origin of abnormal pressure, it is evident that a reservoir containing high pressure must be effectively isolated from any other porous formation which contains normal hydrostatic pressure, otherwise the pressure would be dissipated. This requires a suitable porous reservoir sealed in all directions

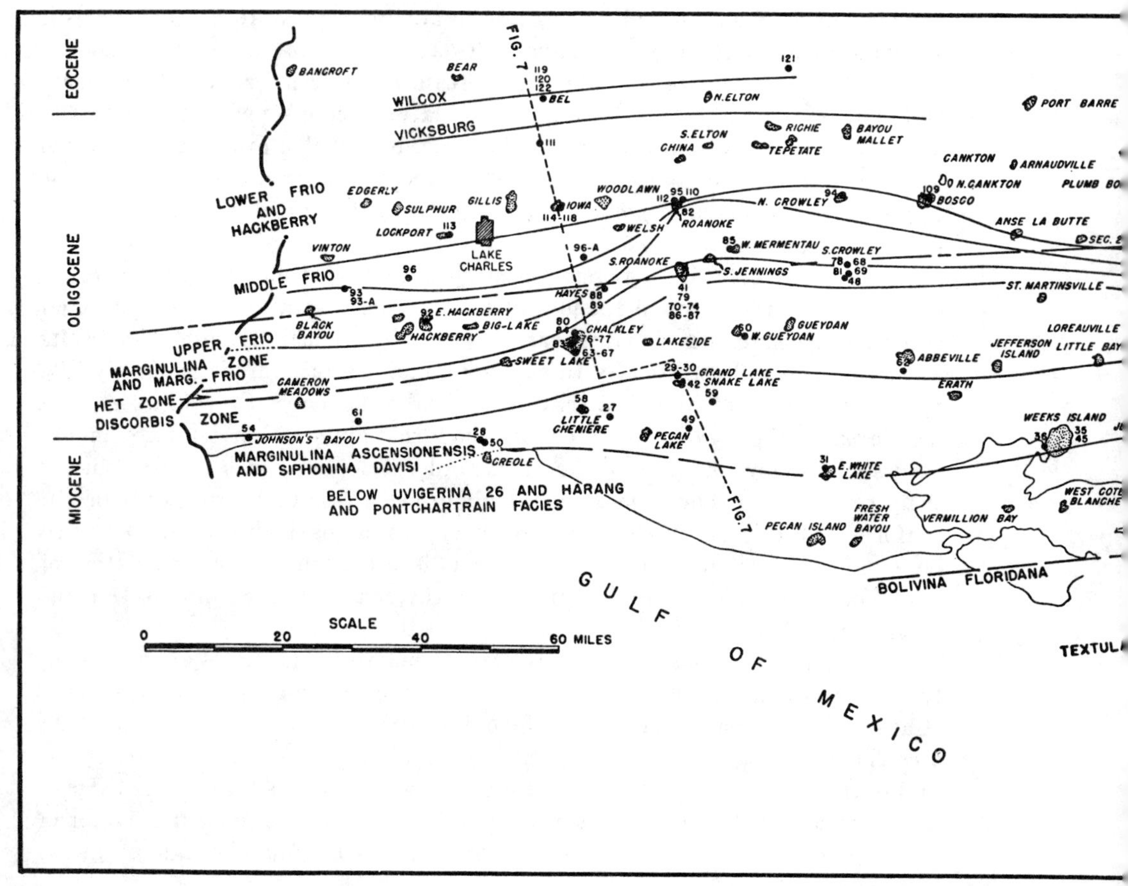

FIG. 3.—Sketch map showing geological age

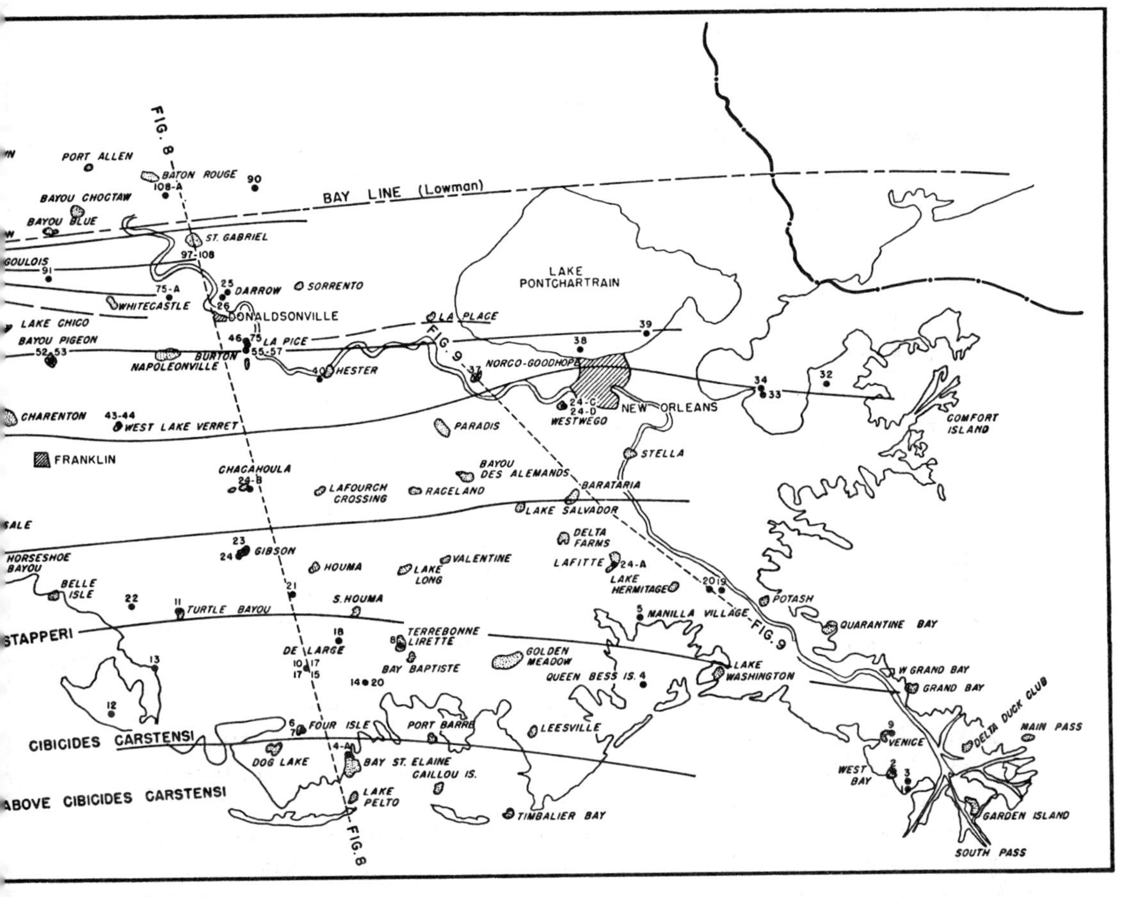

ngest zone in which abnormal pressure occurs.

either by lensing or faulting. However, sand bodies in an essentially shaly series are typically lenticular and erratic so that faulting is not a prerequisite for the preservation of abnormal pressure, even though it is present in nearly all the wells reviewed. Regionally, of course, the downdip seal of all reservoirs can be the change to deep-water facies, but local pinch-out may produce more limited reservoirs. These conditions are shown diagrammatically in sketches a and b, Figure 10. The effect of the relation between the position of the sand body in the

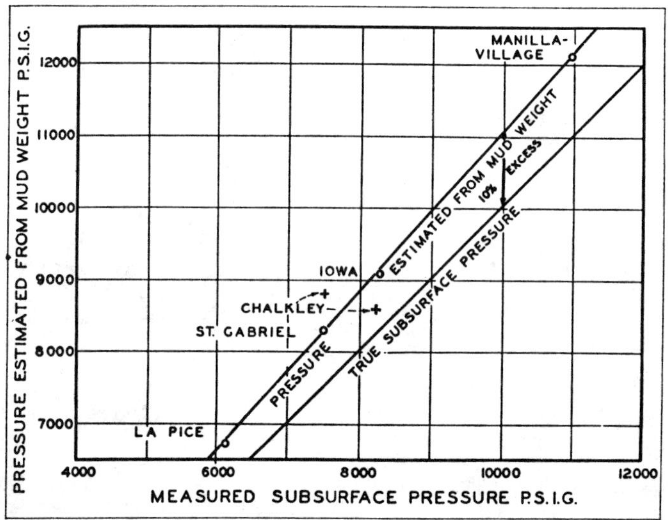

FIG. 4.—Relationship between subsurface pressures measured by pressure bomb and estimated from maximum hydrostatic head of mud required for control during drilling.

shale series and the throw of a fault on the preservation or dissipation of abnormal pressure is shown in sketch c, Figure 10. It is obvious from these diagrams that abnormal pressures can occur near the top of the shale series only if the porous bed is isolated by pinch-out or is faulted down against the shale series as in Chalkley. In the absence of pinch-out of the reservoir abnormal pressures can only be preserved in upthrown blocks at depths below the top of the shale series which are greater in amount than the throw of the fault.

Geological conditions leading to the preservation or dissipation of high pressures are well illustrated in the Chalkley field. Figure 11 is a north-to-south sketch section showing a series of south-dipping normal faults crossing a north-south trending domal structure. The "W" sand in the upper part of a thick shale section contains oil and gas under very high pressures in the south flank but is under normal pressure in the center and north of the structure. The downthrown block of the south flank is effectively sealed updip by being faulted against the thick shale series, whereas in the north flank the "W" sand is faulted against the main sand series and is under normal hydrostatic pressure. In the southernmost

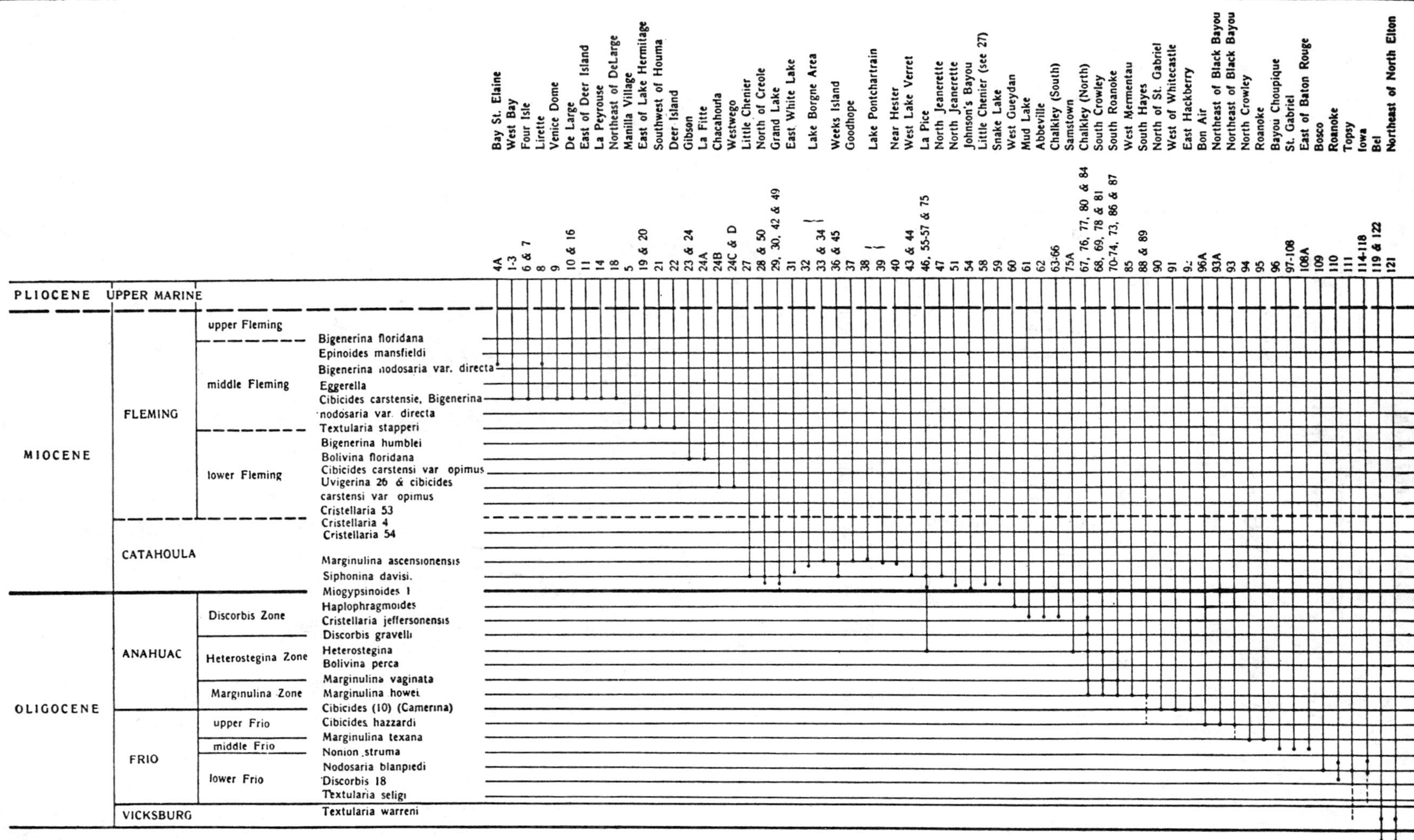

FIG. 5.—Stratigraphic correlation chart of abnormal pressure occurrences.

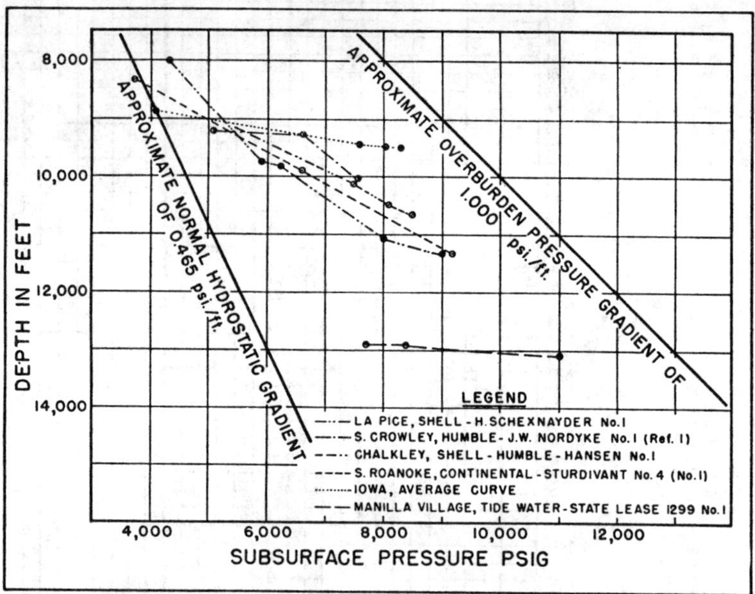

FIG. 6.—Reservoir pressure-depth curves for typical Louisiana Gulf Coast wells.

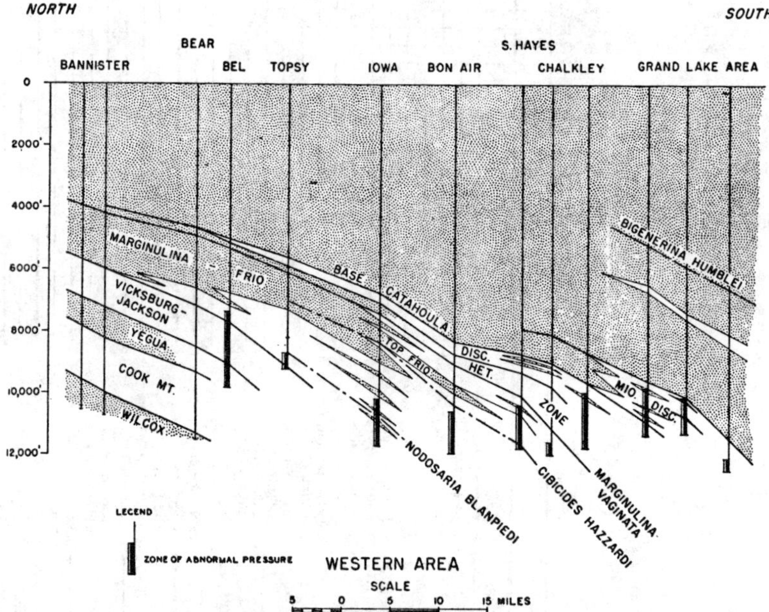

FIG. 7.—Diagrammatic stratigraphic section (western area) illustrating facies control of abnormal reservoir pressure.

of the intermediate blocks sand development is poor and no high-pressure reservoirs were encountered. The "W" sand is present in the other blocks, but it is clear from the section that it is faulted against other sands which have connection to the normally pressured main sand series. However, abnormal pressures were encountered in the two northern blocks at greater depths where the porous zones are sealed by being faulted against higher parts of the thick shale section.

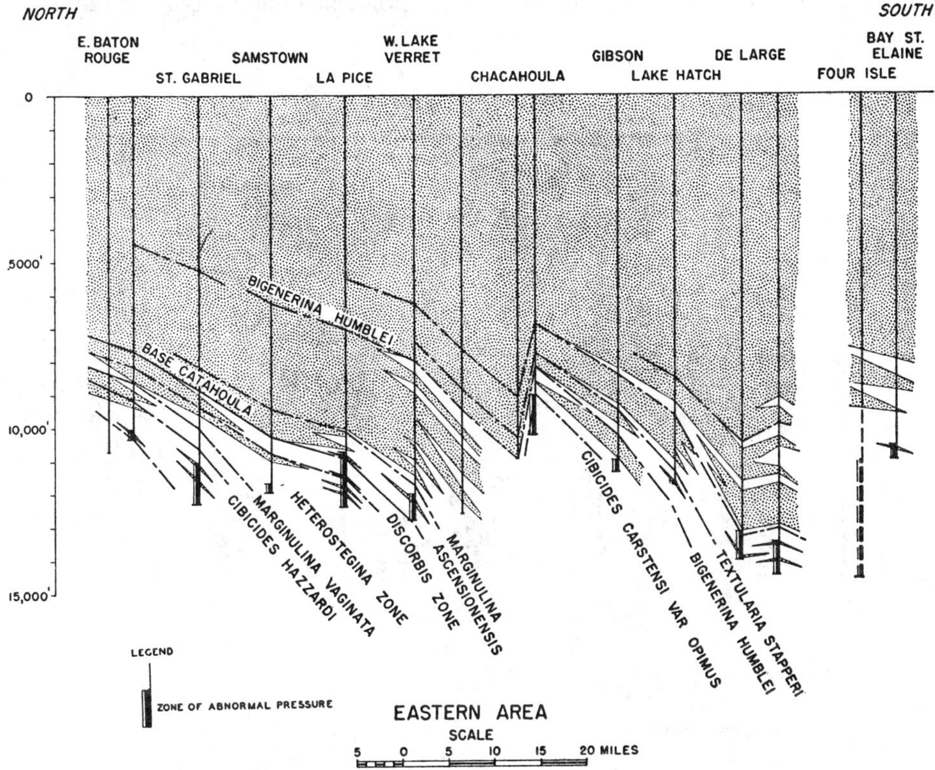

FIG. 8.—Diagrammatic stratigraphic section (eastern area) illustrating facies control of abnormal reservoir pressure.

Abnormal pressure occurrences in upthrown blocks similar to the north flank of Chalkley are numerous in the Gulf Coast region. The amount of uplift above regional is normally relatively small, ranging from about 300 feet in Snake Lake to 1,200 feet in South Crowley and Grand Lake. Uplifts as large as 1,600 feet for the north flank of Chalkley and 3,500 feet in East White Lake are uncommon.

Abnormal pressures below an unsuspected fault are especially difficult to control because of the abrupt change in pressure gradient, such as occurred in several wells in La Pice. Close paleontological control may indicate such a fault and thus permit the mud weight to be increased before a porous zone is penetrated. In some places, where abnormal pressure is encountered unexpectedly, for example, in Shell, Smith A-1, Weeks Island, the fauna may show that the normal reservoir formations have been cut out by a fault.

Little is known about the size of porous zones containing abnormal pressures. Since most of them occur on faulted structures, it is frequently assumed that they are only of limited extent. Most of the abnormal pressure occurrences reviewed are in thin sands containing salt water and in places some gas, although there are also some very high-pressure zones producing oil and gas, for example,

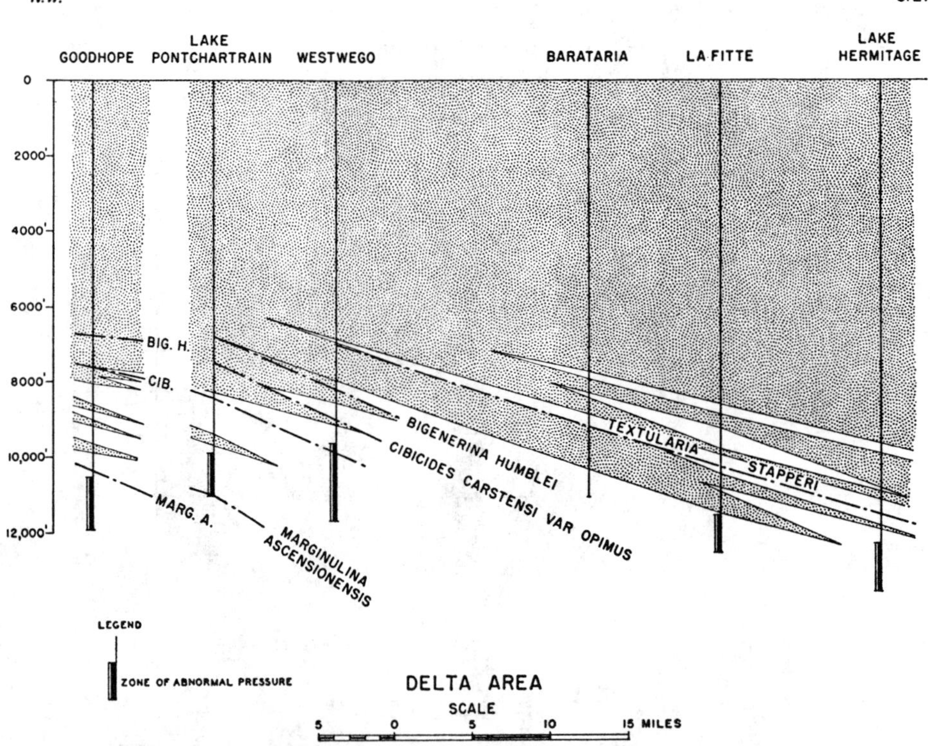

FIG. 9.—Diagrammatic stratigraphic section (delta area) illustrating facies control of abnormal reservoir pressure.

the "FV" and "FX" sands in Iowa, the "W" sand in Chalkley, the "V" sand in St. Gabriel, and a Miocene sand in Manilla Village. Rapid diminution in rate of flow of gas or salt water, or rapid drop in reservoir pressure indicates that some of the high-pressure reservoirs are undoubtedly small in size or poorly permeable, for example, the "V" sand in St. Gabriel appears to have erratic development, and some wells were depleted in a few months. On the other hand, some sand lenses must cover a considerable area as indicated by the large volumes of fluid produced. For example, the Bel crater in Allen Parish produced about 7 million barrels of water without apparent reduction in the rate of

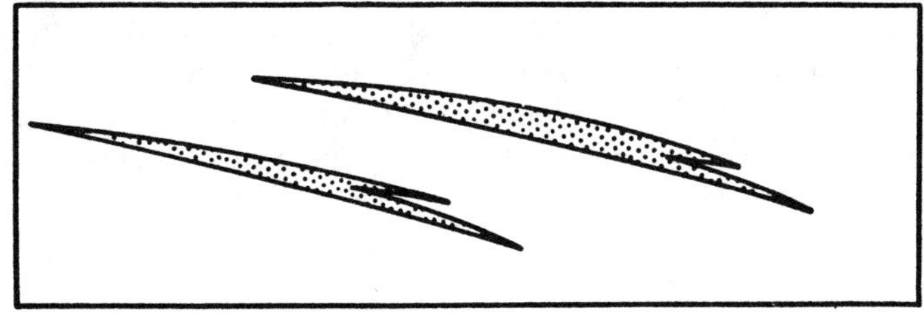

SMALL RESERVOIR SEALED BY PINCHOUT
(a)

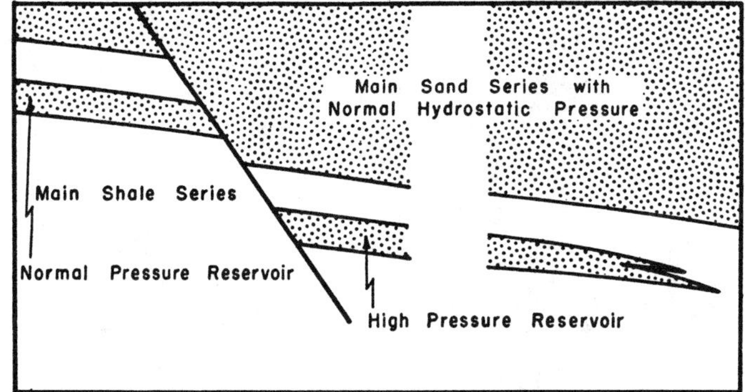

LARGE RESERVOIR SEALED UPDIP BY FAULTING DOWN AGAINST THICK SHALE SERIES, SEALED DOWNDIP BY REGIONAL FACIES CHANGE.
(b)

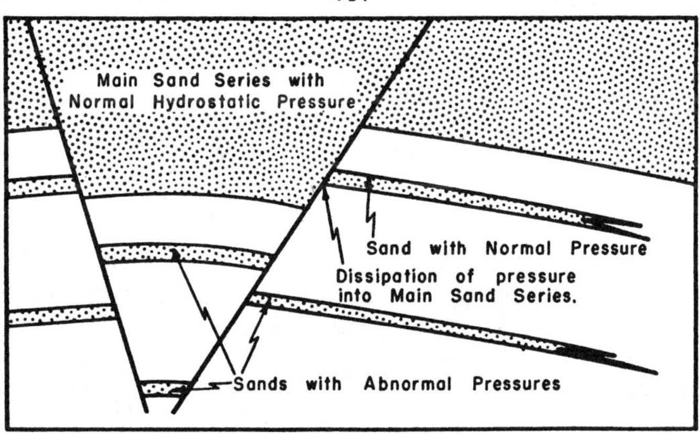

RELATIVE POSITION OF FAULT SEALS IN UPTHROWN AND DOWNTHROWN BLOCKS.
(c)

FIG. 10.—Types of reservoir seals necessary to preserve abnormal pressures.

flow (12). The rapid decline in reservoir pressure of the "FV" sand in the Shell's Fontenot No. 10 in Iowa seemed to indicate a limited reservoir volume, but after 9 months the rate of decline decreased considerably so that either the reservoir is larger than at first supposed, or there has been failure of a fault seal which separated it from an undeveloped high-pressure reservoir.

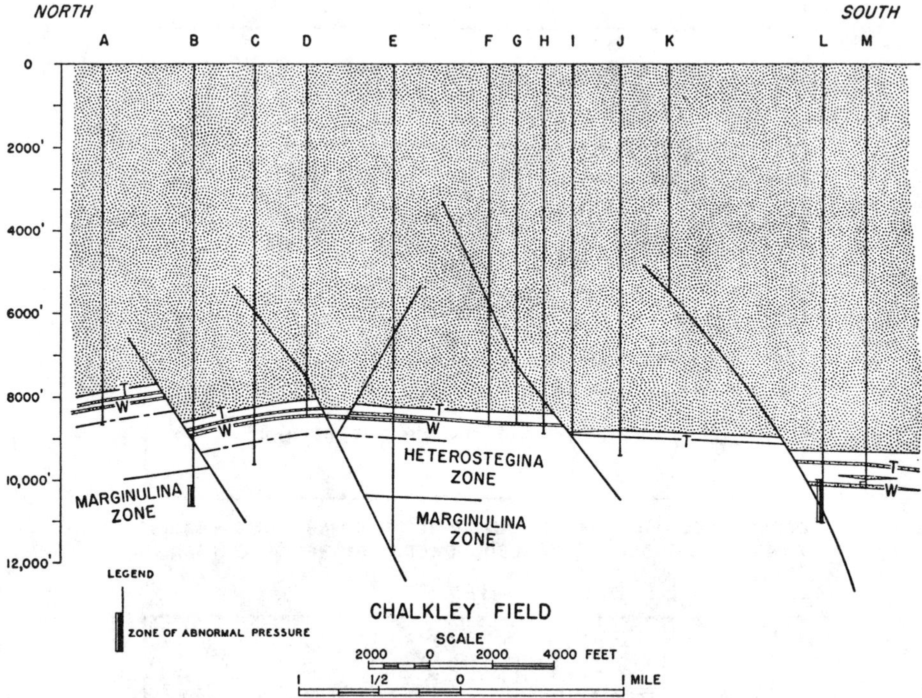

FIG. 11.—Chalkley field, Cameron Parish, Louisiana, north-south sketch section, illustrating conditions controlling abnormal subsurface reservoir pressures.

ORIGIN OF RESERVOIR PRESSURE

The obvious source of normal pressure in a reservoir rock is that due to the head of water filling the pores of the rock and in communication with the surface. In the case of porous beds containing hydrocarbon accumulations above water at normal hydrostatic pressure, the lower density of gas and oil compared with that of water will cause abnormal pressures in proportion to the structural elevation above the water, for example, in the "S" sand on the north flank of Weeks Island where a gas column extends from about 12,150 feet to the gas-oil interface at 13,750 feet subsea (Fig. 12). The original reservoir pressure at 13,800 feet was only slightly above normal hydrostatic at 6,500 pounds per square inch in the Shell's Smith-State Unit 1, wells 1 and 3. If it is assumed that the gas column exerts a pressure of 0.115 pound per square inch per foot depth, the reservoir pressure at the crest of the structure will be about 6,300 pounds per square inch or a pressure gradient of 0.518 pound per square inch per foot depth.

In closed reservoirs different conditions prevail since the fluids are isolated from the normal hydrostatic column and may be under much greater abnormal pressure. The pressure gradient will increase with structural elevation even where a reservoir contains water only as shown diagrammatically in Figure 13. The maximum pressures known approach 90 per cent of geostatic pressure, that is, of the pressure caused by the weight of the overburden. Several hypotheses have been advanced to account for the generation of abnormal pressures in isolated

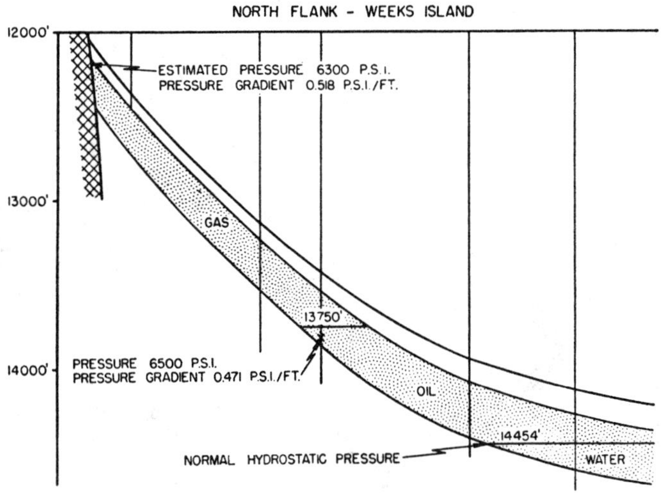

FIG. 12.—Diagrammatic section illustrating abnormal pressure caused by structural elevation in Weeks Island, Iberia Parish, Louisiana.

reservoirs, but the only one that appears to conform to all the conditions prevailing in the Gulf Coast region is that abnormal pressures are caused by the weight of the overburden.

Compression of argillaceous beds during the early stages of sedimentation and the concomitant expulsion of fluid give rise to progressive compaction as additional sediments are added to the overburden. As compaction proceeds the expulsion of fluid becomes more difficult because of decreasing permeability, so that the pressure in the clay will be partly transmitted to the fluids, and thereby to the fluids in any sand body completely enclosed in the compacting mass, even though the sand body itself may not be compressible. So long as the sediments remain plastic, these pressures will ultimately become practically identical and will be determined by the pressure gradient of the combined weight of the sediments and the contained fluids, that is, the geostatic pressure gradient of the overburden.

The pressures in fluids within sediments are dominated by two factors, the compression due to compaction on the one hand and the resistance to expulsion on the other; but as compaction becomes more difficult other factors may become important. Pressure decay at different rates due to variations in the ease of escape of fluid may control the amount of pressure in sands located deep in thick, well compacted shale series. The last stages of compaction of shale are apparently

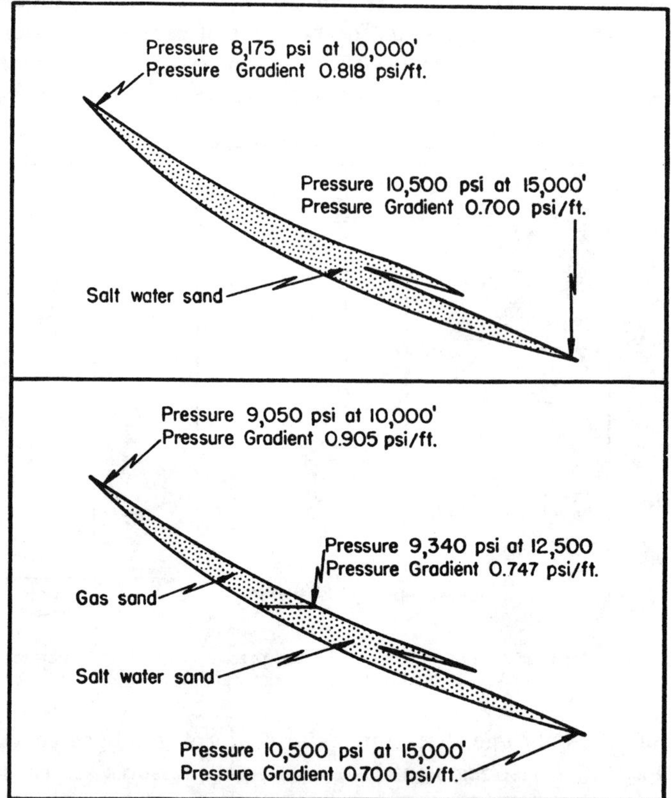

FIG. 13.—Effect of structure on pressure gradients in sands containing fluids under abnormal pressure.

complex. As Thomeer (32) has pointed out, shale resistivity remains relatively constant down to the maximum depth drilled so that the porosity apparently does not decrease to zero. Changes in the mineralogical composition of clay minerals involving absorption of water may also involve reduction in pressure in the later stages of compaction of older shales.

It is well known that newly deposited argillaceous sediments contain 60–90 per cent of pore space which is filled with water, and that shallow clays have a density of about 1.8. A study of the available density measurements and estimates used by geophysicists (6–9, 21, 23) for the Tertiary sediments of the Gulf Coast forms the basis for the depth versus shale density curve in Figure 14.

This curve conforms in a general way with one established by Athy (4) for Permian-Pennsylvanian shales in the Mid-Continent area and seems to confirm that although density *versus* depth relations may follow a well defined law, the relation probably varies according to the age of the formations. For example, the maximum Miocene density measured is 2.51 at about 15,100 feet in Weeks Island compared with 2.58 for Eocene Wilcox at about 8,600 feet in Provident City, Texas, and Athy's 2.59 for Pennsylvanian shales at 4,500 feet in Oklahoma (5). A few shale densities determined on samples from recently drilled deep wells have been incorporated in Figure 14. They appear to show that the curve may be

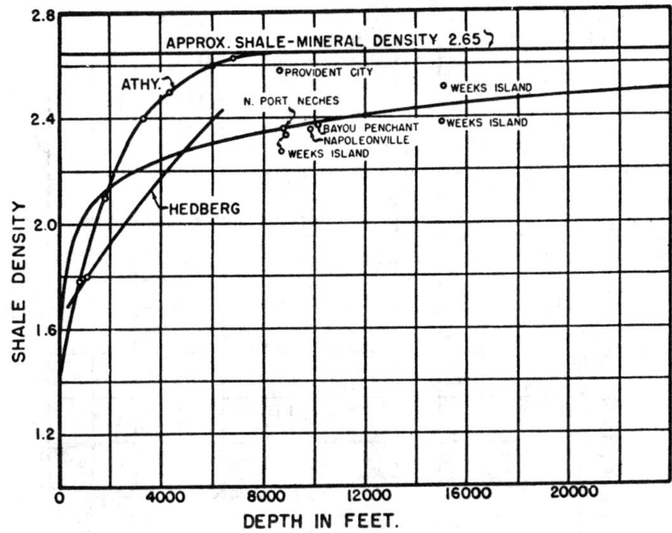

FIG. 14.—Average shale density-depth relationship for Gulf Coast Tertiary formations.

reasonably accurate, at least down to 15,000 feet, although shale densities for Weeks Island seem to indicate that the density-depth curve for the Miocene may be slightly lower.

It is interesting to note that a shale density of 2.4 at 15,000 feet indicates that compaction is probably still taking place, since the ultimate density of shale with little or no porosity must be approximately 2.65 to conform with mineral-density determinations for clays and shales made by many workers (22). Figure 15 shows the percentage of total possible compaction versus depth obtained from the density-depth curve using 80 per cent porosity for mud having a density of 1.4. This curve shows that at 8,000 feet only about 75 per cent of total possible compaction has occurred. Further compaction to a porosity of 5 per cent involves an additional reduction in volume of about 16 per cent; thus it is evident that enormous quantities of water remain to be forced out of the compacting shales by the weight of the overburden.

The compressibility of sand and sandstone has not been extensively investigated. Carpenter and Spencer (14) determined the compressibility of consolidated sand cores from Texas to be about 2.4 per cent under a pressure of 8,000 pounds per square inch, and Athy (4, 5) was able to compress loose, settled sand only 2 per cent with 4,000 pounds per square inch. The very slight compressibility of sand during a limited period of geologic time appears to be confirmed by the general average porosity of 30 per cent prevailing down to at least 10,000 feet for the Miocene, unconsolidated sands of the Gulf Coast region. The lower porosity, averaging about 20 per cent, of some Oligocene, consolidated sandstones in the same area is attributed by Archie[5] to a greater degree of cementation and

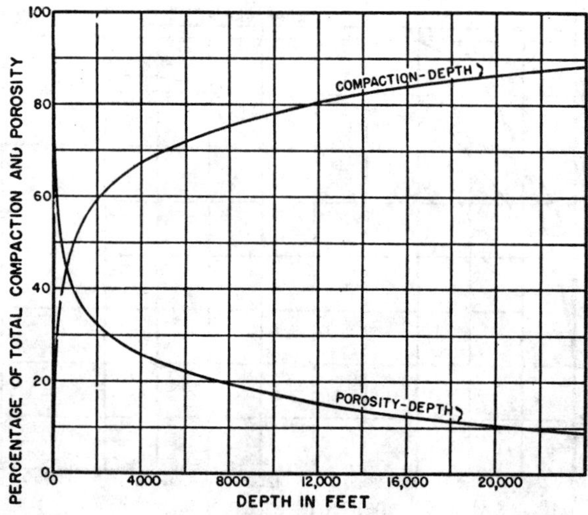

FIG. 15.—Relation of porosity and compaction of shales to depth of burial.

to variations in grain size. Nevertheless, in general, the older a sandstone the more compact and less porous it is, and according to the laws of thermodynamics, there must be solution of silica from the sand grains at points of contact and probably concurrent precipitation around the grains and in the voids. This process has been discussed by Waldschmidt (33) who observed evidence for it in several sandstone members of the Cretaceous and Jurassic in Wyoming. Cogen (11) has reported secondary deposition of silica in the Eocene Wilcox sandstone in the Sheridan field in the Texas Gulf Coast region. This may indicate that there has also been solution of silica at points of contact of the sand grains. At similar depths, about 10,000 feet, the sands and sandstones of the Miocene and Oligocene apparently have not yet been similarly affected, so that age, rather than depth of burial, appears to be the more important factor in this exceedingly slow lithological change in the character of a sand.

[5] Oral communication.

The effects of this process are 2-fold: (1) the solution of silica from sand grains at points of contact results in compaction with consequent decrease in porosity and expulsion of water or rise in fluid pressure; and (2) the precipitation of quartz around the sand grains and in the voids results in a further decrease in porosity and expulsion of water or rise in fluid pressure. The rate of volume reduction from these causes is probably so small compared with that of clays that its effect on the fluid pressure in a sand will be negligible during the greater part of the compaction of the enveloping clays. However, in the later stages of shale compaction, its effect might cause fluid pressures within isolated sand bodies to increase above the residual abnormal pressure generated by the compaction of the clays.

The foregoing hypothesis, that abnormal pressures are caused by the weight of the overburden, appears to conform with known conditions in the Gulf Coast region, whereas the alternative hypotheses discussed in the following paragraphs are not satisfactory in all respects.

P. E. Chaney (15) suggested that progressive degradation of oil and gas in a closed reservoir could give rise to abnormal pressures up to overburden pressure, and that higher pressures would be released by decompaction or fracturing. The disadvantage of this hypothesis is that many of the high-pressure zones in the Gulf Coast region contain salt water with solution gas only. Illing (25) doubts whether changes in the composition of oil and gas occur at so late a stage.

W. E. V. Abraham (1) thought that uplift of sand lenses from great depths might account for abnormal pressures in Trinidad. This hypothesis is untenable for the Gulf Coast since the geological history of the region does not allow postulation of uplift of sufficient magnitude to account for even moderately high abnormal pressures. In addition, as Watts (35) has pointed out, if the uplift is accompanied by the appropriate reduction in temperature, contraction of the confined fluids will decrease the pressure rapidly and under some conditions sufficiently to maintain normal hydrostatic pressure.

Tectonic forces undoubtedly may give rise to very high subsurface pressures in some areas (25, 35) but such forces appear to be absent in the Gulf Coast region, except perhaps locally around salt domes.

ESTIMATION OF OVERBURDEN PRESSURE

A close approximation of overburden pressure based on the shale density-depth relationship is given in Figure 16. It can be seen from this curve that the commonly accepted pressure gradient of one pound per square inch per foot depth is sufficiently accurate for all practical purposes, although its use may lead to underestimation of the overburden pressure at depths greater than about 17,000 feet.

The effect of the great thicknesses of sand in the Gulf Coast section is relatively small. According to Archie (2) Miocene, loosely consolidated sandstone, averages about 30 per cent porosity, and Oligocene, consolidated sandstone, varies between 18 and 35 per cent porosity, with an average of about 25 per cent.

Assuming clean sand with a mineral grain density of 2.65 (quartz) and salt-water density of 1.08, the bulk density of the sandstones will be about 2.18 and 2.26, respectively. Although these densities are lower than the equivalent shale densities at depths greater than about 3,000-4,000 feet, the effect on the overburden pressure is negligible. For example, if the upper 15,000 feet of formation is assumed to be all sand with 30 per cent porosity, the overburden pressure will be about 14,300 pounds per square inch, compared with 14,900 pounds per square inch for an all shale section. These pressures represent the two extremes, so that normally the overburden pressure may be 2–3 per cent below that shown by

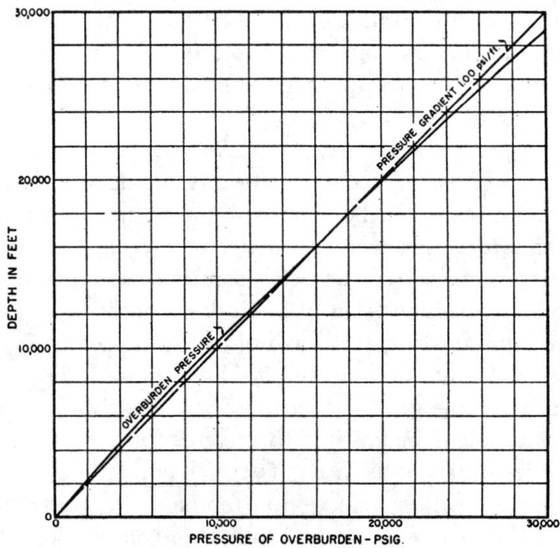

FIG. 16.—Overburden pressure *versus* depth.

the curve in Figure 16 and will be less than one pound per square inch per foot depth to drilling depths at present attainable (20,000 feet).

The current maximum pressure gradients are 0.872 pound per square inch per foot depth for salt water, possible with solution gas, in Johnson's Bayou, and 0.876 pound per square inch per foot depth for gas condensate in the "FV" sand of Iowa. Compared with the maxima of about 0.865 of Cannon and Sullins (13) in 1946, 0.83 of Denton (17) in 1943, and 0.765 of Cannon and Craze (12) in 1938, these gradients appear to indicate that the upper limit of abnormal pressure gradients is being approached, and that it is unlikely that it will exceed about 0.900 pound per square inch per foot depth. Pressures approaching this gradient have been drilled through without excessive trouble by using muds weighing 18 to 18.5 pounds per gallon. The main difficulty with such heavy mud is loss of circulation. Where abnormal pressures have been penetrated successfully, for example, in Iowa, St. Gabriel, and Chalkley, casing was cemented in the top of the shale series before drilling into the high-pressure zones, thus precluding the loss of circulation into the main sand series.

BIBLIOGRAPHY

1. ABRAHAM, W. E. V., "Geological Aspects of Deep Drilling Problems," *Jour. Inst. of Petroleum* (London, 1937), p. 378.
2. ARCHIE, G. E., "Practical Petrophysics," *Shell Oil Company Production Dept. Rept.* (February, 1949), Fig. 20. Published as "Introduction to Petrophysics of Reservoir Rocks," *Bull. Amer. Assoc. Petrol. Geol.*, Vol. 34, No. 5 (May, 1950), Fig. 8.
3. ATHY, L. F., "Compaction and Oil Migration," *Bull. Amer. Assoc. Petrol. Geol.*, Vol. 14, No. 1 (January, 1930), pp. 25–36.
4. ———, "Density, Porosity, and Compaction of Sedimentary Rocks," *ibid.*, pp. 1–24.
5. ———, "Compaction and Its Effect on Local Structure," *Problems of Petroleum Geology*, Amer. Assoc. Petrol. Geol. (1934), p. 814.
6. BARTON, D. C., "Belle Isle Torsion-Balance Survey, St. Mary Parish, Louisiana," *Bull. Amer. Assoc. Petrol. Geol.*, Vol. 15, No. 11 (November, 1931), p. 1342.
7. ———, "Torsion-Balance Survey of Esperson Salt Dome, Liberty County, Texas," *ibid.*, Vol. 14, No. 9 (September, 1930), p. 1135.
8. ———, "Review of Geophysical Prospecting for Petroleum 1929," *ibid.*, pp. 1113–14.
9. ———, "Gravitational Methods of Prospecting," *Science of Petroleum*, Oxford Univ. Press (1938), p. 374.
10. BUGBEE, J. M., "Notes on Drilling and Production, South Pool, Chalkley," *Shell Oil Company Production Dept. Rept.* (September 10, 1945).
11. COGEN, W. M., "Effects of Mud Acid on Cores of Wilcox Sandstone in Sheridan Field, Colorado County, Texas," *Shell Oil Company Exploration Dept. Rept. TG/Misc. No. 315* (September 21, 1942).
12. CANNON, G. E., AND CRAZE, R. C., "Excessive Pressures and Pressure Variations with Depth of Petroleum Reservoirs in the Gulf Coast Region of Texas and Louisiana," *Trans. Am. Inst. Min. Met. Eng.*, Vol. 127 (1938), pp. 31–38.
13. CANNON, G. E., AND SULLINS, R. S., "Problems Encountered in Drilling Abnormal Pressure Formations," *Drilling and Production Practice*, Amer. Petrol. Inst. (1946), pp. 29–33.
14. CARPENTER, C. B., AND SPENCER, G. B., "Measurement of Compressibility of Oil Bearing Sandstones," *U. S. Bur. Mines, R. I. 3540* (October, 1940).
15. CHANEY, P. E., "Abnormal Pressures, Lost Circulation Gulf Coast's Top Drilling Problem," *Oil and Gas Jour.*, Vol. 47, No. 51 (April 21, 1949), pp. 210–15.
16. COLVILL, G. W., "Notes on Deep Well Drilling in Iran," *Jour. Inst. Petroleum* (London, 1937), p. 408.
17. DENTON, H. H., "Abnormal Salt Water Pressures on the Texas and Louisiana Coast," *Field and Laboratory*, Southern Methodist Univ. (January, 1943).
18. EUWER, M. L., "Pressure Maintenance at East Hackberry," *Oil Weekly*, Vol. 119, No. 6 (October 8, 1949), p. 46.
19. GILBERT, C. M., "Cementation of Some California Tertiary Reservoir Sands," *Jour. Geology*, Vol. 57, No. 1 (January, 1949), pp. 1–17.
20. GOLDSTEIN, A., JR., "Cementation of Dakota Sandstones of the Colorado Front Range," *Jour. Sed. Petrology*, Vol. 18, No. 3 (December, 1948), pp. 108–25.
21. GOLDSTONE, F., AND HAFNER, W., "Geophysical Monthly Report," *Shell Explor. Dept. Rept.* (October, 1930).
22. HEDBERG, H. D., "Gravitational Compaction of Clays and Shales," *Amer. Jour. Sci.*, 5th Ser., Vol. 31, No. 184 (April, 1936), p. 279. Includes a good list of references.
23. HEILAND, C. A., *Geophysical Exploration*, pp. 82–84, 278 and 280. Prentice Hall (New York, 1940).
24. HOBSON, G. D., "Compaction and Some Oil Field Features," *Jour. Inst. Petroleum*, Vol. 29, pp. 37–54. London.
25. ILLING, V. C., "The Origin of Pressure in Oil-Pools," *Science of Petroleum*, Oxford Univ. Press (1938), pp. 224–29.
26. KEEP, C. E., AND WARD, H. L., "Drilling against High Rock Pressures with Particular Reference to Operations Conducted in the Khaur Field, Punjab," *Jour. Inst. Petroleum*, Vol. 20 (London, 1934), p. 990.
27. LOWMAN, S. W., "Sedimentary Facies in Gulf Coast," *Bull. Amer. Assoc. Petrol. Geol.*, Vol. 33, No. 12 (December, 1949), pp. 1939–97.
28. NEVIN, C. M., "Porosity, Permeability, Compaction," *Problems of Petroleum Geology*, Amer. Assoc. Petrol. Geol. (1934), pp. 807–10.

29. REED, P., "Trinidad Leaseholds Applies Advanced Methods in Drilling and Production," *Oil and Gas Jour.* (October 5, 1946), p. 45.
30. SLOSS, L. L., AND FERAY, D. E., "Microstylolites in Sandstone," *Jour. Sed. Petrology*, Vol. 18, No. 1 (April, 1948), pp. 3-13.
31. TERZAGHI, K., AND PECK, R. B., *Soil Mechanics in Engineering Practice*, p. 61. John Wiley and Sons, New York.
32. THOMEER, J. H. M. A., "The Electrical Resistivity and Other Physical Properties of a Clay Formation as a Function of Depth and Age," Translation of *P. A. Rept. 5197, Production Dept., Bataafsche Petroleum Mij.* (The Hague, January 26, 1943).
33. WALDSCHMIDT, W. A., "Cementing Materials in Sandstones and Their Probable Influence on Migration and Accumulation of Oil and Gas," *Bull. Amer. Assoc. Petrol. Geol.*, Vol. 25, No. 10 (October, 1941), pp. 1859–63 and 1869–79.
34. WALKER, A. W., "Squeeze Cementing," *Oil World*, Vol. 129, No. 6 (September, 1949), Fig. 1, p. 88.
35. WATTS, E. V., "Some Aspects of High Pressures in the D-7 Zone of the Ventura Avenue Field," *Trans. Amer. Inst. Min. Met. Eng.*, Vol. 174 (1948), pp. 191–205.
36. WESCOTT, B. B., DUNLOP, C. A., AND KEMLER, E. N., "Setting Depths of Casing: Drilling and Production Practice, 1940," *Amer. Petrol. Inst.*, Fig. 10, p. 157.

Reprinted from:
BULLETIN OF THE AMERICAN ASSOCIATION OF PETROLEUM GEOLOGISTS
VOL. 45, NO. 10 (OCTOBER, 1961), PP. 1721–1730

INCREASING OCCURRENCE OF ABNORMALLY HIGH RESERVOIR PRESSURES IN BOREHOLES, AND DRILLING PROBLEMS RESULTING THEREFROM[1]

J. H. M. A. THOMEER[2] AND J. A. BOTTEMA[3]
The Hague, Netherlands

ABSTRACT

Under normal conditions fluid pressures in underground reservoirs will be hydrostatic, i.e., in equilibrium with the weight of a (salt) water column extending from reservoir level to surface. Such hydrostatic pressures can be easily controlled by using a mud flush only slightly heavier than water.

Reservoir pressures higher than hydrostatic are not uncommon, however, and a theory is given explaining the why and when of their occurrence. Composition and previous sedimentation rate of overburden are controlling factors, as are the age and geologic history of formations concerned. Maximum possible reservoir pressure to be encountered is petrostatic, i.e., in equilibrium with weight of overburden. The control of such excessive reservoir pressures in drilling wells requires mud weights, which, in extreme cases, may be equal to or even in excess of overburden weight. These latter pressures will unavoidably cause incurable circulation losses by squeeze action, and under appropriate conditions even less high mud weights may have the same effect. A very serious problem is thus created.

Critical situations of this nature are not uncommon and would appear to require greater attention and more study. A few practical experiences of more recent date (North Germany, Netherlands New Guinea, West Pakistan) are discussed.

From both theory and field experience it appears that with proper care and precaution no unsurmountable difficulties need generally be anticipated in drilling through rocks with reservoir pressures not exceeding 80–90% of overburden pressure. For cases of still higher pressures (90–100%), however, the problem seems still far from solved.

I. INTRODUCTION

It is well known that fluids in porous rocks are normally encountered under hydrostatic pressure, i.e., under a pressure equal to that exerted by a (salt) water column extending from the reservoir level to the surface. With due regard to any smaller deviations from this rule and to normal safety precautions, it follows that such reservoir fluids can be kept from invading the borehole by the use of a mud flush weighing not more than 1.2 kg/litre (10 lbs/gl; 75 lbs/cu. ft).

It is equally well known, however, that this rule is not without exceptions. Many cases are on record where formation fluids were found under far higher than hydrostatic pressures. It will be clear, though, that there must be an upper limit to the value of these "abnormal" or "excessive" reservoir pressures. Evidently, they never can exceed the "petrostatic" level, i.e., the pressure exerted by the weight of the overlying rock column, equalling about 2.2–2.4 times the hydrostatic pressure at the relevant depth. For some time, experience suggested that actual maximum values observed were 0.87 (Gulf Coast, Dickinson, 1951) or 0.90 times petrostatic (world-wide, Kok and Thomeer, 1955). More recent experience, however, appears to indicate that still higher values may actually occur.

In order to keep such abnormally pressured fluids from untimely invading the borehole and causing blow-outs and other complications, mud weights greatly in excess of the aforementioned value of 1.2 kg/litre are required, up to 2.0–2.2 kg/litre (16½–18½ lbs/gl; 125–138 lbs/cu. ft) and, in extreme cases, even of 2.5 kg/litre (21 lbs/gl; 156 lbs/cu. ft) and over.

It will be clear that, quite apart from the problem of preparing and controlling such superheavy muds and the costs involved, their use may easily result in existing joints and fissures in the formations being opened or widened and even in new fissures being formed by "squeeze fracturing" (Bugbee, 1953). Circulation losses resulting from such occurrences can not possibly be controlled by any application of coarse plugging materials in the mud or other known means as long as the mud weight remains excessive. The mud pressure permissible without fracturing taking place may vary with local conditions (Hubbert and Willis, 1957)

[1] Manuscript received, October 8, 1960.

[2] Mining Engineering Department, Technological University at Delft, Netherlands.

[3] Exploration and Production Function of Bataafse Internationale Petroleum Maatschappij, The Hague, Netherlands. The writers express their appreciation to the Gewerkschaft Brigitta; N.V. Nederlandsche Nieuw Guinee Petroleum Maatschappij; and Burmah Oil Company Limited; for their permission to publish the information given in this paper.

and occasionally may even be rather low, but pressures above the petrostatic level are certain to cause fracturing in any case.

It is evident, therefore, that in certain cases the control of reservoir fluid pressure may require the application of a mud that is too heavy to be confined by the walls of the borehole. Formations in which these conditions prevail can not be drilled in the normal manner.

II. Reservoir Pressures and Their Control in Boreholes

Distinction should be made between rock pressure, reservoir pressure, and grain pressure.

Rock pressure denotes the total stress prevailing in a rock. The weight of the overlying rock column is an ever-present component but forces of geotectonic origin may add to and complicate the stress field. Since sedimentary rocks are generally heterogeneous, the rock stresses must be partly carried by the mineral fabric of the rock and partly by the fluid in its pores.

Fluid or reservoir pressure is the pressure prevailing in the pore fluid within a reservoir rock.

Grain pressure is the part of the rock pressure carried by the solid mineral component of the rock. It is generally given as an average value per unit of surface in a section through the rock; actually it is concentrated of course at the contact points between mineral grains where its value will be much higher than this average.

Now let us consider the top layer of a freshly deposited sediment at the bottom of the sea. The fluid in this sediment is continuous with the overlying sea water and its pressure therefore hydrostatic. Its value will depend on the depth of the sediment below the surface of the water. The solid grains in this sediment will only just touch each other. Contact pressure between them is governed by their weight minus buoyancy and, in a thin shallow sediment as considered here, will be very small. In some sediments, i.e., those consisting of clay minerals (sea-bottom muds), the mineral flakes may not even have any direct contact (adsorption of water mantles).

If such a sediment is covered by a thin sheet of impermeable material (e.g., rubber) and sedimentation thereafter allowed to continue, the pressure in it will gradually rise due to the growing weight of the overlying rock column. This rock pressure, however, will be carried nearly exclusively by the water in the sediment. The mineral grains can hardly alleviate this burden because, to carry their part of the weight, they would have to be forced together and compressed, which would only be possible if the water volume between them could be reduced, and this latter can only take place by fluid compression (hardly measurable in this early stage of the process) or by escape (prevented by the rubber sheet). Consequently, in this hypothetical case, the reservoir pressure will practically equal the rock pressure and the loose mineral structure will be preserved.

In actual practice, the impervious rubber sheet is not there and as soon as the column weight increases, the water in the sediment will be able to move out through the continuous pore volume of the overlying younger sediments, thus always preserving its hydrostatic pressure. A further consequence is, however, that now only a part (less than half) of the overlying rock weight can be carried by the water pressure, so that the remainder is transferred to the intertouching solid grains, thus resulting in compaction of the rock (Athy, 1930; Von Terzaghi and Froehlich, 1936; Hedberg, 1936). This very compaction, by the way, will cause decreasing pore volume and consequently further expulsion of water, the pressure of which, however, will still tend to remain hydrostatic. Under normal conditions in a sedimentary column we will therefore encounter hydrostatic reservoir pressures and, as depth increases, growing compaction with reduced porosity. The degree of compaction will also depend on the "compactability" of the minerals: it will be large in the clayey rocks consisting of flexible and brittle mineral flakes, and hardly perceptible—except to some extent at great depths—in sands that consist of hard, rounded grains.

This tendency of fluid pressure in reservoir rocks to remain about hydrostatic is confirmed by general experience in oil well drilling, but exceptions to this rule are also known to occur, more particularly in the form of "over-pressured" reservoirs. These latter can only be satisfactorily explained by assuming that under certain conditions something comparable with the imaginary rubber sheet actually is present in the sedimentary column. Generally there are two possibilities for this situation to occur.

1. *Rock salt or similar deposits* of very great lateral extent, such as present in the North German Zechstein basin. Rock salt is completely impervious to fluid flow and even under high tectonic

stress will not break, but rather be transformed pseudo-plastically (recrystallization effect). If a round-about way of escape for compressed fluid is likewise excluded by the horizontal extent of the deposit, all fluids in all porous formations below this salt deposit should be found under a reservoir pressure equalling the full value of the rock pressure ("petrostatic" pressure). Also these porous formations can not be compacted and therefore may have retained their original soft constitution regardless of age. This situation moreover is permanent as long as the salt deposits are not removed by erosion or otherwise. Examples of such admittedly exceptional conditions have actually been encountered in drilling practice.

2. *Clay or shale deposits of large extent and thickness.*—Clay and shales are not strictly impermeable but their permeability is generally very low. As a result, the escape of water out of the sediments through an overlying shaly column is retarded though not completely impeded. Reservoir pressures below such a shaly cover will normally be hydrostatic, but occasionally higher pressures are found to exist. The chances of this latter condition occurring will increase with the thickness of the shaly deposit (increased resistance to water flow by length of path) and with depth (increased resistance due to compaction having further reduced the permeability). Also the phenomenon is more likely to be encountered in the younger deposits: the older a sediment the greater the possibility that excess reservoir pressure will have been dissipated with time through the low-permeability channels that are always present, and also through smaller cracks and fissures that are not uncommon in the older and more brittle formations. Favorable conditions for overpressured reservoirs below shale columns may be found in some of the younger sedimentary basins, in parts of which clay masses several thousand feet thick may have been relatively rapidly deposited over considerable areas, time thus being too short for hydrostatic reservoir equilibrium to be reached. Any reservoir rock below a clayey or shaly section of this type should be watched for excess pressure conditions. A classic example of this situation is provided by the Tertiary section of the Texas-Louisiana Gulf Coast basin (Dickinson, 1951).

It should also be realized that in cases of this nature not only the fluid in sands below the shale cover, but also that in the pores of the shales themselves, may be found overpressured. Fluid pressures in shale are less easy to measure, but their abnormally high value, necessarily coupled with a lower than normal grain pressure, might be an important factor in explaining the "heaving" and "crumbling" shale character often encountered in these very same basin sediments (Thomeer, 1955).

The question might finally be asked: What happens at very great depth and age when compaction and diagenesis have advanced so far that the permeability not only of shales, but of all rocks will necessarily approach the vanishing point? Would all these rocks be found overpressured? There seem to be two possibilities. If the permeability of all rocks has indeed grown extremely small, so too will the porosity, and the terms reservoir rock, fluid flow, and fluid pressure will have lost their meaning. There is, however, another possibility not infrequently encountered; the compacted and hardened rocks may have acquired a secondary porosity and permeability which may make some of them suitable reservoirs again, while at the same time providing channels regulating the reservoir pressure to normal hydrostatic level.

Summarizing, we should conclude that, normally, reservoir rocks should display *hydrostatic* reservoir pressures, but that occasionally under well defined conditions reservoir pressures may be higher, up to "*petrostatic.*" The latter value is governed by the weight of the overburden. For a completely non-porous overburden, the pressure would amount to about 2.7 times hydrostatic (2.7 being an average value for the density of sediment-building minerals such as quartz and the clay-silicates). Actually the weight is lower as a result of water-filled porosity in the sedimentary column. An average pressure between 2.2 times hydrostatic at shallow depth and 2.4 times at greater depth appears to be the practical maximum.

Tectonic stresses will not affect this maximum. Admittedly, they may occasionally be higher than the vertical overburden pressure, but they can not possibly load up the fluid pressure in rock pores to a value exceeding overburden pressure, since any excess above the latter would be automatically released by a slight lifting of the overburden.

As for the *mud weight* required for drilling through overpressured reservoirs, it stands to reason that this should always be higher than the reservoir pressure, preferably some 10–20% higher, as a safeguard against contingencies such as swabbing effect during upward tool movements, gas adsorption from formations drilled, negligence of drilling personnel in keeping the hole filled, etc. This means that under normal conditions, mud specific weight should preferably not be less than 1.15 ($9\frac{1}{2}$ lbs/gl; 72 lbs/cu. ft), and occasionally much higher values, up to 2.4 (20 lbs/gl; 150 lbs/cu. ft) and even over, may be required.

III. Effect of Mud Pressure on Borehole Wall

It will be clear that there must be a maximum limit to the inside pressure a borehole can sustain without its wall giving way, i.e., without fissures being formed and mud fluid being lost into them. It will be equally clear that fissures formed under such conditions can not possibly be closed by any conventional plugging means, nor mud circulation regained, as long as the fluid pressure in the hole is maintained at the high level: any stoppage of the fluid loss, e.g., by cementing the fissures, will only lead to other, fresh, fissures being formed as a result of the excessive mud pressure.

The point is: what is this maximum pressure limit within a borehole or, in other words: what maximum mud weight is permissible without sudden, complete, and uncontrollable loss of circulation occurring. This is a somewhat complex problem that has attracted considerable attention in recent years, more particularly in connection with the theory underlying hydraulic fracturing techniques (see, e.g., Scott, Bearden, and Howard, 1953; and Hubbert and Willis, 1957). Although there is still some diversity of opinion in some respects, the following aspects will be clear and generally acceptable.

1. All formations are in a state of triaxial stress. One of the three main components is the weight of the overburden, previously referred to as the petrostatic rock pressure, and therefore vertical. The other two components must be horizontal and have their origin partly in the reaction of the rock to vertical compression (Poisson effect) and partly in tectonic pressures, and, within certain limits, the latter may be larger or smaller than the vertical rock-pressure component.

2. Around a borehole, the virgin state of stress is disturbed, which results in local stress deviations (sometimes to the extent of making the hole collapse).

3. By the introduction of a mud column in the borehole, another stress field originating in the weight of the mud, is superimposed on the above, locally disturbed, field of the original rock pressure.

4. The effective stresses to be considered in studying the possibilities of rock fracturing, borehole collapse, etc., are not the rock stresses as such, but rather the differences between these rock stresses and the fluid stresses existing in their pores (reservoir pressures), or, in other words, the grain stresses.

5. The presence of a mud column in a borehole, apart from its direct effects on the stress distribution around this hole, may seriously affect the grain stresses in the vicinity, since, by the penetration of its fluid phase into permeable formations, the reservoir pressure will be locally increased and the grain pressure correspondingly reduced.

6. Under all conditions tension fractures occurring should be perpendicular to the direction of least stress.

It will be clear from the foregoing that a quantitative calculation of stress distribution and fracture possibilities around a borehole is somewhat complex. This has been carried out exhaustively by Hubbert and Willis (1957) and by others to whose work the reader is referred for further detail. For the purpose of the present study, however, the following qualitative conclusions that are obvious without further detailed calculations, may suffice.

a. If vertical overburden pressure is smaller than either horizontal pressure component.

Any fissure caused or opened by excess fluid column pressure should be horizontal. The mud pressure causing this should be at least equal to the overburden pressure. Some excess may be required to overcome the tensile strength of the formation in effecting the first break. In stratified sediments this tensile strength, however, is generally small.

b. If both main horizontal pressures are smaller than the overburden pressure.

Any fissures caused by mud pressure should be vertical and perpendicular to the lesser main horizontal stress component in the disturbed stress field around the borehole. It is far more difficult to calculate exactly the critical mud fluid pressure, but it will be clear that this may be smaller than in the previous case or, in other words, vertical fissuring may occur under these conditions at mud weights less than overburden weight. This latter fact is amply confirmed by field experience (Howard and Fast, 1950).

CONCLUSION

From the foregoing it follows that when the mud column pressure is above the overburden pressure (i.e., when using a mud weight above a specific gravity of about 2.3, equivalent to 19.2 lbs/gl or 144 lbs/cu. ft) formation fracturing (lifting of overburden) *must* unavoidably occur with resultant sudden and complete loss of circulation. Under certain conditions, viz., in tectonically relaxed areas, fracturing with all its resulting complications *may* occur with lower mud weight.

As discussed in the previous chapter, high mud weights are required in all cases where the reservoir pressures are above the normal hydrostatic level. If these reservoir pressures approach the petrostatic level the required weight of the mud may even run up to 2.4 and over.

It follows that in rock sections where reservoir pressures are abnormally high there is a constant danger of running into the pernicious kind of circulation losses that can not be controlled by any sealing means and that, if not stopped, will unavoidably lead to blow-outs, sticking of tools, and other mishaps leading to the eventual loss of the hole.

IV. Case Histories

A. PERMIAN BELOW NORTH GERMAN PLAINS

On a structure in Northwest Germany, three wells were drilled in order to explore the possibilities of gas production from Permian dolomites situated below large masses of rock salt.

The first hole, C-1, encountered the top of the Upper Permian Zechstein at 8,190 ft. From this point down the formation consisted of shale, salty shale, rock salt, and some anhydrite. Much apparent caving was experienced in the lower part of the salt, possibly caused by a plastic deformation of this salt. Although with considerable trouble, drilling could be continued after raising the mud weight from 10/10½ lbs/gl to 12 lbs/gl. Between 9,355 ft. and 9,395 ft. the formation appeared to consist of a very soft and plastic shale which continually tended to fill the hole almost as fast as it was drilled and reamed.

These experiences seem to offer a very clear example of plastic deformation of incompetent rocks under pressure, activated by the drilling of a hole in which the pressure is below that in the surrounding rock. Probably this explanation applies to the salt squeezes, but most certainly to the behavior of the shale. In the writers' opinion this may have been a local lenticular body of clay, completely enclosed in dense, impermeable rock salt. It thus never had an opportunity of losing any of its water content and of compacting under overburden pressure. In spite of its Permian age, it had therefore retained its fresh, highly porous, and plastic condition, and, naturally, began to flow (heave) as soon as it was penetrated with a comparatively light mud flush. Of course, the old and persistent "bentonite swelling" theory, though in the writers' opinion obsolete as a general explanation (Thomeer, 1953), was also brought forward. Subsequent laboratory investigation of formation samples, however, showed them to be free of any montmorillonite.

An attempt to cement a 7-in. casing string below these troublesome formations failed as the string got frozen at 9,100 ft. Many more technical complications had to be overcome but it proved possible to drill deeper after further raising the mud weight to 15½ lbs/gl.

The formation consisted of pure rock salt, locally interbedded with some anhydrite down to 10,300 ft., where the "Platten" dolomite was entered. The mud weight was 16 lbs/gl. This anhydritic-dolomitic complex is locally known as a prospective reservoir rock and when a depth of 10,375 ft. (final depth of the well) was reached, it attempted to blow out. After several attempts to kill and test the well it had to be abandoned by cutting the drill pipe and cementing under highly critical conditions. Mud weights used during these operations were up to 18½ lbs/gl, and from several calculations based on static weight of mud column plus pressures registered against blow-out preventers, estimates could be made of bottom-hole reservoir pressure. Minimum value thus derived was 2.06 times hydrostatic value, maximum 2.38 times hydrostatic.

At any rate this "Platten" dolomite reservoir was highly overpressured, which can be explained by its thick salt cover (at this location about 1,400 ft.) which is moreover known to have a very great lateral extent (hundreds of miles). Under existing conditions it is highly dubious whether this pressure can be controlled by heavy mud without the latter continually squeezing the formation open and becoming lost.

The second hole, C-2, did not reach deeper than 8,512 ft., where it had to be abandoned for technical reasons. It thus advanced less than 1,000 ft. into the Permian and so could not supply much information for our purposes. The only indication of interest was a strong salt water flow at 8,448 ft., which required a $13\frac{3}{4}$-lbs/gl mud to control. The reservoir pressure at this depth could be estimated at 1.6 times hydrostatic, i.e., already far above normal, but less than could be expected at greater depth after penetration of the main salt bodies.

The drilling program (mud, casing scheme, etc.) for *the third hole, C-3,* was drawn up with the experience from the two former wells in mind and thus the caprock of the "Platten" dolomite could be reached without encountering any serious trouble.

A $10\frac{3}{4}$-in. string was safely cemented above the plastic shale zone known from C-1, and the latter penetrated after increasing the mud weight to $18\frac{1}{2}$ lbs/gl. Thereafter this difficult formation was cased off behind a $7\frac{5}{8}$-in. string so that the "Platten" dolomite could be tackled under safest possible conditions.

Subsequently, just after having topped the dolomite at 10,430 ft., a drill-stem test was carried out yielding some salt water. The closed-in bottom-hole pressure amounted to 8,825 psi after 2 hours, from which value a pressure gradient of 1.95 times hydrostatic was calculated. There were indications, however, that the maximum pressure build-up had not yet been reached in the short closed-in time.

After drilling-in to 10,600 ft., a second test was made. The result this time was a strongly saturated solution of mainly Ca and Mg salts, with some gas. During this test the drill pipe gradually became clogged by crystallizing salt and after this phenomenon had completely stopped the flow, bottom-hole pressure rose to 9,135 psi in 15 hours. This corresponds with a minimum pressure gradient of about 2.07 times hydrostatic.

Thus it had proved possible to drill safely through the various trouble zones and even into the really high-pressured "Platten" dolomite, applying a mud weight of not more than $18\frac{1}{2}$ lbs/gl. This was apparently sufficient in this case to control the reservoir pressure and not yet high enough to cause squeeze type circulation losses. It should be noted, however, that at this location the reservoir produced mainly water. Had it been gas, as was presumably the case in the first well, a higher safety margin would probably have been required, thus entailing a danger of circulation losses and a consequent danger of blow-outs.

Meanwhile the test result as such, viz., a relatively small salt-water flow with only traces of gas, was disappointing and, the well being in such a good state, it was decided to explore deeper, penetrating another long rock salt series of great, but locally not quite known thickness, and trying to reach a second dolomite formation ("Haupt" dolomite) below this. During this deepening, the high mud weight had to be maintained because the high-pressure "Platten" dolomite could not be cased off, seeing that sufficient hole size had to be maintained for eventually cementing a 5-in. or $5\frac{1}{2}$-in. string above the "Haupt" (Main) dolomite.

All went well except for some salt water inflow mainly during round-trips. As it was only water, further weighting of the mud was not considered advisable in view of squeeze danger.

The salt section was found to extend to a depth of no less than 14,750 ft., followed by a few meters of basal anhydrite and thereafter a mainly bituminous shale facies with intercalations of limestone and dolomite. Although in the cores some gas traces were observed, as were local indications of fissuring in the dolomitic stringers, this development of the objective formation was not considered worth the high additional expense of casing and testing. It was decided to plug and abandon hole (final depth, 14,863 ft.).

Abandoning operations proved rather complicated and time-consuming. It was found that with a mud weight varying between 18.1 and 18.8 lbs/gl there was a slight but continuous mud flow while circulation was stopped. Once, during circulation with an 18.6-lbs/gl mud the well even threatened to blow out. This seems to indicate that the lower dolomite, although badly developed and possibly of low permeability, carried fluid under a pressure equivalent to at least 2.26 times hydrostatic, i.e., very close to petrostatic (particularly when realizing that in view of the large salt masses in the column, the local petrostatic gradient must have been on the low side). Attempts to control the well with heavier mud failed, since with a weight of 19.3-lbs/gl mud losses began.

The section explored by these three wells provides an excellent example of the effect of salt accumulations on subsurface pressure conditions, beginning in the upper salt measures of the Zechstein and getting progressively worse with increasing depth.

The first signs of abnormal conditions were encountered in salt-enclosed shales (C-1), a couple of hundred feet below the top of the Zechstein salt. They were easily controlled (C-3) by application of an 18½-lbs/gl mud.

In the "Platten" dolomite zone (more than 10,000 ft. deep) the reservoir pressure appeared to have reached a dangerously high level, certainly as high as 2.06 times hydrostatic, and possibly more. In C-1 (gas) it could not be properly controlled, but in C-3 (water) satisfactory control proved possible by using an 18.4-lbs/gl mud. This latter proves that the reservoir pressure can not have been over 2.21 times hydrostatic.

The deepest reservoir level reached, viz., the "Haupt" dolomite at about 15,000 ft. under an enormous additional salt section, was found to contain a pressure of at least 2.26 times hydrostatic. Against an 18.8-lbs/gl mud the well (C-3) flowed, whereas with a mud of 19 lbs/gl and more, squeeze losses started.

B. YOUNGER TERTIARY OF NETHERLANDS NEW GUINEA

The present case refers to the attempted drilling of two exploration wells in the delta area of the Mamberamo River on the north coast of New Guinea. The objects of this campagni were possible Tertiary reservoir formations expected to be encountered below 8,000–10,000 ft. The formations drilled down to about 3,600 ft. consisted of clay alternating with sands and a few coal streaks; down to 6,060 ft., mainly of sticky clay with a few sand intercalations. The sticky nature of these young, unconsolidated clays gave rise to much trouble from bit-balling and swabbing effect, thus seriously aggravating the difficulties experienced after reaching the excessively pressured reservoir formations at a greater depth.

In *the first well, G-1*, some trouble with gas entering the mud occurred from 4,245 ft. down, but was controlled by increasing the mud weight from 10.4 lbs/gl to 12.1 lbs/gl, i.e., already above normal.

Between 5,985 and 6,065 ft. sticky clay trouble occurred to such an extent that a side-track had to be made. In the new hole gas trouble was much more severe, supposedly because deeper gas from the first hole was loading up some shallower sands. An increased mud density of 15.8 lbs/gl proved effective in controlling the gas, but the general condition of the hole proved such that it had to be abandoned.

For *the second well, G-2*, all possible precautions were taken. Subsequent drilling proceeded uneventfully down to a depth of 4,774 ft., beginning with a mud density of 10.3 lbs/gl and loading in steps to 12.5 lbs/gl. From this depth down to 5,738 ft., however, the mud repeatedly became gas-cut. Although the gas was believed to originate from the cuttings only, mud density was increased to 13.3 lbs/gl as a safeguard against swabbing-in fresh gas from the formations during round-trips.

At a depth of 5,738 ft., after 8 hours of observation prior to electric logging, the well began to flow as a result of gas and salt water entering the hole. The mud had to be loaded to 16.3 lbs/gl in order to control this inflow.

While drilling deeper with mud weight 16.7 lbs/gl, a sudden increase of penetration rate occurred at 5,995 ft. At the same time the mud became seriously gas-cut and the well again began to flow. Thereupon the tools were pulled to 5,738 ft. and an attempt was made to weight the mud further while circulating under back-pressure. As a result of the back-pressure nozzles being clogged by cuttings, the bottom pressure became excessive with mud losses as an unavoidable result. As tight hole conditions also manifested themselves, it was decided to pull the bit into the shoe of the 13⅜-in. string (at 2,977 ft.). Thereupon the well remained quiet while circulating an 18.3-lbs/gl mud. When a depth of 4,865 ft. was reached again, the well once more attempted to flow. Drilling was continued, however, with the constant addition of barytes, but only when a mud density of 19.1 lbs/gl had been reached, was the inflow of gas definitely stopped.

Evidently drilling with such a heavy mud, particularly in combination with frequent balling-up of the bit, swabbing effect, etc., entailed an ever-present danger of circulation loss. Soon, while circulating the mud with bit at 3,935 ft. the well again began to flow, mud weight then being checked to be 19.3 lbs/gl. After closing the BOP, within one hour a pressure of 216 psi developed in both drill pipe and annulus. Half an hour later, however, formation fracturing occurred.

The well was eventually plugged and abandoned.

The fact that this well threatened to blow out against the weight of a mud column of 19.3 lbs/gl would seem to indicate a reservoir pressure at total depth very close to petrostatic value. However, this observation was made while circulating through a bit hanging at 3,935 ft. It may well have been that the mud in the lower part of the hole was badly gas-cut and partly replaced by inflowing salt water. The average weight of the fluid column in the hole may have been as low as 17.2 lbs/gl, i.e., still over twice the hydrostatic value. The reservoir pressure could not have been much less than that, however, seeing that on a previous occasion the well threatened to blow out while circulating with a 16.7-lbs/gl mud at bottom.

The following conclusions may be drawn from this New Guinea experience:

a. Reservoir pressures from about 4,000 ft. down in the wells were abnormal and at 5,997 ft. in well G-2 may be estimated to have been between 2.32 and 2.06 times hydrostatic value. Taking into account the weakly consolidated condition of overlying rocks, even the latter value was probably at least 0.9 times petrostatic value.

b. The occurrence of such high pressures at relatively shallow depth should be ascribed to the thick cover and young age of overlying clayey sediments.

C. TERTIARY IN NORTH PART OF WEST PAKISTAN

Severe drilling difficulties were also encountered in relatively shallow formations in the north part of West Pakistan near the western foothills of the Great Himalayan mountain ranges. The objectives were Eocene limestones at an anticipated depth of about 8,000 ft., but these were never reached owing to persistent inflow of extremely pressured water from thin sandstone streaks embedded in the younger Tertiary shales.

In *the first hole, P-1,* the first water inflow occurred at 885 ft. but was controlled by raising the mud density to 13.8 lbs/gl. Several more high-pressure water layers followed, requiring still heavier mud until at a depth of no more than 1,340 ft., the mud weight had already reached a level of 18.2 lbs/gl. Meanwhile, and apparently as a result of this heavy fluid weight, severe mud losses were experienced from 1,180 ft. down.

A further inflow at 1,836 ft. required a mud weight of 18.4 lbs/gl to control it and the next at 1,902 ft. one of 18.7 lbs/gl. When water again broke in at 2,062 ft. the well was shut-in with the latter mud loading the hole, whereupon the surface pressure rose to 315 psi. This would suggest a reservoir pressure of 2,430 psi at 2,062 ft., i.e., 2.5 times hydrostatic.

At this stage, water broke to the surface at six different spots in the vicinity and, when further technical complications added to the difficulties, the well had to be abandoned. The freshly formed water sources in the vicinity are still flowing. The flow appeared to follow a preexisting fault plane which, surprisingly enough, had remained well sealed against the available reservoir pressures (possibly up to petrostatic) and only was opened when pressure in the well exceeded this level.

In *the second well, P-2,* the first inflow of water occurred at a depth of 1,138 ft. against a mud column of 18.8 lbs/gl weight. To control this flow, mud had to be weighted further to no less than 19.7 lbs/gl. Reservoir pressures in this well seemed to be even higher than those in the first.

While drilling at 1,934 ft. severe mud losses set in at a mud weight of 19.5 lbs/gl, but these were stopped by pumping in a plug of highly viscous mud.

An 11¾-in. intermediate string was squeeze-cemented at 1,967 ft. according to plan. While drilling out the cement in the hole, a strong inflow of water occurred, but it could not be ascertained whether this flow originated from behind the casing or from below. With 19.3 lbs/gl, mud fluid was lost while circulating, but it returned as soon as circulation was stopped. After squeeze-cementing the casing once again, it was possible to resume drilling with a 19.2-lbs/gl mud.

Drilling was continued to 2,111 ft. when suddenly the mud level fell out of sight. After some manipulation and observation, it then appeared possible to deepen the hole with a mud of 14.5 lbs/gl without further inflow of water. This seemed to indicate that reservoir pressures were not so extremely high after all, unless one accepted that they were lower below the casing depth than higher up—which seems rather improbable.[4] A better explanation might be that the

[4] According to information subsequently received from Burmah Oil Company the Eocene limestones in West Pakistan (which were also the objective level of

formation below the casing shoe did not contain any reservoir sands and that subsequent inflow of water originated from around this shoe.

Subsequently repeated water flows were experienced, e.g., at 2,570 ft. against 13.8-lbs/gl mud, at 2,575 ft. against 14.4-lbs/gl mud, and at 2,610 ft. against 14.9-lbs/gl mud.

In order to control the last inflow, the mud had to be further weighted to 16.0 lbs/gl, whereupon, however, the mud level dropped out of sight as a result of apparent formation fracturing. The hole was further deepened with repeated inflows, fluid losses and squeeze cementations until at a depth of 3,450 ft. a sandstone was penetrated with strong water flow against an 18.5-lbs/gl mud. Attempts to kill the well by applying mud weights up to 19.3 lbs/gl were unsuccessful owing to heavy circulation losses. Squeeze cementations no longer improved the situation and when heavy caving added to the trouble it was decided to abandon the well.

In conclusion it is apparent that extremely high reservoir pressures occur in this area even at shallow depth. They might possibly be explained by assuming large tectonic stresses in this sub-Himalayan region. The covering clayey sediments under these conditions may have proved insufficiently permeable to allow the escape of water at a rate sufficient to maintain hydrostatic balance in the pores (see also Keep and Ward, 1934, and for similar experiences also Tainsh, 1950). A typical complication of these shallow conditions was that in well No. 1 the squeeze effect of overweighted mud established a connection between the high-pressure (water) sands and the surface.

V. Drilling Problems and Suggestions for Handling Them

The afore-cited experiences are by no means the only instances on record. Similar, though possibly somewhat less severe, cases have been encountered, e.g., in North Borneo, Trinidad, Iran, and France (Berger, 1955).

Reservoir pressures close to petrostatic are theoretically possible and actually seem to occur under appropriate geological conditions. In our normal drilling technique they require the use of mud fluid exerting a pressure on the borehole wall that is close to or above the value of overburden pressure, and this must inevitably result in persistent fluid loss with all its pernicious consequences.

This then constitutes a very serious problem. Actual occurrences might be conveniently subdivided as follows.

1. Reservoir fluid pressures are high but not higher than 80–90% of overburden pressure. At the same time tectonic conditions are such that the direction of least rock stress is vertical, so that any rock rupture will be horizontal and can not occur unless the mud weight exceeds the value of the overburden pressure. Under these conditions the hole can probably be drilled and completed by using a mud flush with a weight carefully adjusted between the values of expected reservoir fluid pressure and overburden pressure.

2. Reservoir pressure in the projected zone of production is very close or even equal to overburden pressure, but pressures higher up in the hole, though also abnormal, are lower. Most probably they will gradually increase with depth. Under these conditions the well may be drilled just as in the previous case down to a safe point in the caprock of the projected reservoir, where a casing string will have to be cemented (cf. case of well C-3 as described in IV. A). The situation will become critical only when the reservoir is subsequently drilled in. It may be necessary at this juncture to close the BOP around the drill stem and keep the latter in the hole for use as a production string until such time as the closed-in bottom-hole pressure drops sufficiently as a result of drainage to allow killing the well and normal completion with liner and tubing.

3. Really serious cases are those where the pressure, not only in the projected zone of production, but also in one or more zones above this, is very close to petrostatic. It might be visualized as a super-pressured oil reservoir covered by a few thousand feet of shale which, in its lower 200 or 300 feet, is interbedded with sandy streaks yielding water or gas at a pressure equally as high as that in the oil sand below. It appears that no general solution can at present be proposed for these cases.

the wells discussed) are generally encountered at normal hydrostatic pressure in spite of petrostatic pressures passed higher in the hole. This is a very remarkable fact, the cause of which would certainly appear to deserve further study.

LITERATURE CITED

Athy, L. F., 1930, Density, porosity, and compaction of sedimentary rocks: Am. Assoc. Petroleum Geologists Bull., v. 14, no. 1.

Berger, Y., 1955, Lacq profond; gisement de gaz a pression anormalement élevee; problèmes de forage: Revue de l'Institut Français du Pétrole, Nov.

Bugbee, J. M., 1953, Lost circulation, a major problem in exploration and development: A.P.I. drilling and prod. practice.

Cannon, G. E., and Craze, R. C., 1938, Excessive pressures and pressure variations with depth of petroleum reservoirs in the Gulf Coast region of Texas and Louisiana: A.I.M.E. petroleum dev. and technology.

Dickinson, G., 1951, Geological aspect of abnormal reservoir pressures in the Gulf Coast region of Louisiana, U.S.A.: 3d World Petroleum Congress Proc., pt. 1, The Hague.

Hedberg, H. D., 1936, Gravitational compaction of clays and shales: Am. Jour. Science, April.

Howard, G. C., and Fast, C. R., 1950, Squeeze cementing operations: A.I.M.E. Tech. Pub. 2795—Trans. A.I.M.E.

Hubbert, M. K., and Willis, D. G., 1957, Mechanics of hydraulic fracturing: A.I.M.E. Tech. Pub. 4597—Jour. Petroleum Technol., June.

Keep, C. L., and Ward, H. L., 1934, Drilling against high rock pressures with particular reference to operations conducted in the Khaur field, Punjab: Jour. Inst. Petroleum.

Kok, P. C., and Thomeer, J. H. M. A., 1955, Abnormal pressures in oil and gas reservoirs: Geologie en Mijnbouw, Aug.

Scott, P. P., Bearden, W. G., and Howard, G. C., 1953, Rock rupture as affected by fluid properties: A.I.M.E. Tech. Pub. 3540—Trans. A.I.M.E.

Tainsh, H. R., 1950, Tertiary geology and principal oil fields of Burmah: Am. Assoc. Petroleum Geologists Bull., v. 34, no. 5, p. 840, 852.

Thomeer, J. H. M. A., 1953, Problems arising from shale unstability in oil well drilling: Geologie en Mijnbouw, May.

——— 1955, The unstable behavior of shale formations in bore holes and its control by properly adjusted mud-flush quality: 4th World Petroleum Congress Proc., Rome.

Von Terzaghi, K., and Froehlich, O. K., 1936, Theorie der Setzung von Tonschichten, ed. Deuticke, Leipzig and Wien.

Reprinted from:
BULLETIN OF THE AMERICAN ASSOCIATION OF PETROLEUM GEOLOGISTS
VOL. 49, NO. 9 (SEPTEMBER, 1965), PP. 1502–1511, 8 FIGS.

QUANTITATIVE DETERMINATION OF RESERVOIR PRESSURES FROM CONDUCTIVITY LOG[1]

J. R. MACGREGOR[2]
New Orleans, Louisiana

ABSTRACT

Use of the conductivity curve of the induction-electrical log to define the depth at which a bore hole enters an abnormally pressured section, as outlined by Wallace (1964), prompted the writer to attempt-to develop a method for making quantitative pressure determinations from the conductivity curve.

A method of plotting conductivity readings of clean shales vs. depth was devised. This plot shows that a sloping straight line can be drawn through the points in the known normally pressured section, whereas the points in the abnormally pressured zone deviate toward higher conductivity. Causes for the straight line and the deviating portions of the plot were analyzed and it was concluded that the magnitude of the conductivity deviation is primarily a function of the pressure abnormality.

Plotting of pressure gradients (calculated from measured reservoir pressure and depths) vs. conductivity deviation in 41 abnormally pressured reservoirs in coastal and offshore Louisiana resulted in a basic curve which can be used to determine approximate pressures in future wells where electric logs can be made available but actual pressure measurements are lacking.

An example of the use of this curve is given for the case where a well passes out of an abnormally pressured section into a normally pressured zone. The lower pressures in the normally pressured zone are masked by the high mud weights required to hold the upper, higher pressures but can be detected by the proposed method.

INTRODUCTION

A need has long been recognized for some method of anticipating and evaluating abnormal pressures in southern and offshore Louisiana wells. One approach to a qualitative determination of abnormal pressure was suggested recently by Wallace (1964), using the conductivity curve of the induction-electrical log. The writer thought that if a quantitative pressure determination could be made from the conductivity curve, it would be of great aid in designing proper protection casing, oil strings, well-head equipment, and in determining the proper packer fluid weight. A method of obtaining a good estimate of reservoir pressure was developed. Continued studies of the factors involved should lead to refinement of the method and reduction of the uncertainty in the estimated pressure.

METHOD

Several plots were made of the conductivity readings in clean shales against depth of hole. The type of graph paper which was found most suitable was semi-logarithmic, with conductivity in millimhos as the abscissa on the logarithmic scale and depth as the ordinate on the linear scale.

Twenty-six wells were plotted in this manner (Fig. 1). These wells had a combined total of 41 abnormally high-pressured reservoirs measured by a pressure bomb. The depths where abnormal pressures were first encountered ranged from 6,500 to 13,000 feet. Four of the wells were in coastal Louisiana fields and 22 were offshore. Their locations are shown on Figure 2. Geologic ages of the plotted parts of the well bores range from Pleistocene to the lower Frio part of the Oligocene.

Several prominent characteristics are noted on a typical plot (Fig. 3). First, the conductivity of the massive sandstone section above the known abnormally high-pressured zone can be represented with reasonable accuracy by a straight line through the scattering of points. This line slopes toward lower conductivity with greater depth and approximately parallels a similar plot of sonic travel time. Second, near the base of the massive sandstone section, both the conductivity and sonic curves deviate sharply toward the right (increasing conductivity and travel time). This deviation coincides with the point at which drilling problems such as gas cutting, heaving shale, and salt-water influx are experienced.

Therefore, it appeared likely that the amount of deviation of conductivity might be related to the degree of pressure abnormality. In developing such a relationship, it is necessary to analyze various factors which should affect the behavior of the conductivity log. The first factor considered was geologic age of the logged sediments.

[1] Manuscript received, February 15, 1965.

[2] Production manager, Ocean Drilling and Exploration Company.

COMPANY	WELL NAME	FIELD	GRAPH SYMBOL	GEOLOGIC AGE OF PLOTTED SECTION
Ocean Drilling & Exploration Co.	OCS 1040 #1	Ship Shoal - Block 275	1	Pleistocene
Kerr Mc Gee Oil Industries	OCS 1025 #1	Ship Shoal - Block 239	2	Pleistocene-Upper Pliocene
Forst Oil Corpo ration	OCS 0900 #1	West Cameron - Block 225	3	Upper Miocene
Union Producing Co.	OCS 0765 #1	West Cameron - Block 215	4	Upper Miocene
Ocean Drilling & Exploration Co.	S.L. 3842 #1	Vermilion - Block 4	5	Middle-Lower Miocene
Ocean Drilling & Exploration Co.	S.L. 3624 #2	Vermilion - Block 17	6	Middle-Lower Miocene
Ocean Drilling & Exploration Co.	S.L. 3762 #1	Vermilion - Block 5	7	Middle-Lower Miocene
Ocean Drilling & Exploration Co.	S.L. 3763 #1	Vermilion - Block 16	8	Middle-Lower Miocene
Ocean Drilling & Exploration Co.	S.L. 3763 #2	Vermilion - Block 16	9	Middle-Lower Miocene
Ocean Drilling & Exploration Co.	S.L. 3763 #3	Vermilion - Block 16	10	Middle-Lower Miocene
Ocean Drilling & Exploration Co.	S.L. 3763 #4	Vermilion - Block 16	11	Middle-Lower Miocene
Ocean Drilling & Exploration Co.	S.L. 3843 #2	Vermilion - Block 16	12	Middle-Lower Miocene
Ocean Drilling & Exploration Co.	OCS 0474 #4	Eugene Island - Block 129-A	13	Upper Miocene
Ocean Drilling & Exploration Co.	OCS 0474 #10	Eugene Island - Block 129-A	14	Upper Miocene
Ocean Drilling & Exploration Co.	S.L. 770 #1	Ship Shoal- Block 93	15	Lower Pliocene-Upper Miocene
Ocean Drilling & Exploration Co.	S.L. 778 #1	Ship Shoal - Block 114	16	Lower Pliocene
Ocean Drilling & Exploration Co.	S.L. 778 #2	Ship Shoal - Block 114	17	Lower Pliocene
Ocean Drilling & Exploration Co.	S.L. 778 #3	Ship Shoal - Block 114	18	Lower Pliocene
Ocean Drilling & Exploration Co.	S.L. 778 #5	Ship Shoal - Block 114	19	Lower Pliocene
Ocean Drilling & Exploration Co.	S.L. 778 #7	Ship Shoal - Block 114	20	Lower Pliocene
Ocean Drilling & Exploration Co.	OCS 0585 #1	Ship Shoal - Block 138	21	Pliocene
Ocean Drilling & Exploration Co.	OCS 0602 #1	South Timbalier - Block 76	22	Pliocene
Exchange Oil & Gas Co.	S.L. 3907 #1	East Lake Sand	23	Lower Miocene
Forest Oil Corporation	Kerr #4	Ellis Field	24	Lower Miocene-Lower Frio
Forest Oil Corporation	Meaux #2	West Ridge Field	25	Lower Miocene-Upper Frio
Sohio Oil Company	ShirleyUnit 1 #1	South Jennings Field	26	Lower Miocene-Lower Frio

FIG. 1.—Tabulation of wells studied for depth-pressure-conductivity relation.

EFFECT OF GEOLOGIC AGE ON CONDUCTIVITY

The normal section straight lines of the 26 wells were plotted on one graph to see the comparison between holes that penetrated sections of different geologic ages (Fig. 4). The slope of all lines is similar, ranging between 2.9° and 6.5° from vertical. With two exceptions, the conductivity at 7,000 feet ranges from 1,100 to 1,800 millimhos; however, no obvious correlation was seen between conductivity variation and geologic age. The two exceptions, however, have values of only 760–800 mmhos at 7,000 feet. Both these wells are located far out in the Gulf and are the only wells studied whose bore holes did not go below the Pleistocene. The question thus arises: why do these two wells have conductivity readings in the normally pressured section which are so much lower than the other 24 wells?

ANALYSIS OF STRAIGHT LINE PORTION OF PLOTS

The probable reason for the net trend toward lower conductivity with greater depth may be explained in part by a theory proposed by Dickinson (1953). As overburden is placed on shale, compaction occurs, causing a porosity reduction with resultant extrusion of saline connate water. In the massive sandstone section, this water is able to escape into a nearby sandstone, leaving the shale with a lower water content than before compaction. Because there is more compaction with greater depth, there is less water with depth; consequently, lower conductivity.

Mitigating this predominant action are at least two other factors which tend to increase conductivity with greater depth. Earth temperature rises with depth drilled and the conductivity of a brine increases as temperature is raised. The second factor, salinity of the connate water of the shale, is less consistent than temperature in that some areas have unusual variations in salinity with depth. However, as a general rule, salinity in wells drilled in south Louisiana and the Gulf of Mexico does increase to a maximum with depth.

In summary, two factors tend to increase conductivity with depth, namely, temperature and salinity. Overriding these factors is the lower water content of the deeper shales, giving the net decrease in conductivity with depth but at a decreasing *rate* of reduction.

Referring now to the question posed by Figure 4: low salinity probably is the factor causing the low conductivity readings in the shales throughout the massive sandstone section of the two wells drilled into the Pleistocene. Shale cores analyzed from bore holes in the vicinity of these two wells show a salinity very nearly that of Gulf water, 20,000–23,000 ppm chlorides. This chloride content is much lower than that found in the massive sandstone section of most other wells in the area studied. Therefore, because the thickness of

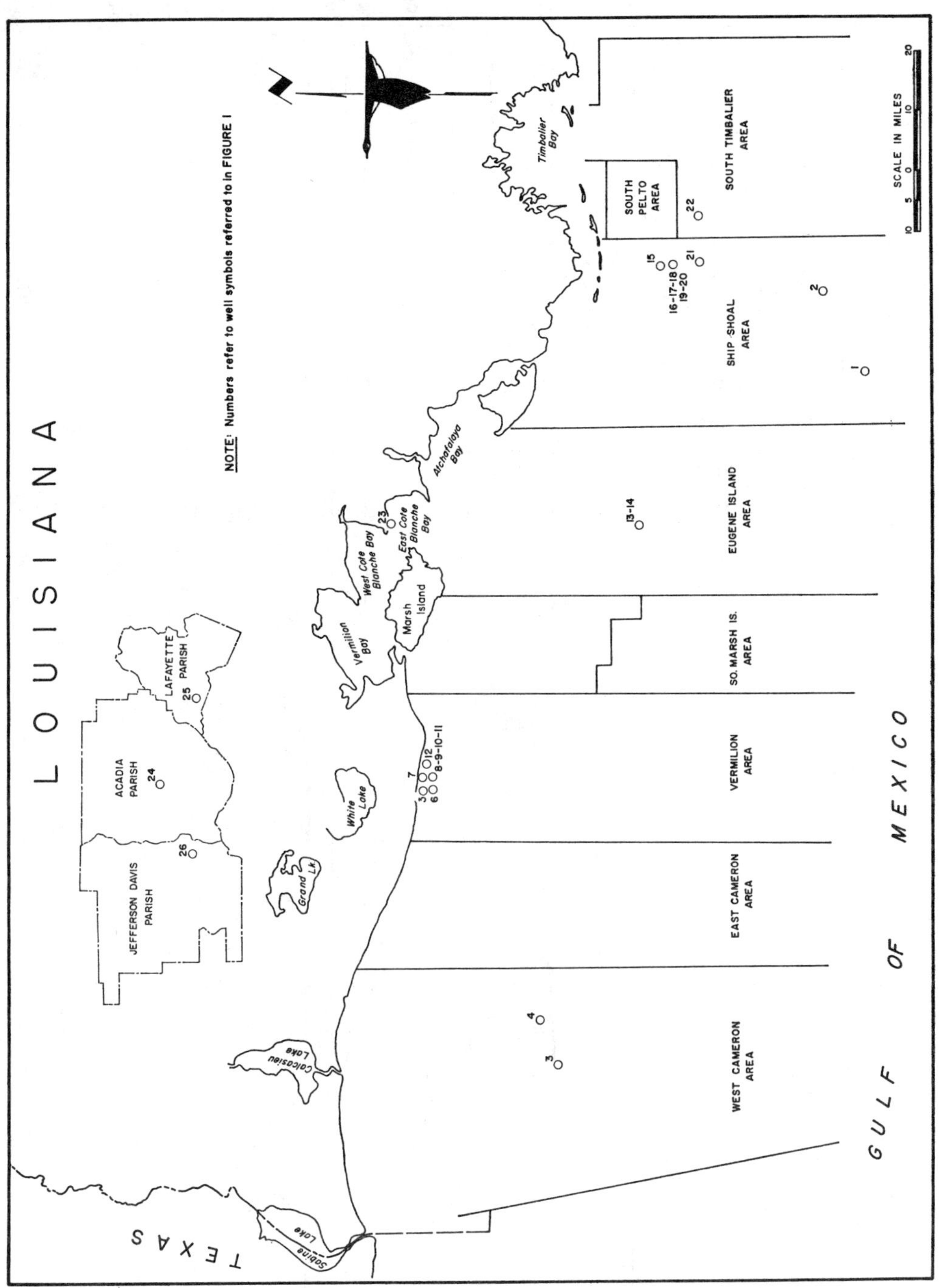

FIG. 2.—Geographical location of wells studied.

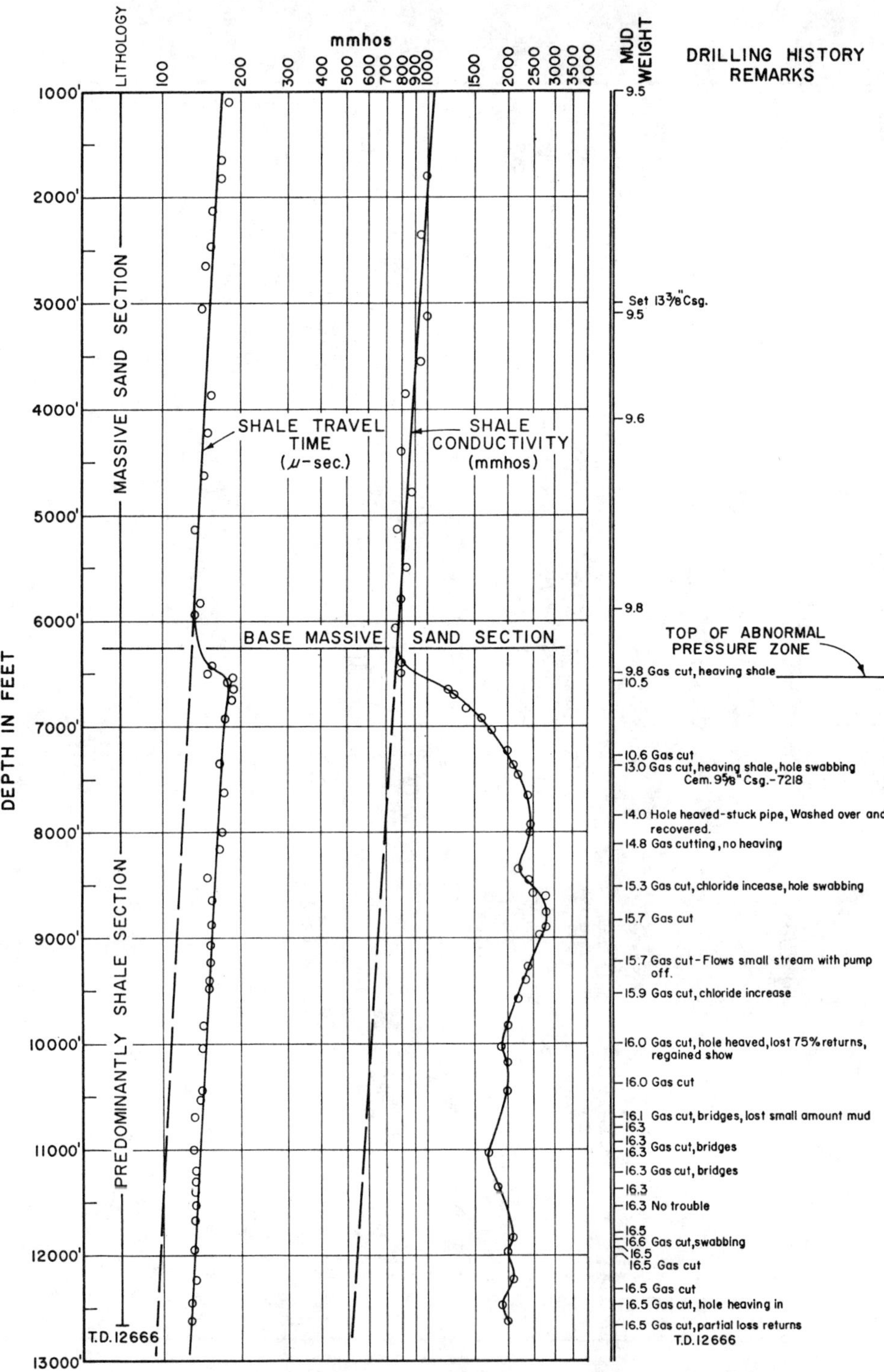

FIG. 3.—Effect of abnormal pressure on sonic and conductivity logs. ODECO, O.C.S.-1040 No. 1, Ship Shoal Block 275.

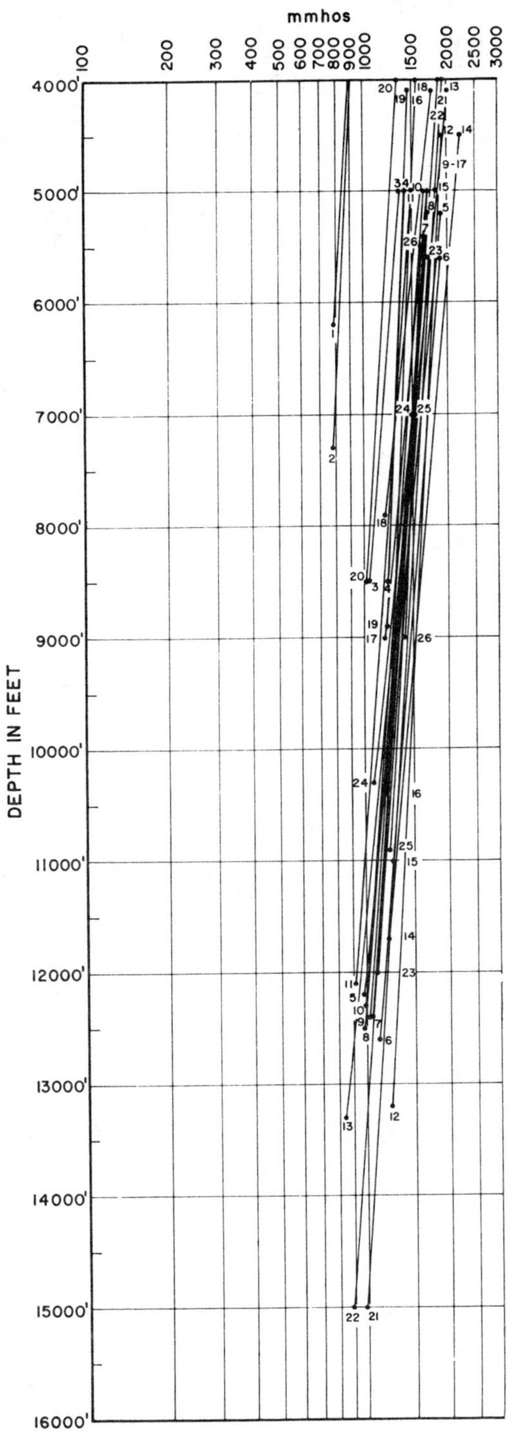

FIG. 4.—Twenty-six wells in 5 offshore areas and 4 onshore fields from West Cameron to South Timbalier areas. Numbers refer to wells shown in Figure 1.

overburden and the gradient of temperature are the same in these wells as in the other 24 wells, it would leave the low salinity of the entrapped brine in the shale as the probable cause of the much lower conductivities.

ANALYSIS OF DEVIATING PORTION OF PLOT

The portion of the conductivity plot which deviated from the "normally pressured" line (or its extrapolation), in all plots examined, was found to lie in abnormally high-pressured sections. By using the reasoning of earlier writers, it appears logical to assume that, as the connate water in the shale can not escape easily as a result of overburden compaction because it is in a predominantly shaly section, the per cent of salt water in a unit volume of shale is greater than in the normally pressured massive sandstone section. This higher salt water content, accordingly, gives higher conductivity. It seemed likely therefore that the amount of deviation might be a measure of the pressure abnormality. Utilizing this thought, measured reservoir pressures were plotted against depths on all graphs, shown in Figures 5 and 6. At these depths, the actual conductivity less the conductivity on the extrapolated part of the "normally pressured" line was noted. This conductivity difference was defined as conductivity abnormality and was represented by the symbol $\triangle c$. The pressure gradient was calculated by dividing the measured reservoir pressure by the depth.

PLOTTING SOLUTION TO PROBLEM

All of the available pressure gradient and $\triangle c$ readings were then plotted on regular coordinate graph paper with gradient as the ordinate and $\triangle c$ as the abscissa. This plot is represented by Figure 7 and is the key to the writer's quantitative approach to determination of degree of pressure abnormality.

It may be noted that, although a solid median line has been drawn through the scattering of points, an envelope is represented by the dotted lines. The variation in this composite plot is probably caused by the use of readings from a large spread of geologic ages and geographical locations.

SUMMARY OF STEPS TO DETERMINE PRESSURE

The method to be used for estimation of reser-

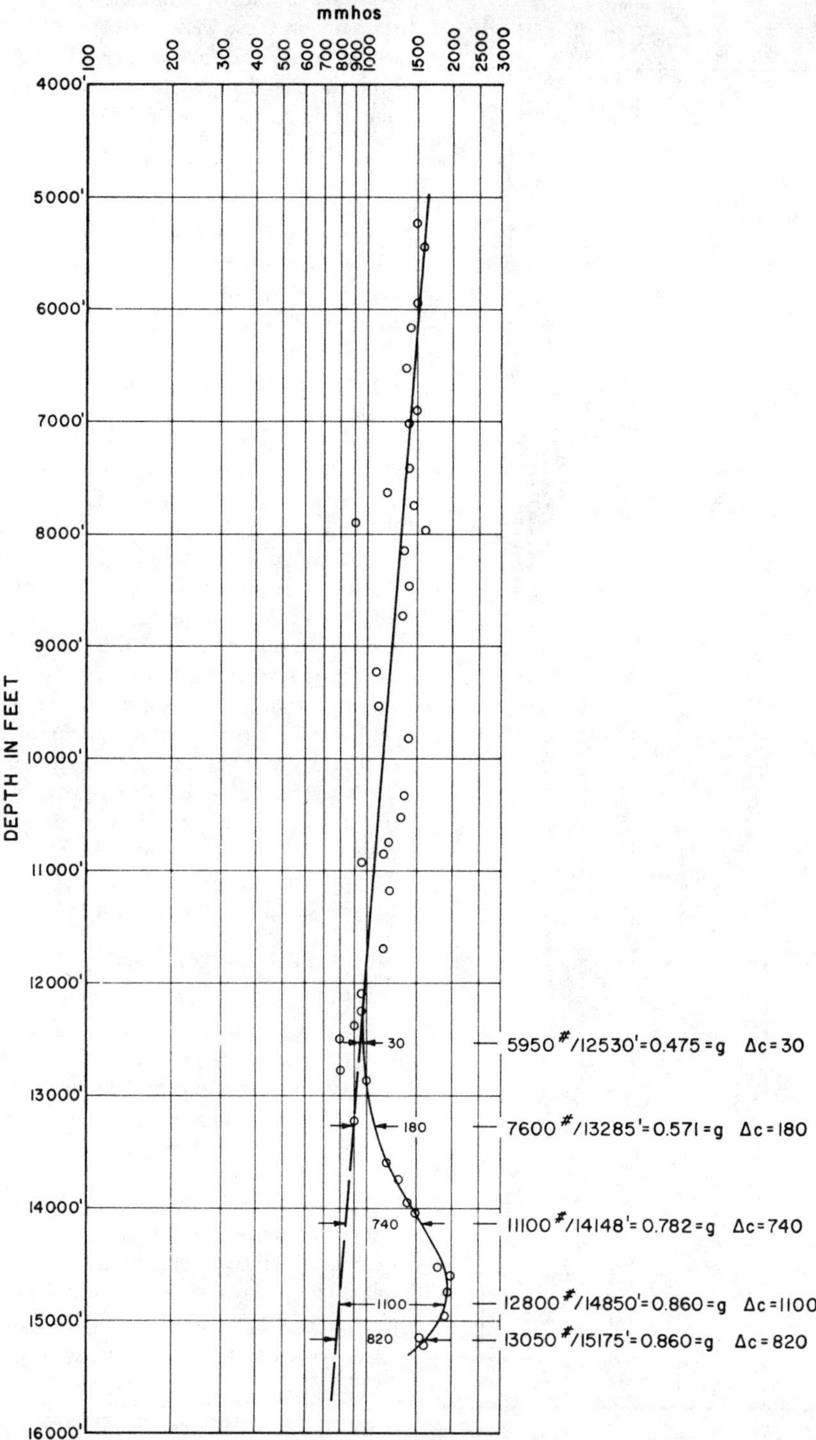

FIG. 5.—Conductivity, depth, pressure relation. ODECO, St. Lse.-3763 No. 3, Block 16, Vermilion area.

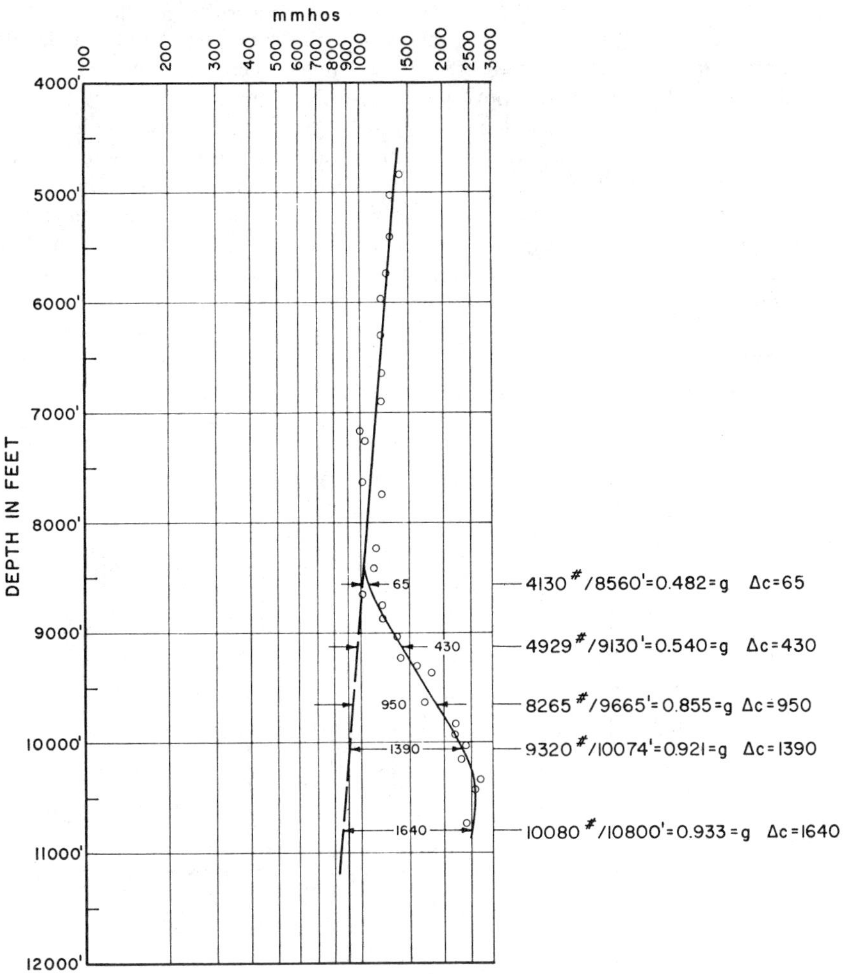

Fig. 6.—Conductivity, depth, pressure relation. Union Producing, O.C.S.-0765 No. 1, Block 215, West Cameron area.

voir pressure from the conductivity log of any well would be as follows.

1. On the induction-electrical log, start below the fresh or brackish-water near-surface sandstones and place a mark on the conductivity curve at the occurrence of the first clean shale. This should be a shale zone of preferably 10 feet or more in thickness with low, flat SP and resistivity readings. Similarly, mark the conductivity log at all clean shales from this point to total depth.

2. Read the conductivity carefully at each marked point and tabulate the results.

3. Make a plot of these conductivity readings vs. depth, on semi-logarithmic paper, with conductivity as the abscissa on the logarithmic scale and depth as the ordinate on the linear scale.

4. If no abnormal pressures are present, it should be possible to draw a straight line through the majority of points. However, if abnormal pressures exist, the conductivity readings in this section will move toward the right of the page. From the point of deviation, an extrapolation should be made of the straight line derived from values higher in the hole.

5. At the depth at which determination of the abnormal pressure is desired, subtract the conductivity of the extrapolated "normal pressure" line from the actual conductivity plotted.

6. Plot this amount on the abscissa of Figure 7, go vertically to the solid curved line, then horizontally to the ordinate, and read the pressure gradient.

7. Multiply this pressure gradient by the depth and get reservoir pressure.

Because this pressure is only approximate, the limits of error also should be noted before putting the results to practical use. This can be done by reading the vertical intercept of conductivity with the two dotted lines and observing the corresponding pressure gradient for each. Again, these gradients, when multiplied by the depth, will give the upper and lower limits of probable pressure at the depth in question.

Special Uses of Method

It is usually possible to obtain a fair idea of actual pressure, while drilling, from calculations using mud-kick data. Negative information also is available from mud-weight history where no kicks are observed with a given mud weight and trips with the drill pipe can be made safely.

However, there are times when this information can be misleading or insufficient to permit proper planning. Some examples are: (1) where high mud weight is necessary to drill through and hold back a sloughing shale section but is not required to hold the pressure of the sandstones encountered below; and (2) where mud weights are increased to control abnormal pressure, following which a re-entry is made into a normally pressured section. In both of these conditions, the actual pressure of the lower sandstones would be masked by the high mud weights.

An example of the latter condition is shown in Figure 8. In this well, an abnormally high-pressured section was entered by the bit at approxi

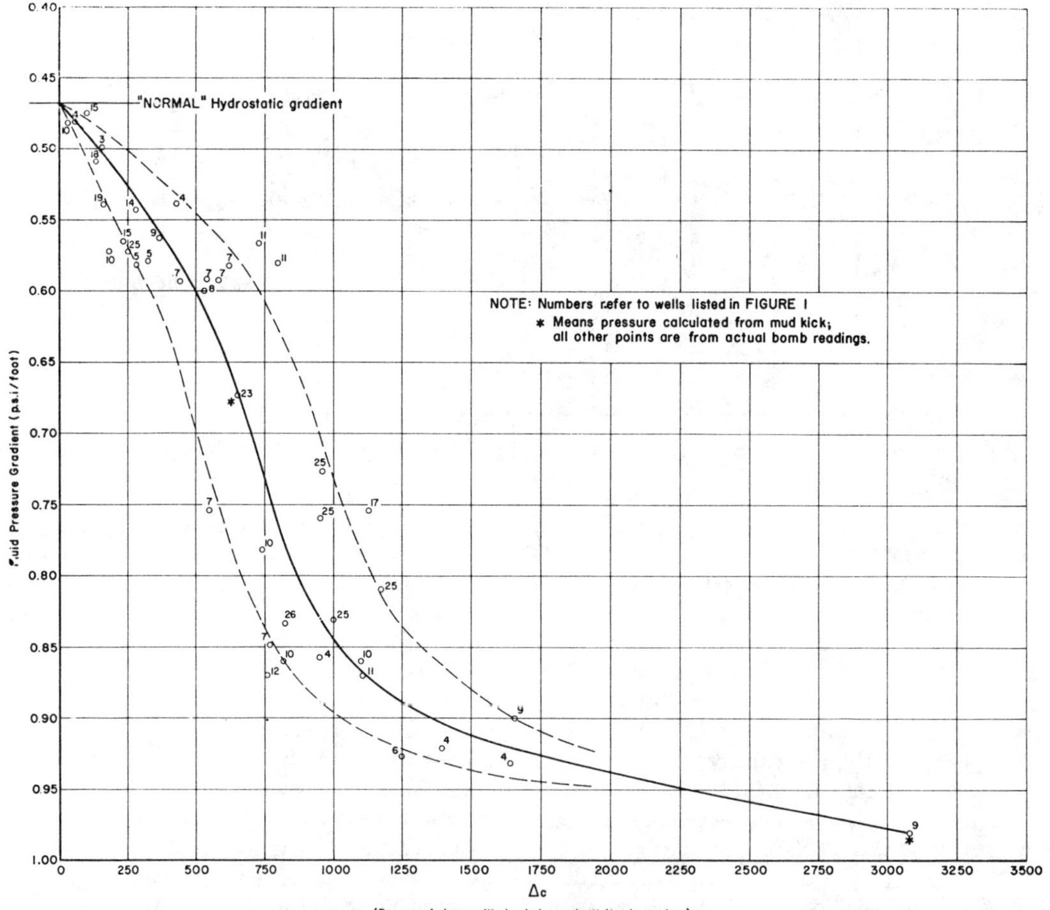

Fig. 7.—Empirical relation of fluid-pressure gradient to degree of abnormality of shale conductivity.

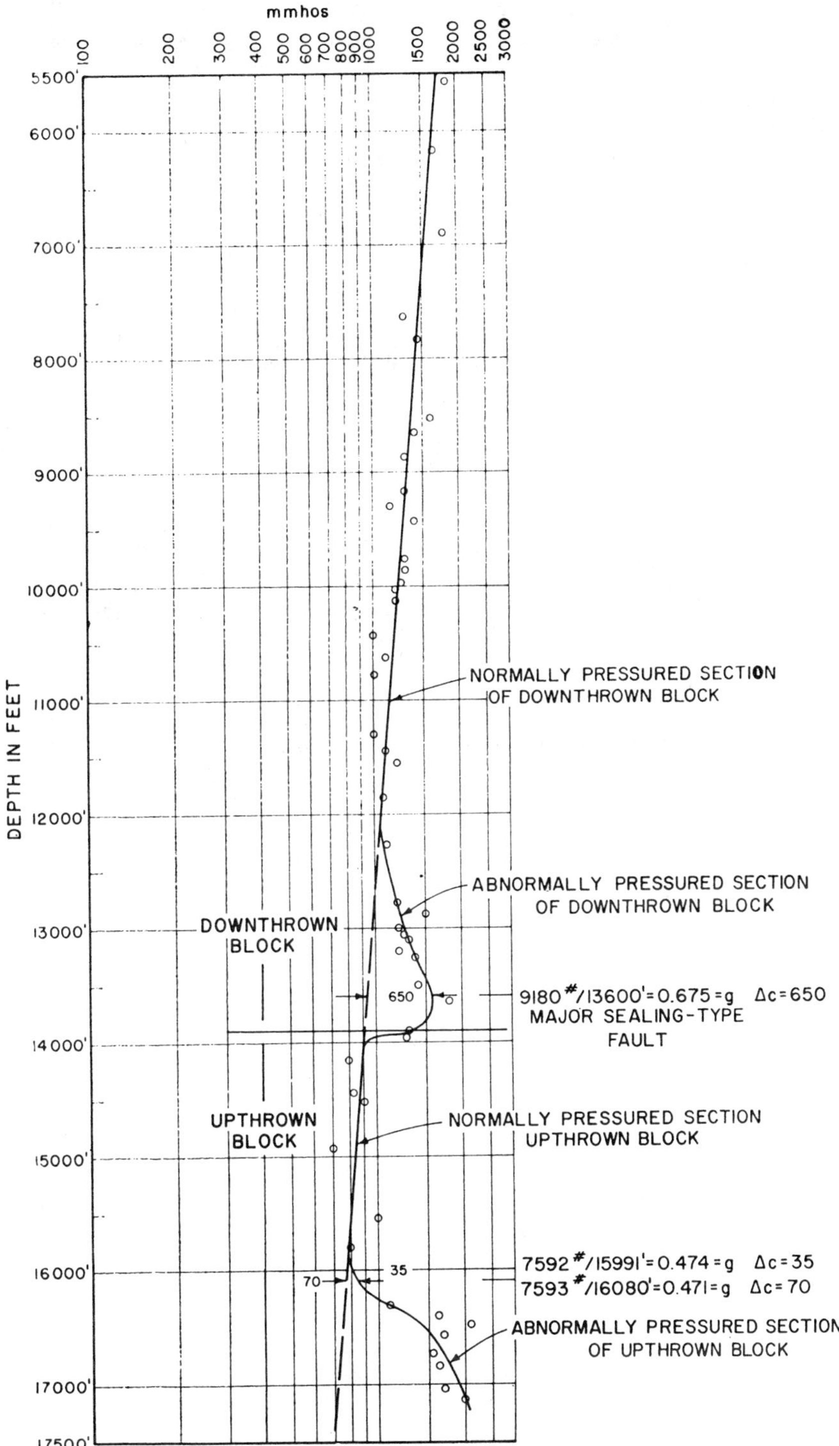

Fig. 8.—Example of repetition of normal pressure, abnormal pressure sequence in same well as result of faulting. EXCHANGE, St. Lse.-3907 No. 1, East Lake Sand field.

mately 12,100 feet. At 13,600 feet, the 12.5 pound/gallon mud being used allowed the well to kick and a shut-in surface pressure of 300 psi was observed. These figures indicate a reservoir pressure of 9,180 psi, or a pressure gradient of 0.673. It was then necessary to set protective casing. Drilling proceeded through about 200 feet more of abnormally high-pressured formation which required high mud weight until, at 13,900 feet, a major, sealing-type normal fault was crossed. The conductivity plot indicates that normally pressured formations were then penetrated for about 2,000 feet before entering another abnormally pressured section. This was confirmed by two drill-stem tests near 16,000 feet which recorded shut-in bottom-hole pressure very close to normal (gradient 0.47).

Conclusions

1. From the wells studied the writer could find little correlation between the attitude of the conductivity curves and geologic age. However, it is probable that examination of a much greater number of wells would permit plotting a gradient-$\triangle c$ curve for each age or for different geographical areas. This would probably result in narrowing the envelope of plotted points which would, in turn, result in more accurate pressure estimation.

2. Of the 26 logs studied, 6 of them were found to adapt poorly to this method of plotting. One reason for this may be that the normally pressured section between the base of the shallow, brackish-water sandstones and the beginning of pressure abnormality was short and extremely sandy, with very few clean shale readings available. This combination of few points and a short distance made drawing of the straight "normally pressured" line and its extrapolation difficult and inaccurate.

3. Particular care should be exercised in the use of Figure 7 when the observed $\triangle c$ is between 500 and 1,000 mmhos. This is true because the slope of the curve is steepest in this range, giving a large increase in pressure gradient for a small increase in $\triangle c$.

Selected Bibliography

Dickinson, George, 1953, Geological aspects of abnormal reservoir pressures in Gulf Coast Louisiana: Am. Assoc. Petroleum Geologists Bull., v. 37, no. 2, p. 410–432.

Hubbert, M. King, and Rubey, Wm. W., 1959, Role of fluid pressure in overthrust faulting: Geol. Soc. America Bull., v. 70, no. 2, p. 115–166.

Myers, Robert L., and Van Siclen, Dewitt C., 1964, Dynamic phenomena of sediment compaction in Matagorda County, Texas: Gulf Coast Assoc. Geol. Societies Trans., v. 14, p. 241–252.

Wallace, W. E., Jr., 1964, Will induction log yield pressure data?: Oil & Gas Jour., Sept. 14, 1964, p. 124–126.

Reprinted from:
BULLETIN OF THE AMERICAN ASSOCIATION OF PETROLEUM GEOLOGISTS
VOL. 51, NO. 7 (JULY, 1967), 1240–1254, 7 FIGS.

FLUID-RELEASE MECHANISMS IN COMPACTING MARINE MUDROCKS AND THEIR IMPORTANCE IN OIL EXPLORATION[1]

MAURICE C. POWERS[2]
Salisbury, North Carolina

ABSTRACT

The application of current knowledge of clay colloid chemistry and mineralogy to the question of how water escapes from muddy sediments suggests two conclusions. (1) The alteration of montmorillonite to illite takes place after deep burial and involves the transfer of large amounts of bound water from montmorillonite surfaces to interparticle areas where it is normal water. This water transfer has an important bearing on the porosity, permeability, abnormal fluid pressure, and initial release of hydrocarbons from mudrocks. (2) In deposits of primary illite, water is compacted out of clay soon after burial, before the formation of hydrocarbons comparable with those found in reservoirs.

These points are important in petroleum exploration in new areas because it appears that the development of a shale source rock requires the initial deposition of a montmorillonitic organic mud, and its subsequent alteration to illite after deep burial. Abnormally high fluid pressures may easily be caused by a volume increase associated with the desorption of the last few monomolecular layers of water from montmorillonite during its diagenesis to illite. This understanding of mineralogical characteristics makes it possible to make meaningful interpretations of data on the bulk properties of compacting mudrocks.

INTRODUCTION

There is no discussion in geologic literature concerning the way in which different clay minerals control the compaction of mudrocks in the deep subsurface. Yet one would probably agree that the role of clay minerals in sediments is the most important and the most complex aspect of sediment compaction. The fact has been known for a long time that overburden pressure alone is not an effective agent in de-watering deep mudrocks (Skempton, 1953). Van Olphen (1963b) showed from theoretical and experimental results that this is what would be expected in view of the present knowledge of colloid chemistry of clays. However, no one has made a concerted effort to define the roles of different types of clay minerals in the subsurface, or the effect of clay diagenesis on the compaction history of sediments. The questions are extremely important to the exploration geologist and petroleum engineer, and they are considered in that light.

A new fluid-release theory has resulted from a synthesis of recent gains in knowledge of clay minerals with data on bulk properties of muddy sediments. The theory offers a logical approach to an understanding of the compaction of sediments. Approaching the problem in this way presents an opportunity to visualize better the interrelationship of many aspects of sediment compaction and it allows one to arrive at a more meaningful interpretation of data found in the literature. Several papers have dealt with the migration of oil from source rock to trap. This paper is concerned only with the initial release of oil from its source rock.

The paper is divided into three parts, serving the following three-fold purpose:

1. A new fluid-release mechanism in mudrocks is presented in which an attempt is made to show that release of water from shale after deep burial is important to several aspects of petroleum geology.

2. The new theory affords explanations for the interrelations of the initial release of fluids from source rocks, abnormal fluid pressures in the subsurface, compaction history of sediments, and the bulk properties (porosity, permeability, and density) of sediments.

3. The theory proposes to set limitations which may be invoked in the assessment of new areas as potential hydrocarbon producers.

Several of the ideas advanced are new, and

[1] Read before the Association at New Orleans, April 29, 1965. Manuscript received, January 1, 1966; accepted, September 30, 1966.

[2] Professor of Geology, Catawba College. The writer is indebted to Roy L. Ingram, University of North Carolina, Donald R. Baker, Marathon Oil Company, and Fred A. F. Berry, University of California (Berkeley) for critically reading the manuscript and offering many valuable suggestions. Gratitude also is extended to Everett Brett for reading the paper before the Association at New Orleans at a time when the writer could not be present.

must be tested carefully by the researcher in the laboratory, and by the exploration geologist in the light of field data.

Four Clues for Development of New Theory

Clue No. 1: Alteration of clays after deep burial.—The alteration of expanding clays to illite and mixed-layer clays in marine sediments, and the associated geochemistry of the host rock, have been described by Powers (1959), who found that the alteration was closely determined by depth of burial. The depth dependency has been discussed and expanded by several investigators (Burst, 1959; Weaver, 1959, 1961) who have substantiated the broad application of the alteration concept. Several geologists and clay mineralogists have recognized the general disappearance of montmorillonite at depth, with a corresponding increase in illite and mixed-layer clays, and it is generally accepted that illite becomes the dominant clay mineral in deeply buried sediments (Boswell, 1961; Keller, 1963; Weaver, 1961; Yaalon, 1962). The general replacement of sodium by potassium in clays after burial, found by Nicholls and Loring (1960), substantiates further the equivalence-level concept advanced by Powers (1959) as an explanation for the general alteration of montmorillonite to illite in the subsurface. Finally, Keller (1963) thoroughly reviewed the subject of diagenesis and wrote (p. 145) that the evidence appears sound that illite forms from montmorillonite by diagenesis. Keller suggested four causes for the diagenesis: depth of burial, time, geothermal gradient, and activities of essential ions. Depth of burial, time, and geothermal gradient are interdependent factors. Depth of burial simply implies that diagenesis is affected by some property which changes with depth, such as temperature, pressure, chemical composition, or combinations of these properties. Time, in the order of a few tens of millions of years, can not be considered as an important factor in the diagenesis of montmorillonite to illite, according to work by Burst (1959) and Powers (1959). Powers (p. 319) found that the alteration of montmorillonite in the Texas Gulf Coast occurred within a small depth range which crossed the entire Tertiary section. It was pointed out that the alteration of montmorillonite to illite is not related to geologic age but in some places gives this appearance because in general the geologically older sediments have been buried deeper than younger ones. It may be that time, in the order of hundreds of millions of years, plays a role in clay diagenesis, but there is no clear-cut evidence for the importance of even this order of magnitude of time.

Burst (1959, p. 328) hinted at the importance of the geothermal gradient as a controlling factor in the alteration of montmorillonite to illite. The weight of evidence definitely points in that direction and suggests that the rate of alteration is temperature-dependent. In this case, the kinetics may be expressed by the Arrhenius equation for the rate constant which expresses the influence of temperature on the rate of reaction (perhaps proportional to e^{-E}/RT) where E is the energy of activation, T is temperature, and R is the gas constant. Considerable additional investigation is necessary, however, before one can depend on this type of relation.

The depth at which montmorillonite alters to illite almost certainly depends on several variables, particularly the geothermal gradient, as well as activities of essential ions, as suggested by Keller (1963, p. 145). Furthermore, montmorillonite may have been buried deeply, altered to illite, and later elevated to a position closer to the surface. The direction of diagenesis is reversed only in the weathering environment.

In summary, the first clue leading to the new theory is the fact that montmorillonite alters to illite and mixed-layer clays after burial. The diagenesis begins at a depth of about 6,000 feet in the Texas Gulf Coast and is fairly complete at a depth of 9,000–12,000 feet. Similar depths seem to apply for most areas studied (Burst, 1959; Powers, 1959; Weaver, 1961).

Clue No. 2: Relations of clay and water.—Clay particles in the natural marine environment settle to the bottom and begin to lose water as they are buried and squeezed by overburden pressures. The particles, or floccules, each consist of "books" (crystals) which in turn consist of individual unit layers of clay. If the clay is a non-expanding type, such as illite or kaolinite, oriented water may occur between parallel "books" within each particle. If the floccule is an expanding clay such as montmorillonite, oriented water will occur not only between particles and books, but also between each of the unit layers (Fig. 1A). There are vastly more unit layers than particles

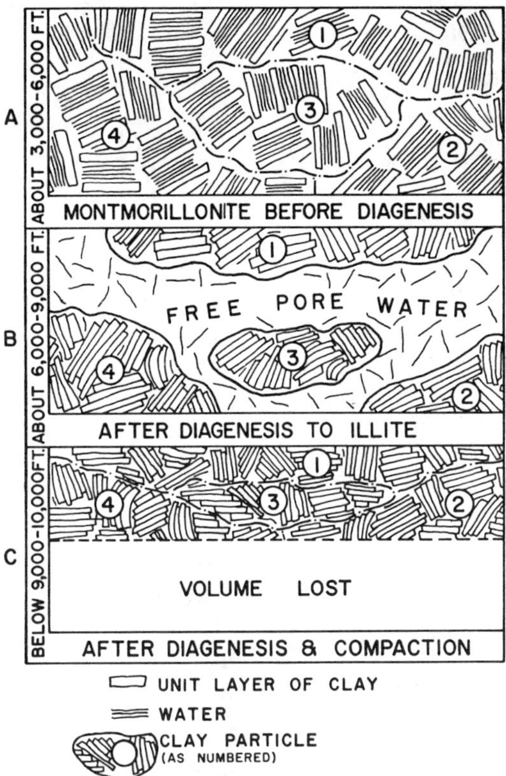

FIG. 1.—Effect of clay diagenesis on compaction of water from mudrocks. It is assumed that same number of particles, "books," and unit layers of clay occur in each compaction stage represented. A.—no effective porosity or permeability; virtually all water is "bound" water. B.—most "bound" water becomes free water; effective porosity and permeability thereby produced. C.—free water squeezed out; effective porosity, permeability, and original volume are greatly reduced.

and floccules combined, and the montmorillonite will therefore adsorb great quantities of oriented water compared with the non-expanding clay minerals.

There are two types of forces which oppose the van der Waals attractive forces. The first is the result of energy from hydrogen bonding between water molecules and oxygen surfaces of the clay mineral. It is important only for the first four water layers. The hydration of interlayer cations also may contribute to these short-range forces (White and Pichler, 1959). Layers of water are oriented by hydration along the basal surfaces between unit layers of clay until at least four monomolecular layers are adsorbed. The oriented water also may occur between the adjacent clay surfaces between "books" and floccules. Ten angstroms may be taken for the appropriate thickness of the unit layer of clay and the same amount for the thickness of four layers of water (each monomolecular layer is 2.5 angstroms thick). Thus the volume of the water of hydration is equal to the volume of the dry particles. The combination of hydration energy and the electrical double layer described so well by van Olphen (1963a) seems to offer a reasonable explanation for most experimental data dealing with the swelling pressures of clays. Data from controlled experiments do not permit extension of the explanation of the double-layer theory much beyond that given by van Olphen.

The second clue for the new theory may be summarized by recognizing that montmorillonite clays may adsorb far more oriented water, primarily as interlayer water, than do non-expanding clays such as illite. The next section reveals that a large part of this water is not squeezed out by overburden pressures after burial.

Clue No. 3: High pressure tests of clay.—Steinfink and Gebhart (1962) and van Olphen (1963a, 1963b) have shown conclusively that the last few monomolecular layers of oriented water adsorbed on expanding clay surfaces are not removed by overburden pressures following deep burial. Van Olphen (1963c) draws the same conclusion for organic molecules adsorbed by expanding clay. Even if a clay is heated to 50°C. while under pressure, it takes approximately 65,000 psi. to remove the last water layer (van Olphen, 1963b). The swelling pressure is the same as the clay's resistance to compaction by confining pressure, and it is equal to the net repulsive pressure between the unit layers of clay and the parallel flat surfaces between "books" and particles. Yong *et al.* (1963) found that swelling pressure in sodium montmorillonite increases with increasing temperature in the range 1 to 23°C.

Van Olphen (1963b) has determined from desorption isotherms the approximate amount of work-of-removal for water layers located between unit layers of montmorillonite. Experimental data from Steinfink and Gebhart (1962) substantiate the work of van Olphen. Figure 2 is a plot of approximate pressures required for the removal of successive monomolecular layers of water from a calcium montmorillonite. The data are calculated from work-of-removal values reported by van Olphen (1963b). Approximate depths of burial corresponding to the desorbtion pressures also are given in the figure, and it would seem that the last four layers of water adsorbed on the clay surfaces are not normally removed by moderately deep overburden pressures. Values used in plotting the curve are only intended to represent the order of magnitude and there will be variation depending on the particular clay used, the water content, and the type of exchange ions adsorbed by the clay.

Beyond four monomolecular layers the hydration energy responsible for the swelling pressure declines to a negligible quantity, and osmotic repulsion assumes the dominant role in accounting for swelling pressures. The swelling pressures at these greater particle distances are approximately equal to the electrical double-layer repulsion, and they probably never exceed 1,500 psi. or an equivalent depth of burial of about 1,500–3,000 feet. At depths shallower than a few thousand feet, water content of the mud may be so high that density values commonly are closer to 1.0 than to 2.0 gm./cc. Consequently, a single density value can not give true values for all depths if converting psi. to equivalent depth of burial.

There appear to be two ways in which the last four monomolecular layers of water may be desorbed from montmorillonite without having to resort to such high pressures. The sample may be heated to somewhat more than 100°C. or one may resort to a large electrostatic force. Clay mineralogists have known for a long time that the repulsion force of clay unit layers may be exceeded by the electrostatic attraction force associated with the exchange of potassium for other ions. Such an exchange causes the expanded clays to collapse to 10 angstroms with accompanying desorption of water. Depending on the lattice deficiencies, the potassium may become fixed and prevent reversal of the collapse. If the lattice charge is such that the potassium becomes permanently fixed, the reaction is termed diagenesis and the mineral illite apparently is formed from montmorillonite.

There are two important clues to the new theory in this section. First, it appears that at least four monomolecular layers of water resist being squeezed out of expanding-type clay by high pressures and they may remain adsorbed even after

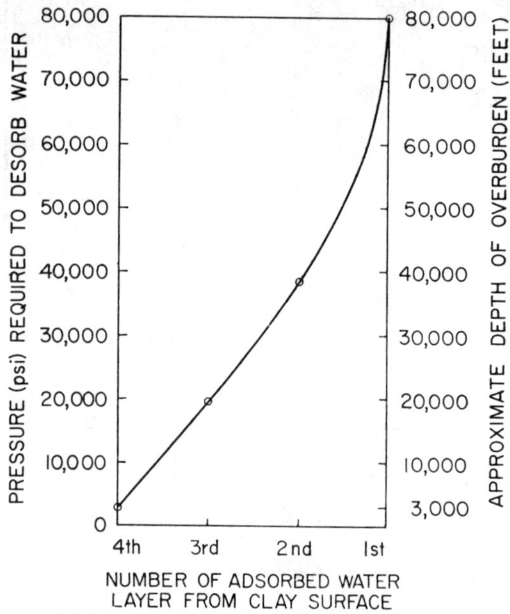

FIG. 2.—Amount of pressure or approximate equivalent depth of burial necessary to remove each of four layers of water from interlayer positions on montmorillonite.

deep burial. Second, it is possible to desorb this water with the electrostatic attraction forces associated with potassium fixation. It was stated previously that potassium fixation is required in the diagenesis of montmorillonite to illite which takes place at depths beginning at about 6,000 feet, and continuing at an increasing rate to about 9,000–12,000 feet. Thus, it is possible to desorb the last four water layers without having to resort to great depths of burial.

Clue No. 4: Density of water bound to clay surfaces.—In a review of the relations between water and clay, Williamson (1951) compared the structure of montmorillonite proposed by Hoffman, Endell, and Wilm with that of Edelman and Favejee, noting the different requirements imposed by the structure for attaching water molecules to basal surfaces. Williamson cited the important structures proposed (to that time) for the interlayer water, including hexagonal networks, two-dimensional liquids, epitaxis (induced oriented growth), and completely rigid (ice) structures. Williamson visualized different arrangements for the water, depending on the degree of hydration; *e.g.*, a hexagonal network arrangement in which additional water molecules are situated in the centers of the water hexagon. He also suggested that this very closely packed arrangement might account for the abnormally high density found for water adsorbed on clays. The basic concepts for visualizing the oriented stacking of hexagonal network layers of water between basal surfaces of clay in relation to the plastic properties of clays are best described by Grim and Cuthbert (1945) and more recently by Grim (1962).

Martin (1962) brought the review up to date and considered the structure of interlayer water in the light of more recent knowledge of clay crystal chemistry. He suggested a minimum density of adsorbed water of about 0.97 gm./cc. at 0.7 gram of water per gram of clay (approximately seven monomolecular layers of water). For fewer monomolecular layers, density of the interlayer water rises abruptly to 1.4, and for more than seven layers it rises until it equals normal water at 6.5 grams of water per gram of clay. The density values greater than 1.0 reported by Martin cast doubt on the Hendricks-Jefferson (1938) hexagonal network structure that has received attention for so long. However, the abnormally high density values may be reconciled by adding water molecules in the centers of the water hexagons, as suggested by Williamson (1951). Whether the hexagonal net structure may be salvaged in this way can not be settled at this time, but the fact remains that data from many sources confirm the high density of the last few monomolecular layers of water, with some determinations extending to 1.7 gm./cc. (Boswell, 1954, p. 20). Thus, the final clue needed for the interpretation of the new theory indicates that the last few layers of water adsorbed on montmorillonite surfaces have a higher-than-normal density.

Explanation of New Theory

In the previous section some theoretical and experimental aspects of clay mineralogy were explored. These have led to the following four-fold premise:

1. Expanding clays, such as montmorillonite, are altered to illite after deep burial.

2. Montmorillonite clays adsorb far more hydrogen-bonded oriented water to the basal surfaces than do illite and other non-expanding clays.

3. At least four monomolecular layers of water stacked between the unit layers of montmorillonite are not desorbed by the overburden pressures associated with deep burial, but the forces of electrostatic attraction during the adsorption of potassium by montmorillonite are great enough to desorb the water from the clay.

4. Abnormally high density values have been established for the last few monomolecular layers of water bound on montmorillonite basal surfaces.

Now it will be seen how this information improves the understanding of sediment compaction and leads to a new hypothesis for fluid release.

If montmorillonite is deposited in a marine or estuarine environment and buried to a few hundred feet, a balance is reached between the water retained in the mud and the water-retaining properties of montmorillonite. Further increase in pressure alone, resulting from deeper burial, is ineffective in squeezing the remaining water out of the plastic mud. At depths greater than about 1,500–3,000 feet, most of the water is water of hydration which is stacked at least four monomolecular layers thick between the unit layers of montmorillonite. A much smaller amount of oriented water occurs between "books" (crystals) and particles (Fig. 1A). As the montmorillonite is altered to illite, the interlayer bound water is desorbed and becomes free pore water (Fig. 1B). In this way, water and space are added to the interparticle area by transfer from a bound interlayer position, and the transfer causes a decrease in clay particle size with a corresponding increase in the effective porosity, permeability, and free pore water (Fig. 1B). At the pressure attained for the depths considered, about 6,000–9,000 feet, most transferred water molecules are located at distances from particle surfaces too great to be held in oriented form by forces arising from the electrical double layer. Consequently, overburden pressures compact the water from the clay until a balance is established once more, this time having the water-retaining properties of the illitic alteration product (Fig. 1C). If it is considered that, for a pure clay sediment, up to 50 per cent by volume of the expanded montmorillonite becomes pore space filled with free water, it is apparent that large volumes of water and porosity are produced. The alteration of montmorillonite to illite, therefore, has a profound effect on the free pore water content, effective porosity, permeability, and other bulk properties of the sediment.

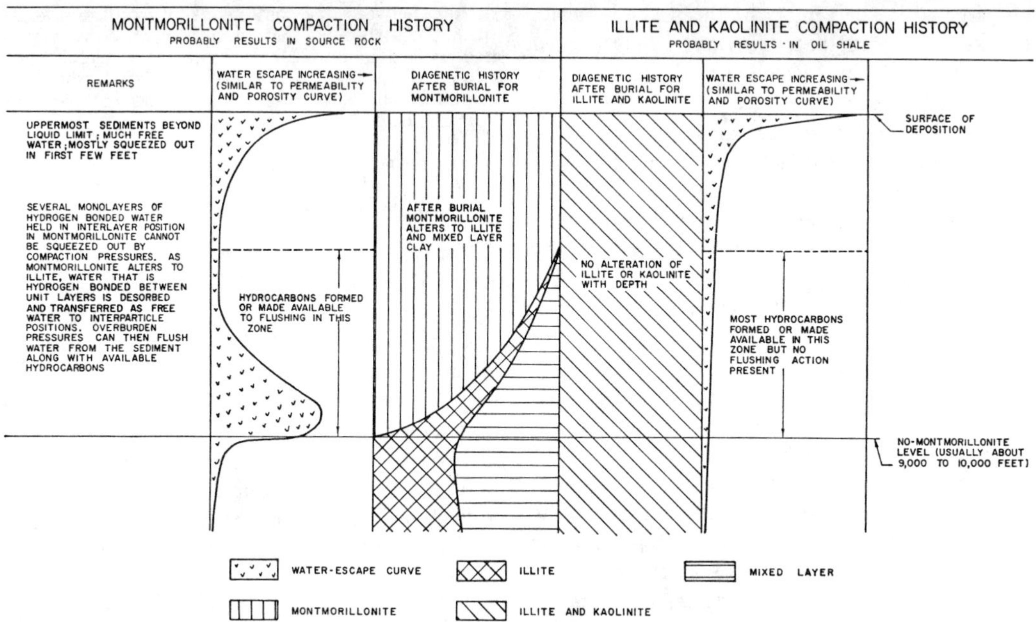

FIG. 3.—Compaction history of different clay minerals when deposited in marine environment and its probable relation to release of hydrocarbons from shale.

The relations between water escape, type of clay, and depth of burial are illustrated diagrammatically in Figure 3, both for a montmorillonitic expanding clay and for non-expanding illitic and kaolinitic deposits. The water-escape curves are diagnostic of the permeability, porosity, and bulk density of compacting mudrocks. It is clear that the compaction history of mudrocks depends largely on their original clay composition and the diagenesis which they undergo after burial.

It was said earlier that interlayer water, especially the last four layers, has a considerably higher density than ordinary water. Therefore the water must increase in volume as it is desorbed from the montmorillonite.

If 1.4 gm./cm.^{3} is used as the average density of the last four layers of water, the volume increase of the water upon transfer is 40 per cent. Thus the water requires 40 per cent more room as ordinary water between the particles than it did as interlayer water. The actual amount of water involved depends on the ratio of expanding clay to other minerals in the sediment, and the figure probably would range from about 2.5 to 20 per cent clay for Texas Gulf Coast mudrocks. Such a mechanism would be expected to produce abnormally high fluid pressures in the mudrock. The compressibility of water for the depths considered here, and for a temperature increase corresponding to these depths, is practically zero. Even where a very small quantity of water is involved in the volume increase, it probably would produce high fluid pressures because it can not be compressed and there never was any extra pore space for it. The problem is like that of trying to cram a little more water into a container that is already full.

There seems to be an exponential increase in the permeability of a rock corresponding to an increase in porosity (Rubey and Hubbert, 1959, p. 178). This exponential relation between porosity and permeability, together with the increase in abnormal fluid pressure referred to above, would result in a pronounced increase in the escape of water from the mudrock, very probably amounting to a flushing action.

The shape of the porosity and permeability curves should be similar to the shape of the water-escape curve, and therefore closely related to the depth and rate of alteration of montmorillonite to illite (Fig. 3). Density, on the other hand, should show a reverse relation to the water-escape curve. As the water is flushed out of the mudrock, it carries along organic constituents in

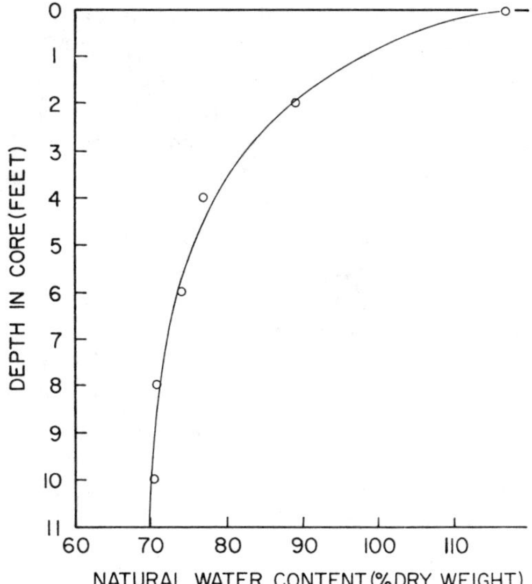

FIG. 4.—Natural water content of muddy sediments in York Estuary. Average values for 12 cores.

suspension, as well as hydrocarbons, either as an emulsion or in solution. Only those mudrocks that contain montmorillonite undergo the flushing action and therefore might become source rocks in this way.

A Fresh Look at Compaction

BULK PROPERTIES

As floccules of clay settle gently to the bottom surface of a basin, they form gels and contain very large amounts of water. The structure of the gel in the first few inches of sediment probably is an edge-to-face arrangement (van Olphen, 1963a; Norrish and Rausell-Colom, 1963), in which the "books" of unit layers form roughly 90° angles with each other. Below the first few inches of soupy bottom sediment, this house-of-cards structure probably gives way to a "hinge" structure as described by Mungan and Jessen (1963, Fig. 7). This structure may persist as a gel to considerable depths.

The retention of water by clay at, and near, the bottom surface of the depositional basin is the result of the electric double-layer repulsion forces which range from less than one to several times ten atmospheres (van Olphen, 1963b, p. 181). These forces are not effective at very short distances from the clay surfaces (Norrish and Rausell-Colom, 1963) and they appear to affect only the water layers at distances beyond the fourth layer. The distribution of the repulsion forces outward from the clay surfaces is such that by far most of the water in this category is lost at a burial depth of rarely more than 3–4 feet (Fig. 4). The influence of electrolyte concentration on the structure of clays in water is considerable for low pressures, but apparently it is not important above 10 atmospheres (or about 300 feet of burial) according to von Engelhardt and Gaida (1963).

Several investigators (Emery, 1960; Ericson *et al.*, 1961) have reported a very marked drop-off in natural water content of sediments in the first few inches below the bottom surface, followed by very little to no apparent decrease from there to the bottom of a cored interval ranging from 10 to 20 feet. Fisk and McClelland (1959) found about the same relation for water content of muds in the Gulf of Mexico as Ericson *et al.* (1961) did for the Pacific and Powers (1957) did for the Chesapeake Bay area. Kerr and Barrington (1961, p. 1698) reported that shale along the Gulf Coast is "water soaked" to a depth of 5,000 feet. Emery (1960, p. 260) discussed the thixotropic properties of surface muds. He referred to the high yield stress of these deposits which commonly permits a 3-inch-diameter core, 1–2 feet long, to be held by one end in a horizontal position without shear.

Organic matter in high concentrations may have a considerable influence on the water-retaining capacity of a sediment near the bottom surface. Barring the effect of organics, illite and kaolinite, because most of the water they contain is free water, are expected to lose water at a faster rate under the light loading conditions of the near-surface zone than does montmorillonite. This free water has less resistance to a small compaction load than does the oriented water of montmorillonite (Grim, 1962).

Most investigators who have measured natural water content have found a marked decrease in it very near the surface followed by very little to no decrease for the next several tens of feet downward. The shapes of the uppermost parts of the water-escape curves in Figure 3 reflect this trend, and it is seen that these are the results expected from the knowledge of clay-water systems.

Kerr and Barrington (1961) point out that muddy sediments are no longer water-soaked below 5,000 feet depth in the Louisiana Gulf Coast, and that significant measurements of compaction and bulk density are feasible. Boswell (1961) says that beds as old as the Jurassic still are plastic and that it has " . . . certainly been possible for sediments to remain in the plastic state for at least a hundred and fifty million years" (p. 87). One must conclude that age alone has little to do with the compaction of muddy sediments. Skempton (1944) found that, upon consolidation, the natural water content of a sediment usually is very near to its plastic limit. This discovery was supported by the work of Grim and Cuthbert (1945) in which they suggested that the rigidity of the water layers on the clay surface was the dominant factor determining the plastic properties of the clay. Furthermore, the number of water layers found on clays at the plastic limit by White (1958) shows a correspondence to the natural water content of compacted clays.

Rittenberg *et al.* (1963, p. 152) found very little change in water contents of clay sediments in the Pacific from the bottom surface to depths of 400 feet. These muds, cored in the Mohole project, contained about 50 per cent water, which is equivalent to approximately 10 layers of oriented water, if one assumes that the sediment is pure clay. Impurities increase the number of water layers required by distribution of the water in the clay fraction of the sediment; thus, 10 layers probably are a minimum value. The outer five or six layers are lost at a decreasing rate as depth is increased, and at about 2,000–3,000 feet only the four innermost oriented layers remain in the clay. This requires a loss of about 30 per cent of the clay water by compaction between depths of 400 and 3,000 feet, and this is the expected amount for a highly montmorillonitic sediment.

Almost 30 years ago, Hedberg (1936) recognized a change in the mechanics of compaction of shale at depths greater than about 6,000 feet. He even suggested that the recrystallization of minerals was in some way responsible for the change. More recently Hedberg (1964) described compaction as a continuous process in which water content of a sediment is reduced at an ever-diminishing rate as overburden is increased. Skempton (1953) noted that the void ratio in clay sediments decreases at an increasing rate at depths greater than about 5,000 feet, corresponding to what would be expected from the water-escape curve for montmorillonite. Yet laboratory experiments, in which water-saturated expanding clays were pressured to equivalent depths much greater than 6,000 feet (Skempton, 1953), do not show any variation in the rate of compaction at depths comparable to those cited above. The answer, to what seems like a paradox, is simply that the absence of diagenesis in the laboratory compression experiments precluded their duplication of the natural situation. Weller (1959, p. 292) recognized that these experiments did not produce the crystallization that he believed was necessary for the production of shale. Thus, several investigators have recognized that a change in the nature of the compaction of mudrocks at about 5,000–6,000 feet depth probably is caused by some sort of recrystallization. The significance of their work to this paper is that the new fluid-release theory proposed here may account for the mineral alteration that they suspected at about the depth which they indicated.

Unfortunately there are very few data published that can be used for testing the water-escape curve for depths greater than a few tens of feet. If the expense of sampling and the lack of interest that the production engineer has in samples from shale sections are considered, the reasons for this lack of data are obvious.

A porosity curve from Russian data for a Mesozoic shale section (Fig. 5) shows a marked decrease in porosity from the surface to about 1,000 feet followed by a decrease of less than 10 per cent from there to 6,000 feet. If the values from 6,000 feet downward are considered, a clear increase in porosity is seen for the next 2,000 feet followed by a distinct decrease for the next 1,500 feet. Correlation of the porosity curve of Figure 5 with equivalent depths on the water-escape curve of Figure 3 suggests that its shape can be most logically explained by the new fluid-release theory.

The three curves in Figure 6 utilize all reliable field data from the literature (Hedberg, 1936; Johnson, 1950) for density values of shale plotted against depth. The sudden increase in density, shown by each curve beginning at about 5,500 feet, correlates closely with the density increase expected from the shape of the water-escape curve for montmorillonite (Fig. 3) at virtually the same depth. This correlation lends added support to the new fluid-release theory.

ABNORMAL FLUID PRESSURES

The fluid pressure in a rock, expressed as

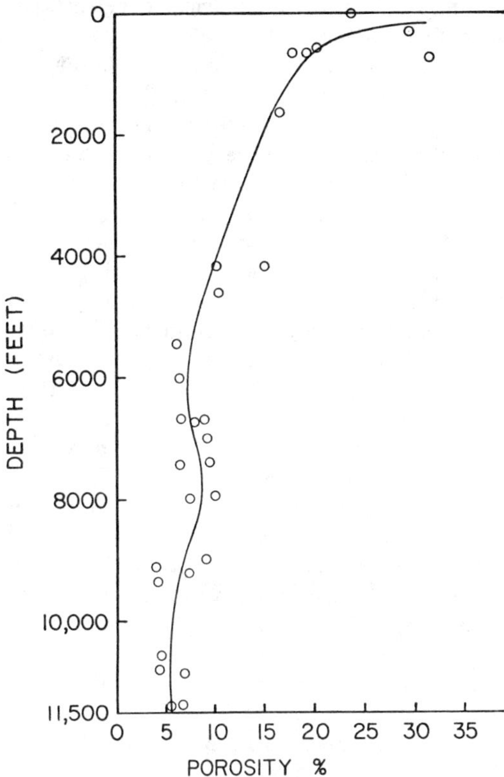

FIG. 5.—Porosity of shale section in Mesozoic of Russia. After B. K. Proshlyakov (*in* Maxwell, 1964).

pounds per square inch (psi.) is said to be abnormal when it corresponds to a value greater than the normal hydrostatic load or about half the burial depth of the rock in feet. Geologists and petroleum engineers have known for a long time about abnormal fluid pressures in deep subsurface rocks. Most knowledge concerning their origin is contained in four excellent papers by Dickinson (1953), Hubbert and Rubey (1959), Rubey and Hubbert (1959), and Handin *et al.* (1963). In their discussion of the origin of abnormal fluid pressures in subsurface sediments, Hubbert and Rubey consider particularly the mechanical compression of water-filled pores. This compression may result either from rapid gravitational loading, which produces deep undercompacted sediments, or from stresses of tectonic origin. The resulting pressures may approach in amount the geostatic load. Handin *et al.* (1963) pointed out that a permeability barrier is required for the development of abnormal pressures in excess of 0.5 of overburden pressure. The rapid gravitational loading of water-soaked sediments has been cited as a probable origin of abnormal fluid pressure in the Gulf Coast sediments by Hubbert and Rubey (1959) and this has been suggested as the most probable cause of high pressure by Handin *et al.* (1963).

Tkhostov (1963) reviewed the occurrence and possible causes of abnormally high pressures. He concluded that there is no single commonly accepted opinion on the subject, and he recognized four principle sources of rock energy which may explain pressures in oil and gas deposits (p. 18): (1) geostatic pressure, (2) hydrostatic pressure, (3) geotectonic pressure, and (4) the availability of routes of communication between beds having different pressures. He stated that these conditions operate with variable degrees of intensity, which has changed in time and in space.

The sources of rock energy cited by Tkhostov must be considered as possible causes of abnormal fluid pressures. In addition, the gravitational loading theory rests on good evidence from soil and rock mechanics experiments, and it may be a common contributing cause of abnormal pressure. In addition, pressures resulting from fluid trans-

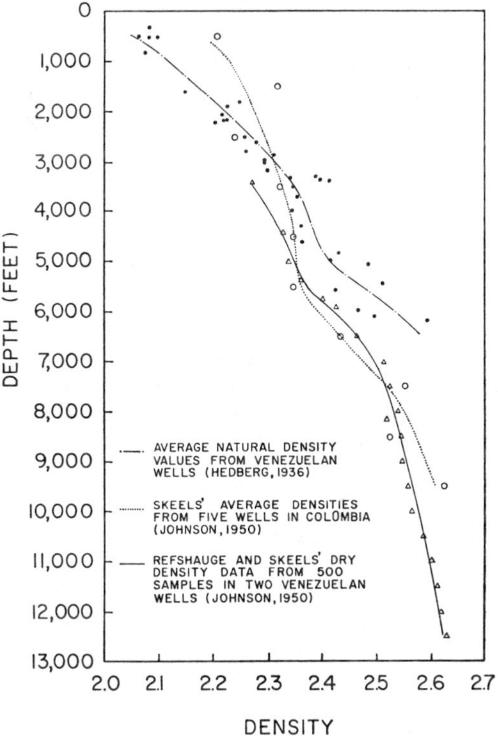

FIG. 6.—Density variations with depth in shale sections.

fer during the alteration of montmorillonite to illite may be considered.

As the last few layers of water of hydration are desorbed from an interlayer position and transferred as free water to interparticle pores, the water undergoes a density decrease with corresponding increase in volume, as described previously. All pores in the mudrock are assumed to be filled with fluid at the time of transfer, and even small amounts of water create a potential "room problem" amounting approximately to

$$Vi = PC,$$

where Vi is the increase in bulk volume of the sediment; P is the difference in density between the last few monomolecular water layers and normal water; and C is the per cent of montmorillonite which collapses to 10 angstroms. Four layers of water probably are desorbed, in which case the volume of water transferred is equal to the volume of dry clay undergoing alteration.

Pursuing this line of reasoning further, the volume increase in mudrocks of the Texas Gulf Coast would range from a probable low of 2.5 per cent for sandy mudrocks to a high of probably 20 per cent for relatively pure clayrocks. The pressure buildup would depend on the rate at which normal density water formed and the simultaneous rate of escape of the normal water from the shale. The increase in volume of transferred water corresponding to a density decrease from 1.4 gm./cc. (Martin, 1962) to 1.0 gm./cc. would be limited only by the compressibility of water which is nearly zero for the temperature-pressure relations considered. Abnormal pressures therefore could easily equal the entire weight of the overburden (geostatic or lithostatic pressure) for the depths considered. In this case, the total overburden of rock would be in a state of flotation as described by Hubbert and Rubey (1959), who have considered the importance of abnormal fluid pressure in fault mechanics.

Dickinson (1953) found that fluid pressures in the Louisiana Gulf Coast are "normal" to a depth of about 7,000 feet, and that a pressure gradient of 0.465 psi. per foot is normal for the Gulf Coast to depths of 16,000 feet. He also found that the change from normal to abnormally high pressures occurs through a very short vertical interval. It is significant that Dickinson usually found abnormally high pressures associated with porous reservoir beds isolated in thick shale sequences. His data suggest that the high-pressure zone is depth-dependent.

Figure 7, taken from Dickinson (1953, Fig. 2), permits one to determine at a glance at what depth abnormal pressures occur, as well as the amount of deviation above the normal gradient. The greatest deviation above normal pressure occurs within a depth interval of about 9,000–12,000 feet. Using the water-escape curve for montmorillonite (Fig. 3) as an explanation for this distribution of Dickinson's data, the pressure gradient curve could be redrawn as indicated by the dashed line in Figure 7. The redrawn curve is parallel with the normal pressure gradient to a depth of about 6,000 feet, where it increases sharply to about 9,000 feet, and from there down is again parallel to the normal gradient.

Two important relations are now clear. The new pressure gradient curve fits the actual field data closely, and the shape-to-depth relations of this curve coincide very closely with the water-escape curve for montmorillonite. These points suggest a causal relation between the abnormal fluid pressures given by Dickinson and the alteration of montmorillonite to illite.

X-ray diffraction analyses of cuttings from deep wells in the Texas and Louisiana Gulf Coast were carried out by the writer while working with the Shell Development Company. The results suggest that there is also a relation between the depth at which montmorillonite is altered to illite and the depth at which abnormal fluid pressure is reached in the Gulf Coast, according to Dickinson and others.

The maximum pressure gradient is first reached at 9,000–10,000 feet (Fig. 7) and, as depths increase below that level, increasing numbers of fluid-pressure values are closer to the normal hydrostatic pressure curve. This trend suggests that the abnormal pressures were formed at about 9,000–10,000 feet and that, in most cases, some pressure leaked off before the porous reservoir trap was buried to a much greater depth. In accordance with the new theory, one would expect pressures below about 9,000 feet to show maximum values which do not exceed that allowed by an increase determined by the normal hydrostatic pressure gradient. It is evident from Figure 7 that all but three points plotted for depths greater than 9,000 feet conform with this restriction by falling on the normal pressure side of the new pressure gradient curve. This agreement further

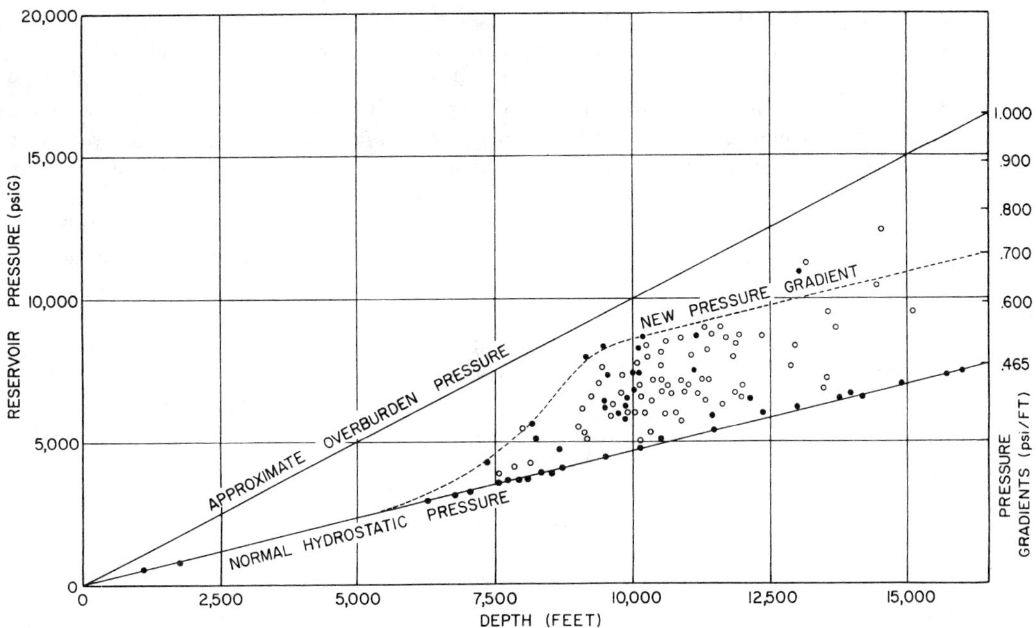

FIG. 7.—Reservoir pressure *versus* depth for Louisiana Gulf Coast wells. Solid circles: measured pressures. Open circles: estimated pressures. After Dickinson (1953).

suggests the likelihood that the new theory may explain abnormal pressures. The distribution of points below the new pressure-gradient curve is explained in part by different degrees of leakage of high pressure from the trap. Handin *et al.* (1963) noted that, once abnormal pressures are formed, they would be retained for a period, even for a period of geologic time, which would depend on the effective permeability both along and across the bedding.

FLUSHING HYROCARBONS FROM SOURCE ROCKS

Rubey and Hubbert (1959) suggested that porosity and fluid pressure are interdependent, and that slight increases in porosity are accompanied by large increases in permeability. According to the new fluid-release theory, alteration of montmorillonite to illite in the deep subsurface initiates rapid increases in porosity, permeability, and fluid pressures in the compacting mudrock. Conditions would seem to be excellent for a fluid-flushing action which may be part of the mechanism by which hydrocarbons are moved out of source rocks (Fig. 3). Most investigators indicate that oil migrates from source rock in water that is squeezed out after considerable burial (Uspenskii, 1962); their reasoning is that source rocks are restricted to those sediments from which water has been expressed (see, for example, Baker, 1959). Weeks (1961) emphasized the interdependency of the origin, migration, and accumulation of oil. He stated (p. 5–29) that ". . . at no time in the history subsequent to its subsidence and deposition is there anything remotely approaching the magnitude of the migration and flushing action that takes place during that early subsidence and receiving of sediments." Weeks expressed a realistic concern for the lack at that time of a known deep flushing mechanism. The new fluid-release theory provides a possible mechanism needed for moderately deep (about 6,000–9,000 feet) flushing, and accounts for the need of early trap formation cited by Weeks (p. 5–30). The optimum depth zone suggested by Weeks for the occurrence of oil (p. 5–42) lies between 2,000 and 9,000 feet, and this depth range is compatible with the new fluid-release theory.

According to the work by Bray and Evans (1961) and Silverman (1965), hydrocarbons comparable with those found in reservoirs form mainly after deep burial. Cox (1946) gave a minimum overburden pressure of 5,000 feet for the formation of petroleum. Hedberg (1964) pointed out the difficulty of establishing the existence of indigenous oil occurrences in beds which do not appear to have had overburden pressures comparable with a burial depth of at least 2,000–3,000 feet. Silverman (1965) suggested that petroleum matures with increasing depth of burial and that secondary migration induces further minor differences in oils. On the other hand, several investigators have put limitations on the maximum depth of formation of oil by restricting the temperature of formation to low values (Dunning and Moore, 1957; Hodgson and Baker, 1957; Weeks, 1961). Still others contend that hydrocarbons form and migrate after only shallow burial. A summation of the information on this subject suggests that hydrocarbons comparable with those found in reservoirs form after burial to depths of a few thousand feet but probably not deeper than 9,000–10,000 feet (Fig. 3). The movement of the deeper fraction of hydrocarbons in this depth range, from source rock to porous reservoirs, could be accounted for by a flushing action associated with the diagenesis of montmorillonite to illite at depths of about 6,000–10,000 feet.

Continuing this line of reasoning, water is squeezed out of illitic or kaolinitic sediments soon after burial, and hydrocarbons that form at greater depth would have no flushing action available to move them toward porous reservoirs. Therefore illite and kaolinite may be characteristic of oil shale, or rocks that do not give up their hydrocarbons.

Conclusions and Geological Implications

A combination of experimental data from the laboratory, field data, and current knowledge of clay mineralogy suggests the following tentative conclusions:

1. Where montmorillonite is buried to a depth of about 3,000 feet, most of the water is expelled from it except the last few bound layers that are oriented along basal surfaces between the unit layers of clay. This bound water may comprise nearly 50 per cent of the volume of the deposit for pure clayrocks, and it apparently can not be squeezed out by pressures associated with increasing depth of burial.

2. At 3,000 feet the effective porosity and permeability are essentially zero for the mudrock because virtually all space is occupied by the solid unit layers of clay and the rigid water layers bound to the clay. The clay and fixed water layers have a "wrap-around" structure on sand and silt particles and the influence that even large amounts of contaminant might have on porosity and permeability is materially reduced.

3. Alteration of montmorillonite to illite begins at a depth of about 6,000 feet and continues at an increasing rate to a depth, usually about 9,000–10,000 feet, where there is no discrete montmorillonite left. The alteration offers a mechanism for desorbing the last few layers of bound water from the unit layers of clay, and transferring it as free water to interparticle locations. In this new position the water literally "produces" effective porosity and permeability.

4. Laboratory experiments support the suggestion that the last few layers of bound water have a considerably greater density than free (normal) water, and this is interpreted to mean that water undergoes a volume increase as it is desorbed from between unit layers. As the water expands to its normal density it increases the pore fluid pressure to abnormally high levels.

5. The combination of increased porosity, increased permeability, and abnormally high fluid pressure suggests that a flushing action may occur in which fluids are released suddenly from montmorillonitic mudrocks in the deep subsurface. It may be at this depth that mudrocks develop fissility.

6. The flushing action apparently would be absent from illite and kaolinite deposits because there is no alteration in the deep subsurface to provide the fluid-release mechanism.

7. The alteration of montmorillonite to illite provides a mechanism for the desorption of interlayer hydrocarbons from montmorillonite, and these hydrocarbons, together with others in the sediment, may be moved out of the mudrock in the course of the flushing action. The mechanism therefore may be an important aspect in the development of shale source rocks. In this light illite and kaolinite deposits rich in the organic raw materials for forming hydrocarbons may be expected to compact to oil shales rather than to source rocks.

8. It is therefore suggested that there may be two primary requisites for the development of most shale source rocks:

a. The original deposit consists in part of montmorillonite.

b. The montmorillonite must be buried to a depth great enough to cause its diagenesis to illite.

If one pursues this line of reasoning, oil in the Gulf Coast Tertiary should not move freely to the reservoir rock before burial to a depth of about 6,000 feet, and probably deeper. Perhaps the low productivity of the updip Catahoula Miocene sediments in southeast Texas can be explained in this way; the montmorillonite there has not been buried deeply enough to have altered to illite and thereby release the fluids it contains. These same beds downdip have greater productivity. Furthermore, the Pliocene sediments in Louisiana are most productive where they are buried to depths great enough that the montmorillonite has been altered to illitic clay.

The writer has suggested that oil may not be released from source rocks much above depths of about 6,000 feet. It is suggested that much of this oil may move out of the source rock at the peak of water escape, which will occur at about 7,000–10,000 feet in the Texas Gulf Coast and probably somewhat deeper in the Louisiana Gulf Coast. The absence of the flushing mechanism described here, below the no-montmorillonite level, may place depth limitations on the movement of oil from shale source rocks. Where oil is released from source rocks at the depths suggested, it may then follow fractures or other avenues of escape to much shallower depths.

If the hydrocarbons are released where there is no trap, they eventually may be lost to the surface if avenues of escape are available. Once the hydrocarbons are trapped, they may be carried to greater depths of burial by subsidence resulting from continued deposition at the surface. Viewed in this way, the importance of the relation between the depth at which the fluids are released and the depth and time of formation of a trap should be of vital interest to exploration geologists.

Below the no-montmorillonite level (Fig. 3), small amounts of water continue to be lost from the shale as the montmorillonite fraction of the mixed-layer clays collapses to form illite. This slow release continues to a depth of about 14,000 feet in the Gulf Coast, below which there is no swelling component remaining in the mixed-layer clay (Burst, 1959).

Finally, there are two additional areas which should be investigated in the light of the new fluid-release theory presented here.

1. The liquid limit and plastic limit for illite and kaolinite are several times less than for montmorillonite (Grim, 1962), and the sudden formation of free water simultaneously with the alteration of montmorillonite to illite may have a profound effect on the plastic properties of mudrock in the deep subsurface. Numerous examples suggest a relation between the new theory and an increase in the plasticity of mudrocks within the depth range of 6,000–10,000 feet. For example, Thomeer and Bottema (1961, p. 1723) found abnormally high fluid pressures within shale sections as well as in porous reservoirs. They found abnormally high pressures at 9,000–10,000 feet in Permian shales of North Germany. These high pressures continued to bottom-hole depths of nearly 15,000 feet. In one well, at 9,355 and 9,395 feet they found (p. 1725) ". . . very soft plastic shale which continually tended to fill the hole almost as fast as it was drilled and reamed." The authors believed that this shale forms a lens that is completely enclosed within dense, impermeable rock salt. Furthermore, they reported that there is no montmorillonite in the shale at these depths. These conditions of no-montmorillonite (it is probably diagenetic illite and mixed-layer clay), the highly plastic nature of the clay, and the abnormally high fluid pressures found are predicted by the new theory for mudrocks at this depth which are completely sealed off by an impermeable barrier. In Tertiary sediments of Netherlands New Guinea, Thomeer and Bottema reported (p. 1727) "sticky clay trouble" which was so bad at about 6,000 feet in one well that a sidetrack had to be drilled. In another New Guinea well, they reported "a sudden increase of penetration rate" at 5,995 feet. In both wells the fluid pressures were abnormally high. They ascribe the high pressures found at these moderately shallow depths to the thick sequence of young clayey sediments, but the alteration of montmorillonite to illite at these levels could account for both the high pressures and the highly plastic nature of the clay.

2. Weeks (1961, p. 45) stated that the adsorp-

tion of ions on clay particles and their later release under high temperatures and pressures ". . . may well account for the common increase in salinity with depth and the changes that occur in the composition of the salt content in basins." The alteration of montmorillonite to illite would almost certainly result in the release of ions from exchange positions on the clay, and this exchange may be related in an important way to the origin of subsurface brines. The data and theory available at this time support the suggestions made. There are several "loose ends" to the ideas advanced here, and only after further testing and investigation will it be possible to gain confidence in the application of the ideas.

References Cited

Baker, E. G., 1959, Origin and migration of oil: Science, v. 129, p. 871–874.

Boswell, P. G. H., 1954, The natural moisture-content, voids, and compaction of unconsolidated clayey rocks: Liverpool and Manchester Geol. Jour., v. 1, p. 1–22.

——— 1961, Muddy sediments: some geochemical studies for geologists, engineers and soil scientists: Cambridge, Eng., Heffer & Sons, 140 p.

Bray, E. E., and E. D. Evans, 1961, Distribution of *n*-paraffins as a clue to recognition of source beds: Geochim. et Cosmochim. Acta, v. 22, p. 2–15.

Burst, J. F., Jr., 1959, Postdiagenetic clay mineral environmental relationships in the Gulf Coast Eocene, *in* Clays and Clay Minerals, v. 2: New York, Pergamon Press, p. 327–341.

Cox, B. B., 1946, Transformation of organic material into petroleum under geologic conditions ("The geological fence"): Am. Assoc. Petroleum Geologists Bull., v. 30, p. 645–659.

Dickinson, George, 1953, Geological aspects of abnormal reservoir pressures in Gulf Coast Louisiana: Am. Assoc. Petroleum Geologists Bull., v. 37, p. 410–432.

Dunning, H. N., and J. W. Moore, 1957, Porphyrin research and origin of petroleum: Am. Assoc. Petroleum Geologists Bull., v. 41, p. 2403–2412.

Emery, K. O., 1960, The sea off southern California: New York, John Wiley and Sons, Inc., 366 p.

Engelhardt, W. von, and K. H. Gaida, 1963, Concentration changes of pore solutions during the compaction of clay sediments: Jour. Sed. Petrology, v. 33, p. 919–930.

Ericson, D. B., M. Ewing, G. Wollin, and B. C. Heezen, 1961, Atlantic deep sea sediment cores: Geol. Soc. America Bull., v. 72, p. 193–286.

Fisk, H. N., and B. McClelland, 1959, Geology of continental shelf off Louisiana: its influence on offshore foundation design: Geol. Soc. America Bull., v. 70, p. 1369–1394.

Grim, R. E., 1962, Applied clay mineralogy: New York, McGraw-Hill Book Co., Inc., 422 p.

——— and F. L. Cuthbert, 1945, Some clay-water properties of certain clay minerals: Am. Ceramic Soc. Jour., v. 28, p. 90–95.

Handin, J., R. V. Hager, Jr., M. Friedman, and J. N. Feather, 1963, Experimental deformation of sedimentary rocks under confining pressure tests: pore pressure tests: Am. Assoc. Petroleum Geologists Bull., v. 47, p. 717–755.

Hedberg, H. D., 1936, Gravitational compaction of clays and shales: Am. Jour. Science, v. 31, p. 241–287.

——— 1964, Geologic aspects of origin of petroleum: Am. Assoc. Petroleum Geologists Bull., v. 48, p. 1755–1803.

Hendricks, S. B., and M. E. Jefferson, 1938, Structures of kaolin and talc-pyrophyllite hydrates and their bearing on water sorption of clays: Am. Mineralogist, v. 23, p. 863–875.

Hodgson, G. W., and B. L. Baker, 1957, Vanadium, nickel, and porphyrins in thermal geochemistry of petroleum: Am. Assoc. Petroleum Geologists Bull., v. 41, p. 2413–2426.

Hubbert, M. K., and W. W. Rubey, 1959, Role of fluid pressure in mechanics of overthrust faulting, I: Geol. Soc. America Bull, v. 70, p. 115–166.

Johnson, F. W., 1950, Shale density analysis, *in* Subsurface geologic methods: Colo. School Mines, p. 329–343.

Keller, W. D., 1963, Diagenesis in clay minerals—a review, *in* Clays and Clay Minerals, v. 13: New York, The Macmillan Co., p. 136–157.

Kerr, P. F., and J. Barrington, 1961, Clays of deep shale zone, Caillou Island, Louisiana: Am. Assoc. Petroleum Geologists Bull., v. 45, p. 1697–1712.

Martin, R. T., 1962, Adsorbed water on clay: a review, *in* Clays and Clay Minerals, v. 9: New York, Pergamon Press, p. 28–70.

Maxwell, J. C., 1964, Influence of depth, temperature, and geologic age on porosity of quartzose sandstone: Am. Assoc. Petroleum Geologists Bull., v. 48, p. 697–709.

Mungan, N., and F. W. Jessen, 1963, Studies in fractionated montmorillonite suspensions, *in* Clays and Clay Minerals, v. 13: New York, The Macmillan Co., p. 282–298.

Nicholls, G. D., and D. H. Loring, 1960, Some chemical data on British Carboniferous sediments and their relationship to the clay mineralogy of those rocks: Clay Minerals Bull., v. 4, p. 196–207.

Norrish, K., and J. A. Rausell-Colom, 1963, Low-angle X-ray diffraction studies of the swelling of montmorillonite and vermiculite, *in* Clays and Clay Minerals, v. 12: New York, The Macmillan Co., p. 123–149.

Powers, M. C., 1957, Adjustment of land derived clays to the marine environment: Jour. Sed. Petrology, v. 27, p. 355–372.

——— 1959, Adjustment of clays to chemical change and the concept of the equivalence level, *in* Clays and Clay Minerals, v. 2: New York, Pergamon Press, p. 309–326.

Rittenberg, S. C., K. O. Emery, Jobst Hulsemann, E. T. Degens, R. C. Fay, J. H. Renter, J. R. Grady, S. H. Richardson, and E. E. Bray, 1963, Biogeochemistry of sediments in experimental Mohole: Jour. Sed. Petrology, v. 33, p. 140–172.

Rubey, W. W., and M. K. Hubbert, 1959, Role of fluid pressure in mechanics of overthrust faulting, II: Geol. Soc. America Bull., v. 70, p. 167–205.

Silverman, S. R., 1965, Migration and segregation of oil and gas, *in* Fluids in subsurface environments:

Am. Assoc. Petroleum Geologists Memoir 4, p. 53–65.

Skempton, A. W., 1944, Notes on the compressibility of clays: Quart. Jour. Geol. Soc. London, v. 100, no. 397–398, pt. 1–2, p. 119–135.

——— 1953, Soil mechanics in relation to geology: Yorkshire Geol. Soc. Proc. v. 29, p. 33–62.

Steinfink, H., and J. E. Gebhart, 1962, Compression apparatus for powder X-ray diffractometry: Rev. Sci. Instruments, v. 33, p. 542–544.

Thomeer, J. H. M. A., and J. A. Bottema, 1961, Increasing occurrence of abnormally high reservoir pressures in boreholes, and drilling problems resulting therefrom: Am. Assoc. Petroleum Geologists Bull., v. 45, p. 1721–1730.

Tkhostov, B. A., 1963, Initial rock pressures in oil and gas deposits: New York, The Macmillan Co.

Uspenskii, V. A., 1962, The geochemistry of processes of primary oil migration: Geochemistry (a translation of *Geokhimiya*), no. 12, p. 1157–1178.

van Olphen, H., 1963a, An introduction to clay colloid chemistry: New York, John Wiley and Sons, Inc.

——— 1963b, Compaction of clay sediments in the range of molecular particle distances, *in* Clays and Clay Minerals, v. 13: New York, The Macmillan Co., p. 178–187.

——— 1963c, Clay-organic complexes and the retention of hydrocarbons by source rocks, *in* International Clay Conference 1963, Proc. v. 1: New York, The Macmillan Co., p. 307–317.

Weaver, C. E., 1959, The clay petrology of sediments, *in* Clays and Clay Minerals, v. 2: New York, Pergamon Press, p. 154–187.

——— 1961, Clay mineralogy of the Late Cretaceous rocks of the Washakie basin: Wyo. Geol. Assoc. 16th Ann. Field Conf. Guidebook, p. 148–154.

Weeks, L. G., 1961, Origin, migration, and occurrence of petroleum, *in* G. B. Moody, ed., Petroleum exploration handbook: New York, McGraw-Hill Book Co., Inc., p. 5-i–5-51

Weller, J. M., 1959, Compaction of sediments: Am. Assoc. Petroleum Geologists Bull., v. 43, p. 273–310.

White, W. A., 1958, Water sorption properties of homoionic clay minerals: Ill. State Geol. Survey Rept. Inv. 208, 46 p.

——— and E. Pichler, 1959, Water sorption characteristics of clay minerals: Ill. State Geol. Survey Circ. 266, 20 p.

Williamson, W. O., 1951, The physical relationships between clay and water: British Ceramic Soc. Trans., v. 50, p. 10–34.

Yaalon, D. H., 1962, Mineral compostiion of the average shale: Clay Minerals Bull., v. 5, p. 31–36.

Yong, R., L. O. Taylor, and B. P. Warkentiss, 1963, Swelling pressures of sodium montmorillonite at depressed temperatures, *in* Clays and Clay Minerals, v. 13; New York, The Macmillan Co., p. 268–281.

Reprinted from:
BULLETIN OF THE AMERICAN ASSOCIATION OF PETROLEUM GEOLOGISTS
VOL. 52, NO. 12 (DECEMBER, 1968), PP. 2466–2501, 27 FIGS., 8 TABLES

COMPACTION AND MIGRATION OF FLUIDS IN MIOCENE MUDSTONE, NAGAOKA PLAIN, JAPAN[1]

KINJI MAGARA[2]
Tokyo, Japan

ABSTRACT

The main hydrocarbon reservoirs in Nagaoka Plain, Japan, are volcanic and pyroclastic rocks of the Nanatanian-Nishiyaman Stages (Neogene). The migration of hydrocarbons seems to have occurred from adjacent sedimentary source beds to the volcanic reservoirs, facilitated by differential compaction. Therefore, it is possible to determine the directions of the compaction currents by using horizontal porosity distributions of covered mudstone bodies. The differential compaction maps of the mudstone source beds overlying the reservoirs may indicate the directions of hydrocarbon migration, and may suggest the location of hydrocarbon accumulations in Nagaoka Plain.

As fluid is expelled from pores of the mudstone, compaction occurs and porosity decreases. However, if fluid expulsion is inhibited because of low permeability, mudstone compaction may not be great and a high porosity-high fluid pressure situation will result. With such a low-permeability mudstone acting as a barrier to fluid expulsion, compaction currents could move downward in the level below the low-permeability zone and upward above that zone.

The volumes of the compaction currents are estimated from the vertical porosity distributions and the porosity differences before and after compaction. The resultant figures are discussed in connection with the possible hydrocarbon accumulations.

There are some abnormal pressures in the reservoirs of the Fujikawa, Kumoide, and Mitsuke gas and oil fields in the plain, whereas normal hydrostatic pressures are observed in the reservoirs of such fields as Katagai, Nishi-Nagaoka, and Higashi-Sanjō. This suggests the presence of permeability barriers in the volcanic reservoir rocks.

The depths of the volcanic reservoirs and the thicknesses of the sealing mudstone beds are greater at Fujikawa, Kumoide, and Mitsuke, where abnormal pressures are found, than at Katagai, Nishi-Nagaoka, and Higashi-Sanjō, where reservoir pressures are hydrostatic. Therefore abnormal pressures in this district may be ascribed to the thick mudstone seals and overburden pressures.

INTRODUCTION

Nagaoka Plain is on the Japan Sea side of Honshu, the main island of Japan, about 230 km northwest of Tokyo (Fig. 1). Nagaoka Plain is one of the most actively explored areas in Japan, and contains such fields as Katagai, Sekihara, Nishi-Nagaoka, Kumoide, Fujikawa, Mitsuke, and Higashi-Sanjō (Fig. 2). A common characteristic of these fields is that the main reservoirs are composed of several kinds of volcanic and pyroclastic rock.[3] There are marked differences in the composition of hydrocarbons in the plain; some reservoirs produce gas with small amounts of condensate or light oil, and others produce large quantities of oil with some gas. Another remarkable feature of the volcanic reservoirs is the unusually wide pressure range; in some the pressure is hydrostatic and in others it is abnormal. The differences in hydrocarbon composition and pressure may be the result of differences in the characteristics of the source rocks and the sealing rocks.

The volcanic material in Nagaoka Plain seems to have formed banks or synsedimentary highs in the geologic past. The migration of hydrocarbon is thought to have occurred from the overlying and underlying mudstone source beds into the volcanic masses, facilitated by differential compaction. The overlying mudstone beds may be more important than the underlying ones, because they may serve as cap rock.

Hydrocarbons seem to be generated in early stages after the deposition of their source materials and to move with water as a compaction current (Mitsuchi, 1960). Therefore, it is very important and useful in petroleum exploration to detect the directions and the amount of compaction current which occurred in the geologic past.

[1] Manuscript received, February 8, 1967; accepted, July 25, 1967.

[2] Japan Petroleum Development Corporation. Present address: Institute of Sedimentary and Petroleum Geology, Geological Survey of Canada, Calgary, Alberta.

The writer thanks the management of Japan Petroleum Exploration Co. Ltd. (JAPEX) and Teikoku Oil Co. Ltd. for permission to use data gathered in their explorations. He is greatly indebted to K. Kawai, University of Tokyo; K. Nakazawa, Kyoto University; and A. Kujiraoka, JAPEX, for their suggestions. He also thanks Y. Ikebe, S. Juge, and M. Katō of JAPEX and K. Hoshino of the Geological Survey of Japan for encouragement and valuable advice.

[3] Herein, volcanic reservoirs include lava, tuff breccia, tuff, and agglomerate.

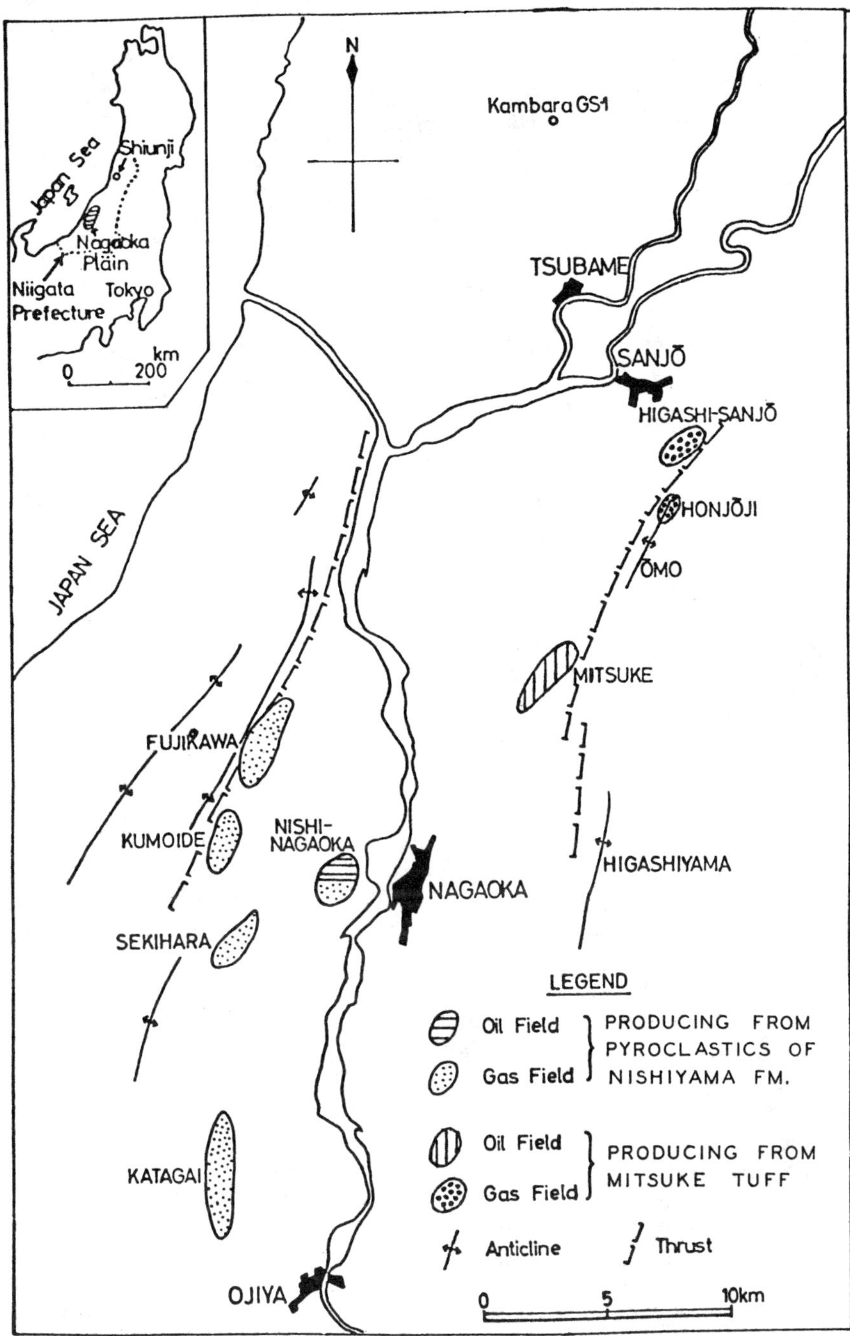

Fig. 1.—Distribution of oil and gas fields in Nagaoka Plain, Japan.

Lithofacies, biofacies, isopach and subsurface structure, and other maps have been prepared and used to determine the migration and accumulation of hydrocarbons in basins. However, no method which directly indicates the direction of the migration and the amount of compaction current has been developed. The writer proposes two new methods to detect the directions and the

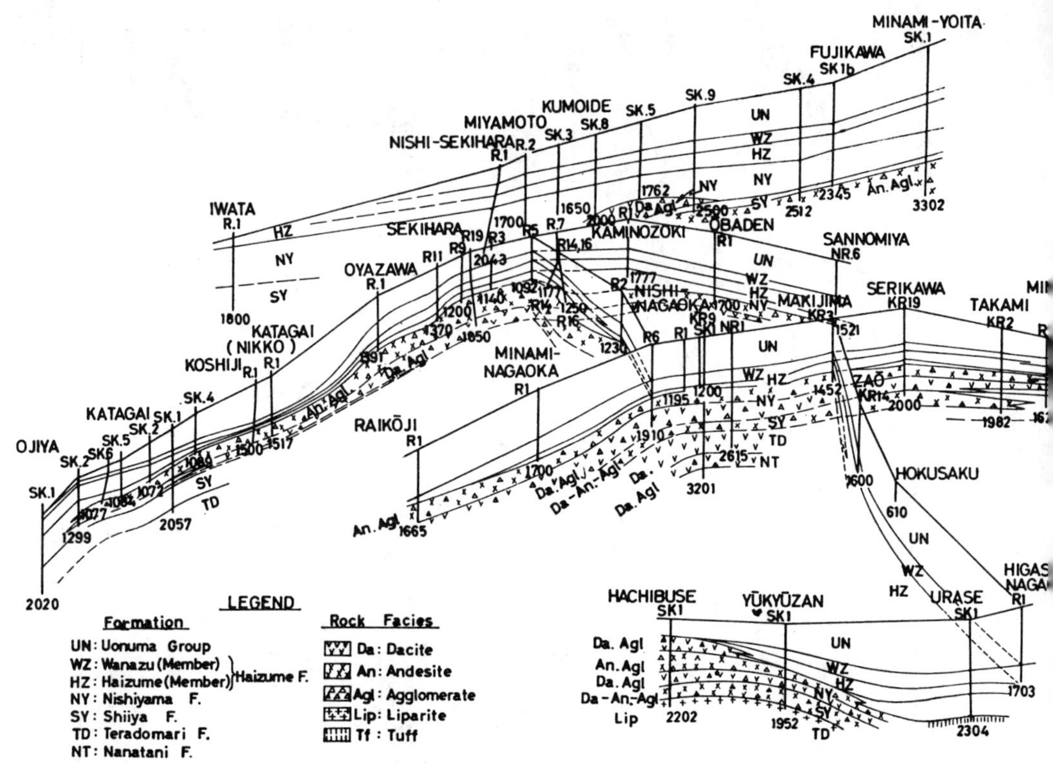

amount of compaction current, based on the horizontal and vertical porosity distributions of mudstone bodies.

Water expelled from mudstone may move in a carrier bed from the point of greater expulsion to the point of lesser expulsion. Therefore the horizontal porosity distributions of mudstone beds above and below the carrier bed suggest directions of the compaction currents. If water expulsion is inhibited, compaction of mudstone may not be great and high porosity–high fluid pressure will result. Because such a low-permeability mudstone seems to seal water movement, and fluid pressure in it seems to become relatively higher, compaction currents could move downward to the level below the low-permeability zone and upward above that zone. The amount of the compaction currents is estimated from the vertical porosity distributions and the porosity differences, before and after compaction.

General Geology

Table I shows the general stratigraphy and producing zones in Nagaoka Plain and the surrounding area. The basement rocks of the oil-bearing Tertiary System consist of Paleozoic strata and intrusive granite bodies, which crop out in the eastern mountainous district. The Paleozoic rocks are composed mainly of slate, sandstone, and quartzite.

Tsugawa Formation, basal Neogene, lies unconformably on the basement, and consists mainly of plagioliparite lava and tuff, and intercalated thin sandstone, conglomerate, and shale beds.

Mitsuke SK-36 (Fig. 2), one of the deepest wells in the region, penetrated liparite from 2,769 to 2,875 m (T.D.), similar in lithology to the Tsugawa Formation which crops out in nearby hills.

"It is supposed that submarine volcanisms, characterized chiefly by plagioliparite and partly by andesite, were active in the Tsugawan Stage. But the relationship between these volcanisms and sedimentation or tectonic movements is not clear, because of poor data" (Kujiraoka, 1965).

The Mitsuke Tuff (the Koguriyama Tuff), one of the important objectives for exploration in Nagaoka Plain, is composed of light-green to gray

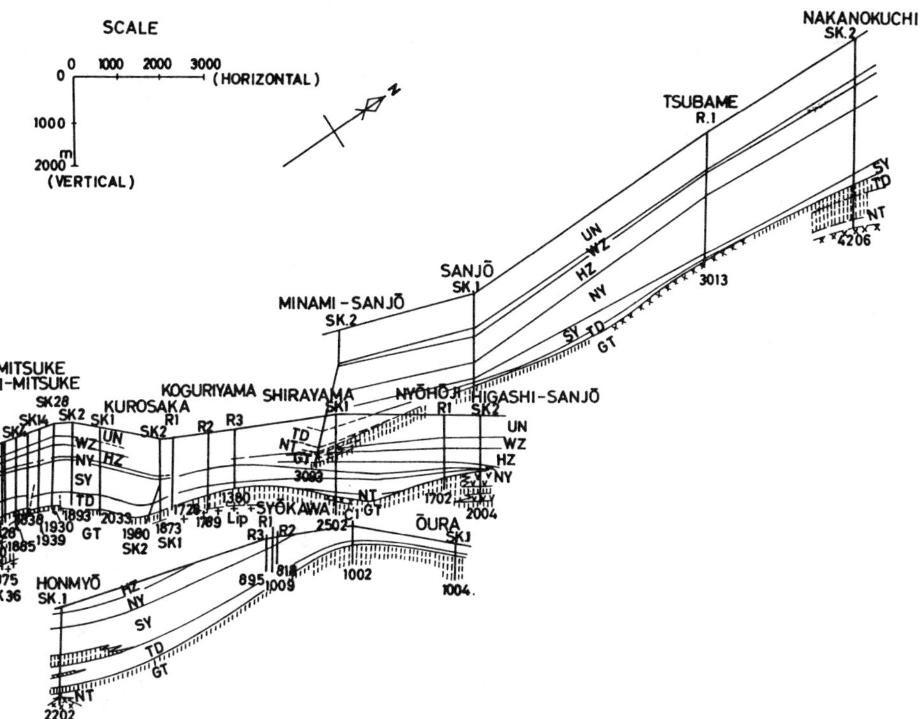

Fig. 2.—Panel diagram of Nagaoka Plain, Japan.

tuff and tuff breccia, intercalated with dacite lava. Fractured parts of the lava form the main hydrocarbon reservoirs of the Mitsuke oil field and Higashi-Sanjō gas field. The Mitsuke Tuff region, including these fields, is the northeastern part of the plain.

The Mitsuke Tuff is overlain in places by the Nanatani hard mudstone, and elsewhere by the Teradomari black mudstone. These mudstone bodies seem to have been the major source rocks in the Mitsuke Tuff region because of their rich organic content. Figure 3, an isopach map of the Teradomari and Nanatani clastic rocks, illustrates the synsedimentary highs at Mitsuke, Koguriyama, Ōmo, Honjōji, Higashi-Sanjō, and Ōura. This area is called the Mitsuke Tuff region. Hydrocarbons may have moved toward the highs, and some of them may have accumulated in commercial quantity, aided by the differential compaction of the mudstone source beds. After volcanic materials had accumulated in parts of the area, continuing compaction may have caused the masses of volcanic rock to remain higher than the surrounding mudstones and, therefore, to become reservoirs for the accumulation of hydrocarbons.

The Teradomari Formation is overlain conformably by the Shiiya Formation composed mainly of dark-gray mudstone, with some sandstone and volcanic rocks.

At the Fujikawa gas field on the western margin of the plain, andesite agglomerates, which are the gas reservoirs of this field, are overlain by mudstone of the Shiiyan Stage, hence are assigned to the Shiiya Formation. In the Kumoide gas field, about 4 km south of Fujikawa, the main reservoirs are dacite agglomerate and pyroclastic sandstone correlated with the Nishiyama Formation. Other examples of volcanic and pyroclastic rocks in the Nishiyaman Stage are found at Sekihara, Katagai, and Nishi-Nagaoka fields.

Herein, the region including Fujikawa, Kumoide, Sekihara, Katagai, and Nishi-Nagaoka fields is called the "Nagaoka agglomerate region" for convenience of description.

Sekihara SK-2, the deepest exploratory well at the Sekihara gas field, penetrated volcanic rocks from 1,192 to 1,743 m. The lowest part consists of a dacite agglomerate similar to the agglomer-

TABLE I. GENERAL STRATIGRAPHY AND PRODUCING ZONES IN NAGAOKA PLAIN AND SURROUNDING AREA, JAPAN (Redrawn from Kujiraoka, 1965)

Age	Formation		Max. Thick(m)	Lithology	Characteristic Megafossils	Volcanic Rocks	Producing Zones
Pleistocene	Jingamine F.		150	Sand, Gravel & Clay			
Pleistocene	Uonuma Group		2,400	Sand, Gravel & Clay			
Pliocene	Haizume F.	Wanazu M.; Haizume M.	1,000	massive Sandstone; bluish gray Siltstone			
Pliocene	Nishiyama F.	Nishiyama M.; Hama-tsuda	1,700	gray Mudstone; Alternation of Sandstone & Mudstone			
Miocene	Shiiya F.	Shiiya Facies; Araya Facies	1,300	Alternation of Sandstone & dark gray Mudstone; dark gray Mudstone			
Miocene	Teradomari F.		1,000	black Mudstone			
Miocene	Nanatani F.	Mitsuke Tuff	1,500	hard Mudstone; green Tuff			
Miocene	Tsugawa F.		1,600	green Tuff, Sandstone, Conglomerate & Mudstone			
Basement	Paleozoic Complex and Granite				Daijima flora; Chlamys kaneharai, Patinopecten kimurai; Sagarites chitanii; Omma-Manganji fauna; Palliolum peckhami; Parastegodon cf. akashiensis; Palaeoloxodon namadicus naumanni	Liparite; Dacite; Andesite; Basalt	Higashi-Sanjō; Ōmo; Mitsuke; Fujikawa; Kumoide; Sekihara; Nishi-Nagaoka; Katagai

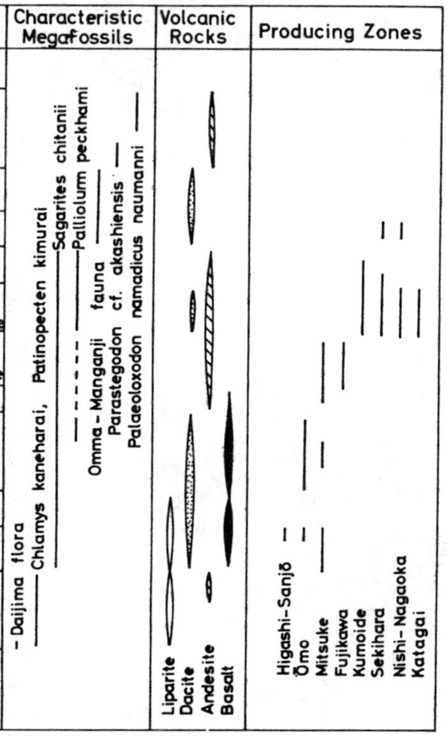

ate at Kumoide; the upper part, composed of andesite agglomerate, may be lithologically equivalent to the volcanic rocks found from 1,025 to 1,440 m at Katagai SK-1. Nishi-Nagaoka SK-1, drilled to 3,201.5 m, penetrated volcanic and pyroclastic rocks from 1,162 to 2,574 m. The upper part (from 1,162 to 1,865 m) consists of agglomerates of andesite and dacite. At Kumoide, Sekihara, Katagai, and Nishi-Nagaoka the andesitic and dacitic rocks are overlapped by the Nishiyama mudstone and are correlated with the Nishiyama Formation.

The middle Nishi-Nagaoka volcanic rocks (from 1,865 to 2,127 m) are andesite agglomerate and tuff of the Shiiyan Stage, and the lower part (from 2,127 to 2,574 m) consists of dacite lava and agglomerate of the Teradomarian Stage. The main hydrocarbon pools of the Nagaoka agglomerate region are in the uppermost pyroclastic rocks of the volcanic sequence.

The isopach map of the Nishiyama and Haizume clastic rocks overlying the volcanic reservoirs in Nagaoka Plain (Fig. 4) shows the synsedimentation highs at Katagai, Sekihara, Kumoide, Fujikawa, Nishi-Nagaoka, and elsewhere. Hydrocarbons may have migrated and accumulated in these highs.

Deposition of the Uonuma Group followed that of the Haizume Formation. Pronounced tectonic movements at the end of the Uonuman Stage produced many of the present structures. The strata in the hilly lands are folded strongly and in places overturned toward the Nagaoka Plain, with some thrusting. Below the plain, however, the folding is rather weak. The Highashi-Sanjō, Kumoide, and Fujikawa gas fields underlie the thrusts.

DIRECTIONS OF COMPACTION CURRENT

Several empirical relations between depth of burial and degree of compaction of shale and mudstone have been proposed. Figure 5, a comparison of the depth-porosity relations in several regions, shows (1) Athy's (1930) curve based on bulk densities of numerous samples of Pennsylvanian and Permian shale beds in northern Oklahoma; (2) Hedberg's (1936) curve based on bulk

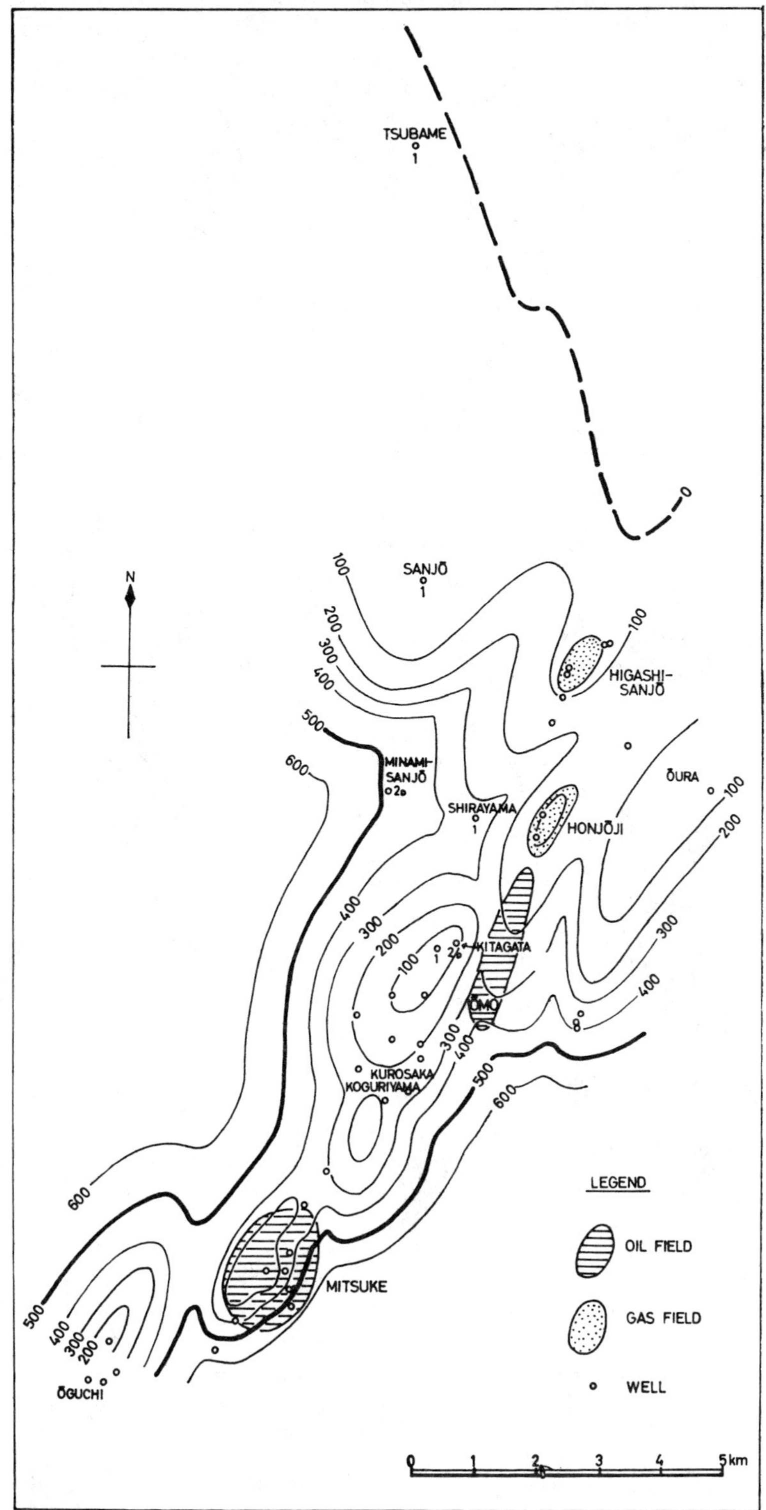

FIG 3.—Isopach map of Teradomari Formation and Nanatani clastic sediment in Mitsuke Tuff region.

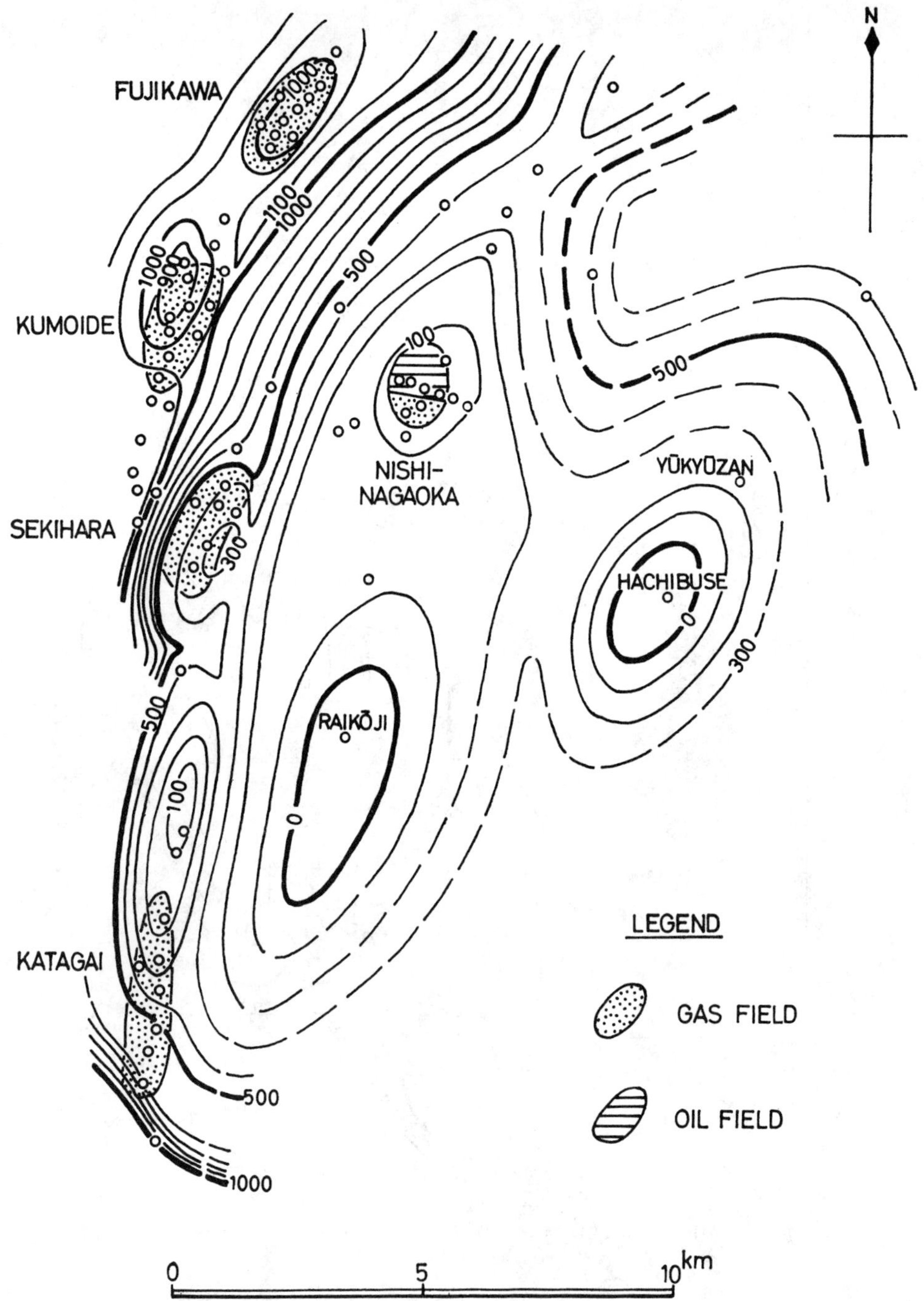

FIG. 4.—Isopach map of Haizume Formation and clastic rock of Nishiyama Formation in Nagaoka agglomerate region.

densities of samples of Tertiary shale in Venezuela; (3) an average relation for Tertiary sediments in the Gulf Coast region based on Dickinson's (1951) regional curve of bulk densities; and (4) Hosoi's (1963) curve based on bulk densities of many samples of Tertiary mudstone in Akita and Yamagata Prefectures, Japan.

These curves indicate a marked decrease in po-

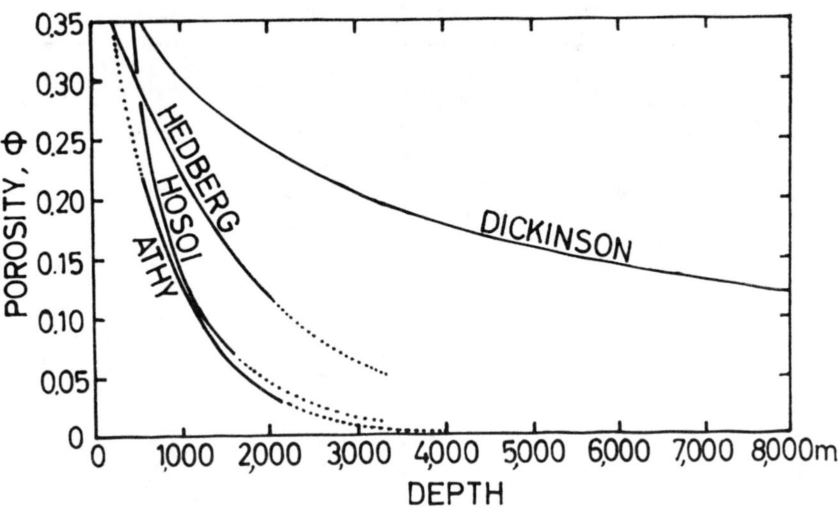

Fig. 5.—Comparison of depth-porosity relations in several regions: Athy (1930), Oklahoma; Hedberg (1936), Venezuela; Dickinson (1951), Gulf Coast; Hosoi (1963), Japan.

rosity at shallower depths; therefore, mechanical compaction of clay may occur in a geologically short time if water expulsion occurs. In addition, migration of fluid may depend on rapid mechanical compaction (Gussow, 1954; Levorsen, 1954; Schwade *et al.*, 1958).

As fluid escapes from pores of clayey rocks, compaction occurs and porosity decreases. However, if escape is inhibited, compaction may not be great and high porosity–high fluid pressure can result. More water accompanied by hydrocarbon may be expelled from clayey rocks to carrier beds in the former case than in the latter case.

As sediment accumulates on the sea bottom, lower layers are compacted to support the weight of the overburden. If the water contained in the lower layers can escape easily, it mingles with the water in the overlying rock and ultimately with the seawater. Hence, the fluid pressure in the lower layer would be nearly equal to the hydrostatic pressure corresponding to that depth. If, on the contrary, water within a layer were not free to move under the overburden, the water would not be squeezed out and the matrix could not be packed more closely because water is almost incompressible. Therefore, the water in the layer must support the added weight of the sediments. Fluid in pores would have greater pressure than the normal hydrostatic head, and abnormal pressure would be established.

Water expelled from mudstone beds may move in a carrier bed from the point of greater expulsion to the point of lesser expulsion. Therefore it is possible to determine directions of compaction currents by using the horizontal porosity distributions of shale or mudstone bodies above and below the carrier bed.

Data on rock density or porosity formerly depended on laboratory study of cores, but recently sonic and gamma-gamma logs (formation density logs) have been introduced and can be used to determine density or porosity of rocks.

After numerous laboratory tests, Wyllie *et al.* (1956, 1958) concluded that in consolidated strata with uniformly distributed small pores, there is a linear relation between porosity and transit time:

$$\Delta t \text{ log} = \Phi \Delta t \text{ liquid} + (1 - \Phi)\Delta t \text{ matrix} \quad (1)$$

or

$$\Delta t \text{ log} = (\Delta t \text{ liquid} - \Delta t \text{ matrix})\Phi + \Delta t \text{ matrix} \quad (1')$$

where Δt log is transit time on sonic log, Δt liquid and Δt matrix are transit times in formation liquid and matrix, respectively, and Φ is porosity of formation. This equation means that transit time increases with the porosity increase in rock of uniform lithology.

The relation between mudstone porosity values and acoustic transit time can be found by using the data derived from spot cores and the sonic log at Kambara GS-1, one of the deepest stratigraphic

Table II. Data from Core Analysis and Sonic Log of Kambara GS-1

depth(m)	core analysis		Sonic log
	density(gr/cm³)	porosity (%)	transit time([illegible])
1,029.08-1,029.23	2.00	39.00	145
1,609.40-1,609.60	2.11	33.15	127
1,808.63-1,808.76	2.13	26.56	130
2,150.65-2,150.85	2.27	24.60	109
2,296.00-2,296.20	2.22	24.26	114
2,443.46-2,443.66	2.24	23.08	110
2,607.16-2,607.33	2.26	21.82	102
3,062.77-3,062.98	2.28	19.60	99
3,205.36-3,205.53	2.32	18.80	95
3,505.25-3,505.46	2.35	15.90	104
3,701.29-3,701.49	2.42	14.60	92 (assumed)

test wells in Niigata Prefecture (Fig. 1). The result of the core analysis and transit times is shown in Table II, and the relation between mudstone porosity values, Φ, derived from the core analysis, and the transit times, Δt log, is indicated in Figure 6. It should be noted that there is a linear relation between Φ and Δt log which coincides with the Wyllie *et al.* equation.

Figure 7 shows sonic logs of the volcanic reservoirs of the Nishiyaman and Shiiyan Stages and the overlying and underlying mudstone beds of the main wells in the Nagaoka agglomerate region. The acoustic transit times of the overlying mudstone beds, which correspond to their porosity values, are considerably different from place to place. For example, at Minami-Nagaoka R-1 well the mudstone transit time, Δt, is 103 μsec/ft; at Sekihara R-5, $\Delta t = 125$ μsec/ft; and at Kumoide SK-8, $\Delta t = 107$ μsec/ft. Figure 7 also indicates the porosity values of the overlying mudstone beds in these wells, which are determined from the acoustic data by use of Figure 6.

Similarly, the acoustic data and the porosity values of the mudstone beds which overlie the volcanic reservoirs in the Nagaoka agglomerate region are listed in Table III. Figure 8 shows the porosity distributions of the overlying mudstone beds, which indicate a differential compaction in the region. As has been noted, compaction currents may move in the reservoir rocks from a place of greater water expulsion from the overlying mudstone, with resultant lowering of porosity, to a place of lesser water expulsion. The directions of migration in the volcanic reservoirs are shown by arrows in Figure 8.

The arrows suggest the locations where oil or gas has accumulated. In addition, these porosity distributions correspond closely to the present structural pattern of the volcanic reservoirs at the Nishi-Nagaoka, Sekihara, and Katagai oil and gas fields. In the Kumoide-Fujikawa trend, the overlying mudstone porosity increases northward, although the present structure plunges north. The compaction current may have flowed along this

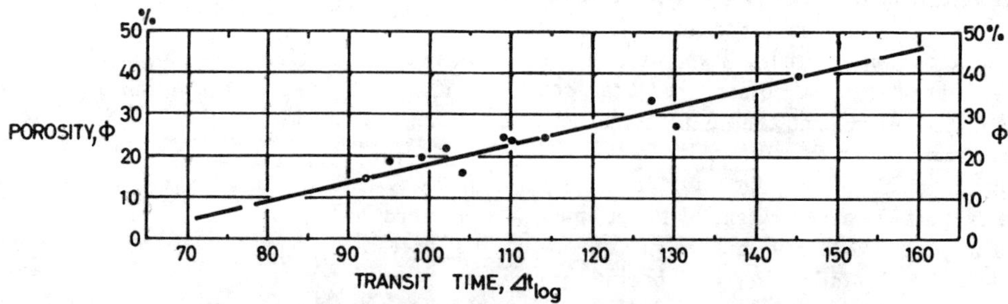

Fig. 6.—Relation between mudstone porosity, Φ, and transit time, Δt log (μ sec/ft).

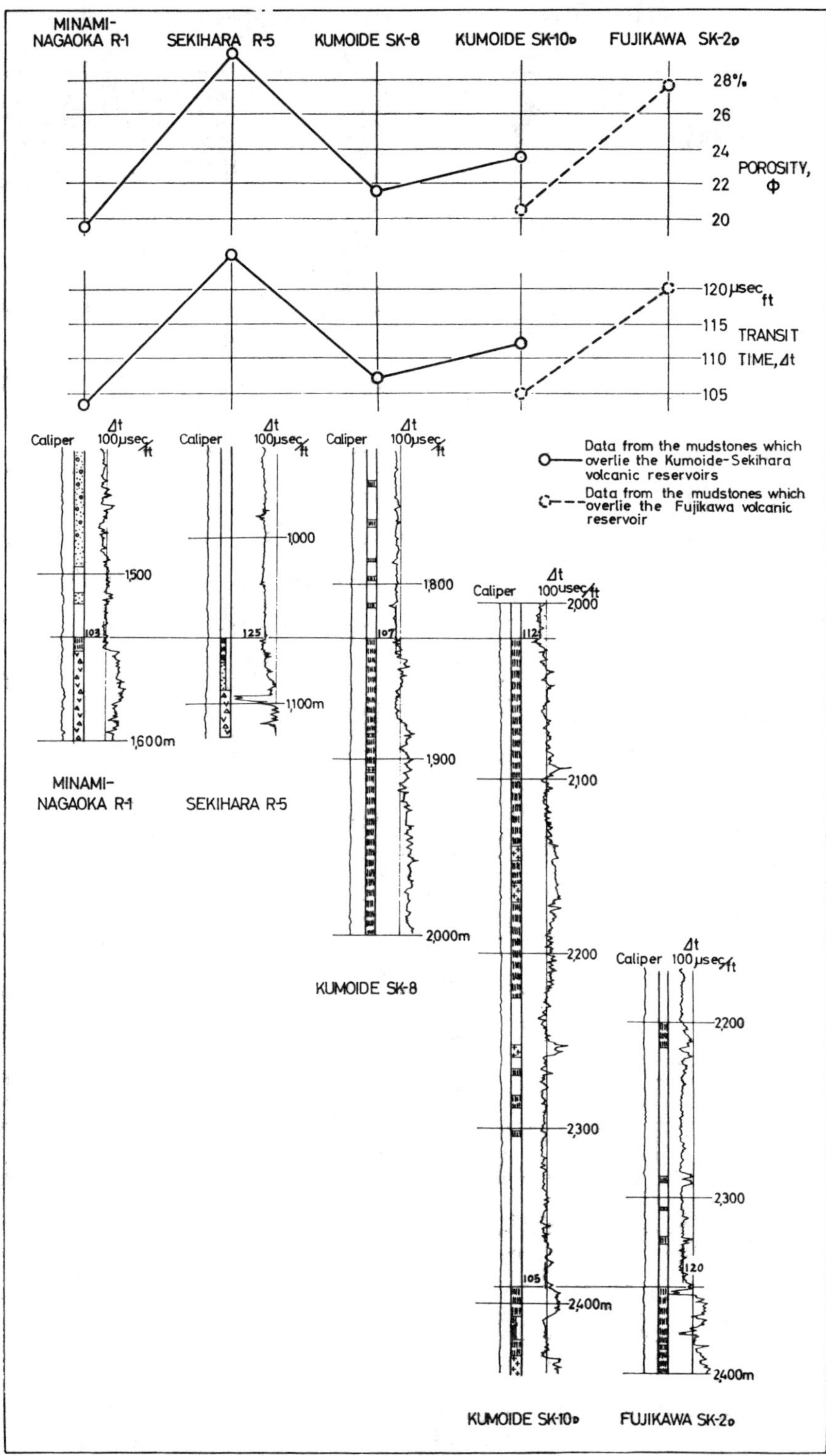

Fig. 7.—Sonic logs of volcanic reservoirs and overlying and underlying mudstone beds of main wells in Nagaoka Plain.

Table III. Transit Times and Porosities of Overlying Mudstones in Nagaoka Agglomerate Region

well		depth(m)	transit time(μs/ft)	porosity(%)	well		depth(m)	transit time(μs/ft)	porosity(%)
Fujikawa	SK-1b	2,300	123	28.5	Sekihara	R-8	1,085	125	29.5
	2_D	2,330	120	27.5		9	1,085	125	29.5
	3_D	2,330	120	27.5		11	1,290	115	25.0
	4	2,315	109	22.5		19	1,100	130	32.0
	5_D	2,310	122	28.5		20	1,140	120	27.5
	6_D	2,420	115	25.0	Oyazawa	R-1	1,350	112	23.5
	8_D	2,300	115	25.0	Kaminozoki	R-2a	1,250	120	27.5
	9_D	2,445	115	25.0	Ōbaden	R-1	1,435	116	25.5
	10_D	2,500	115	25.0	Minami-Nagaoka	R-1	1,540	103	19.5
Kumoide	SK-8	1,820	107	21.5	Koshiji	R-1	1,230	123	28.5
	9	2,100	123	28.5	Katagai	SK-2	1,000	133	33.5
		2,350	111	23.0		3	1,090	135	34.5
	10_D	2,020	112	23.5		4	1,065	140	36.5
		2,380	105	20.5	Nishi-Nagaoka	KR12-2	1,180	132	33.0
	13_D	1,995	100	18.0		16-1	1,170	123	28.5
	14_D	1,800	105	20.5		16-4	1,200	123	28.5
	16	1,830	105	20.5		16-5	1,220	130	32.0
	17	2,255	116	25.5		15-1	1,155	130	32.0
		2,480	95	16.0		15-2	1,185	129	31.5
	18_D	2,520	105	20.5		15-3	1,180	130	32.0
Miyamoto	R-1	1,780	107	21.5	Minami-Ōguchi	R-1	880	133	33.5
Nishi-Sekihara	R-1	2,000	105	20.5	Yūkyūzan	SK-1	1,040	122	28.5
Sekihara	R-5	1,075	125	29.5	Urase	SK-1	2,160	106	21.0
	7	1,300	115	25.0	Minami-Yoita	SK-1	2,520	113	24.0

trend from south to north, and the discordance between the porosity distributions and locations of present structures are explained in the following paragraphs.

In the Kumoide-Fujikawa trend, the overlying Haizume and Nishiyama Formations thicken markedly southward. Therefore, it seems that these formations subsided more deeply and consequently were more compacted in the south than in the north. Because major compaction may occur in the early stage of subsidence, early compaction may have been the main cause of present porosity distribution of the mudstone. However, this explanation does not entirely solve the problem, because the present structure plunges northward in this trend.

Philipp *et al.* (1964) stated that there is a relation between the porosity values of the Aalenian (Early Jurassic) water-filled sandstone beds in northwestern Germany and the maximum depths of burial. In addition, these authors assumed that the decrease in porosity with depth was interrupted by hydrocarbon accumulation. If this assumption can be applied also to overlying mudstone beds, the compaction situation of the Kumoide-Fujikawa trend seems to be solved. Pressure, however, may be a major problem.

Hubbert and Rubey (1959) have demonstrated that the load, S, is supported jointly by the fluid pressure, p, and the effective stress, σ, of the clay aggregates.

$$\sigma = S - p, \tag{2}$$

and load, S, can be expressed

$$S = \overline{\rho_{bw}}\ g\ Z \tag{3}$$

where $\overline{\rho_{bw}}$ is the mean value of the water-saturated bulk density of the overlying sediments, g is the acceleration of gravity, and Z is the depth of burial.

The effective stress σ exerted by the porous clay depends solely on the degree of compaction of the clay. The stress σ increases continuously as the porosity decreases. Hence, this can be expressed as follows (Rubey and Hubbert, 1959),

$$\sigma = f(\Phi). \tag{4}$$

It also may be stated that for a specified clay there is for each value of porosity Φ a maximum value of effective stress σ which the clay can support without further compaction. Consequently, the effective stress σ or porosity Φ at a particular burial depth is dependent on the fluid pressure p.

If the fluid pressure is abnormally high, the effective stress σ will be abnormally low, or the porosity will be abnormally high for a specified

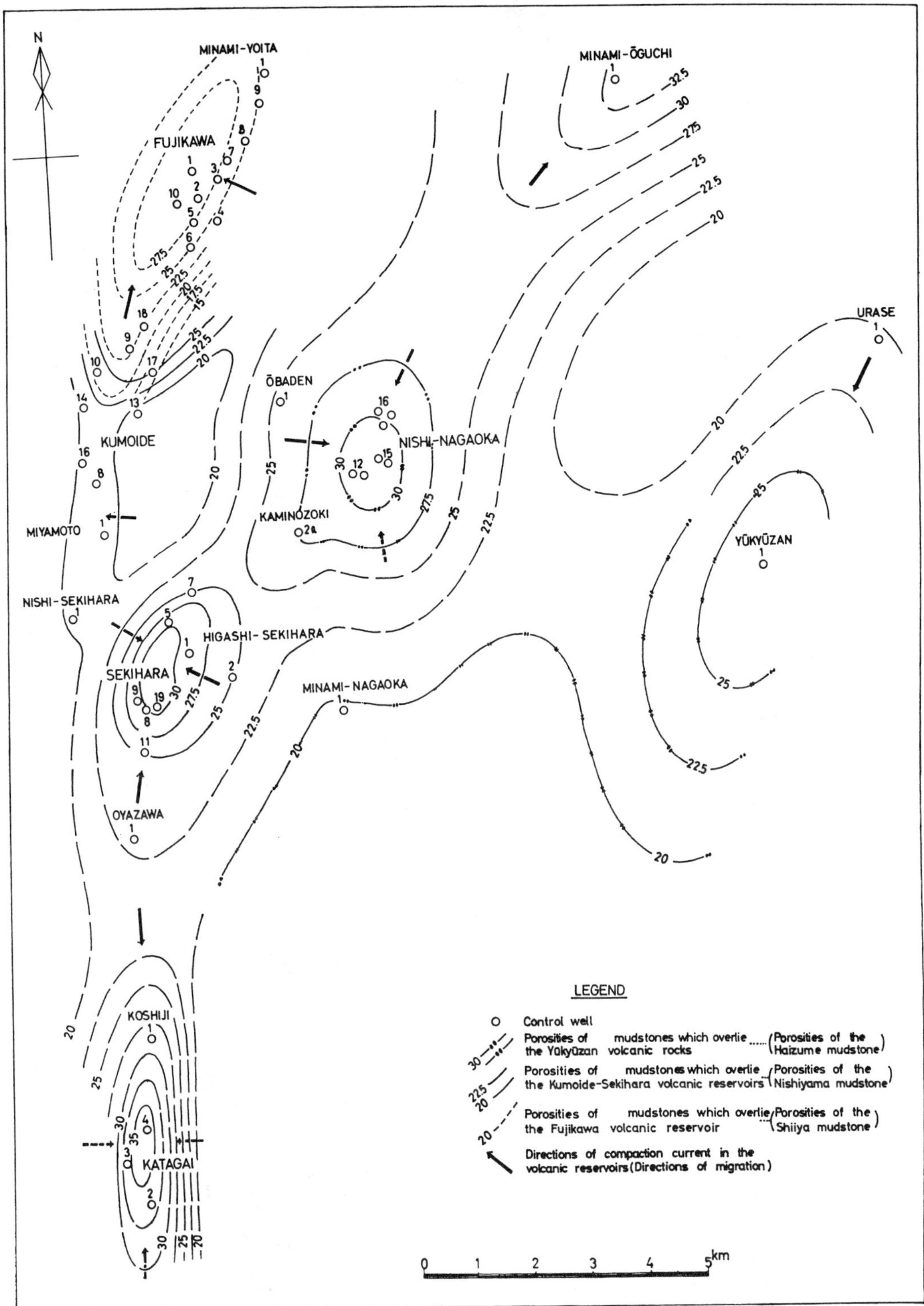

FIG. 8.—Differential compaction map of overlying, mudstone beds in Nagaoka agglomerate region.

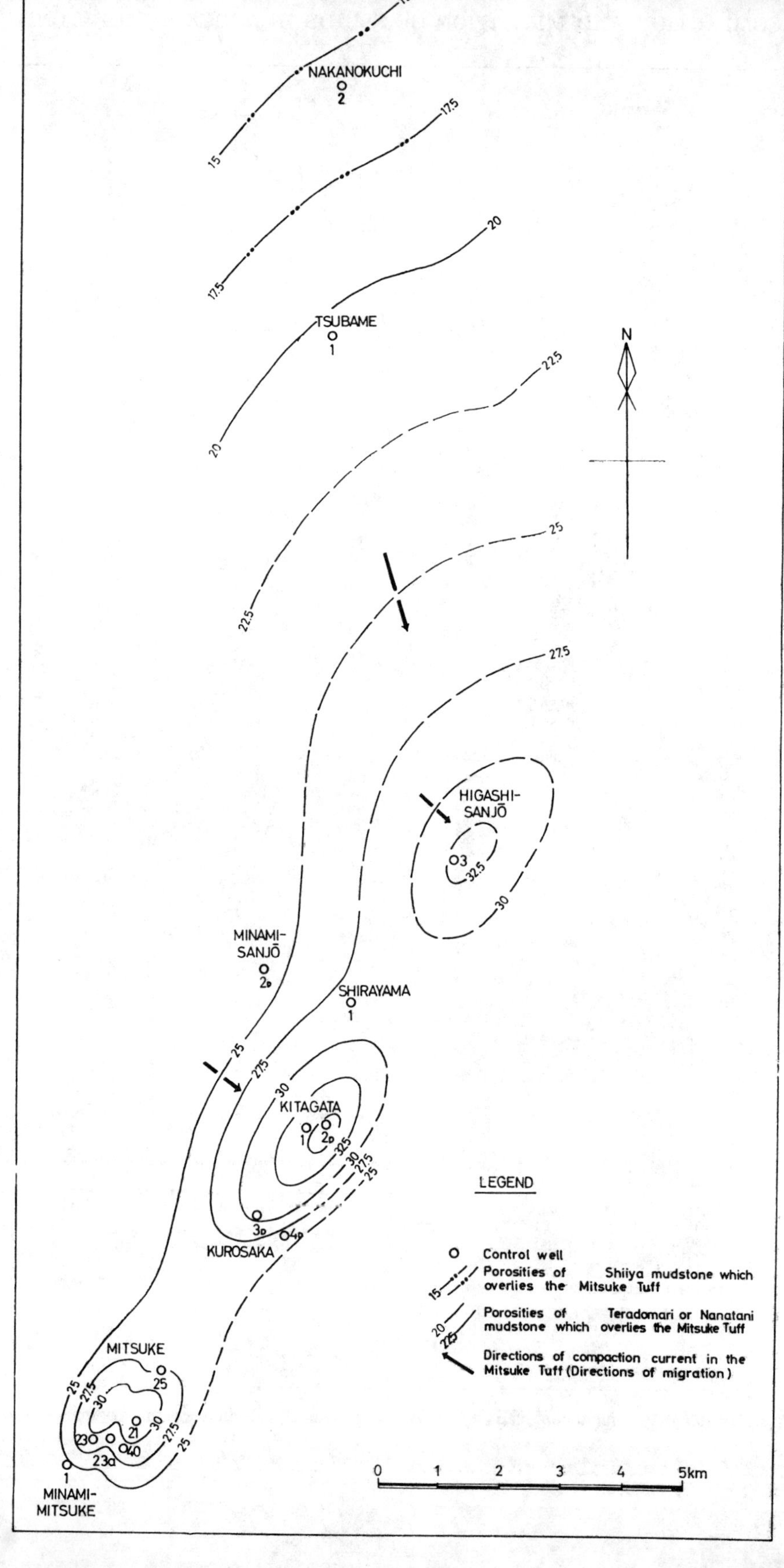
NAKANOKUCHI
2
TSUBAME
1
N
HIGASHI-
SANJŌ
3
32.5
30
MINAMI-
SANJŌ
2D
SHIRAYAMA
1
25
27.5
30
KITAGATA
1
2D
32.5
3D
4D
KUROSAKA
MITSUKE
25
27.5
30
21
23
23a
40
1
MINAMI-
MITSUKE
15
17.5
20
22.5
LEGEND
Control well
Porosities of Shiiya mudstone which overlies the Mitsuke Tuff
Porosities of Teradomari or Nanatani mudstone which overlies the Mitsuke Tuff
Directions of compaction current in the Mitsuke Tuff (Directions of migration)
0
1
2
3
4
5km

burial depth Z. Therefore, it would be reasonable to say that the abnormal reservoir pressure in the Fujikawa gas field caused the high porosity of the mudstone bodies.

As a result of greater subsidence, the formations were more compacted in the southern part than in the northern part in the early stages of compaction, and compaction current then moved northward in the volcanic reservoirs. In geologic time, abnormally higher pressure was generated in the northern part, and compaction of the overlying mudstone beds may have been arrested.

Another example, shown in Figure 9, indicates the porosity distribution of the overlying mudstone beds in the Mitsuke Tuff region. The transit times and the porosity values of the mudstone beds are also listed in Table IV. The data available for preparing a differential compaction map are insufficient, although the mapped area contains the Mitsuke oil field and Higashi-Sanjō gas field. The general direction of the compaction current in the Mitsuke Tuff seems to be southeast.

TABLE IV. TRANSIT TIMES AND POROSITIES OF OVERLYING MUDSTONES IN MITSUKE TUFF REGION

well		depth(m)	transit time(μs/ft)	porosity(%)
Mitsuke	SK-21D	1,610	130	32.0
	23	1,720	125	29.5
	23a	1,690	122	28.5
	25D	1,800	120	27.5
	40a	1,550	125	29.5
Minami-Mitsuke	SK-1	2,160	112	23.5
Kurosaka	SK-3D	1,620	128	31.0
	4D	1,715	118	26.5
Higashi-Sanjō	SK-3	1,310	132	33.0
Minami-Sanjo	SK-2D	2,720	110	23.0
Kitagata	SK-1	1,195	135	34.5
	2D	1,170	138	35.5
Shirayama	SK-1	1,815	121	28.0
Nakanokuchi	SK-2	3,180	95	16.0
Tsubame	R-1	2,840	105	20.5

QUANTITATIVE EVALUATION OF COMPACTION CURRENTS

Even more useful to petroleum exploration than the directions of compaction currents in the volcanic reservoirs in Nagaoka Plain is the evaluation of the amount of compaction current in each field and its relation to hydrocarbon accumulation.

The Shiunji gas field provides an excellent example to demonstrate the calculation of the amount of compaction currents in the geologic past. The Shiunji gas field is in the northern part of Niigata Prefecture (Fig. 1) and the pay zones are the lenticular sandstone bodies in the Nishiyama-Shiiya Formations and the Shuinji tuff bed in the Nanatani Formation, which is approximately equivalent to the Mitsuke Tuff. The Shiunji tuff bed directly overlies Paleozoic basement rock composed mainly of metamorphosed sandstone. Therefore, source rocks would not be expected below the tuff bed and it would be reasonable to postulate some downward or lateral migration from overlying source beds. Landes *et al.* (1960) summarized the possibility of oil accumulation in basement rocks:

> Basement rock accumulations obtain their oil from one of three possible sources: (1) overlying organic rock, from which the oil was expelled downward during compaction, (2) lateral, off-the-basement but topographically lower, organic rock from which oil was squeezed into an underlying carrier bed through which it migrated updip into the basement rock, and (3) lower lateral reservoirs from which earlier trapped oil was spilled due to tilting, or to overfilling.

Landes's last statement (3) seems to be inadequate to explain the accumulation in the Shiunji tuff. The deep exploratory well at Shiunji, Shuinji SK-21, was drilled and successfully completed in two pay zones in the tuff. Figure 10 indicates conductivity and acoustic transit time of mudstone, drilling rate, and specific gravity of drilling mud derived from Shiunji SK-21. The "normal trends" in the conductivity and transit-time plots have been established from all Shuinji well-log data. Following Hottman and Johnson (1965), the writer defines "normal trend" as the linear approximation of the relation between the logarithm of mudstone conductivity or transit time and depth in hydrostatic condition. The maximum positive deviation from the normal trend in the conductivity or transit-time plots seen in Teradomari mudstone seems to correspond qualitatively

←

FIG. 9.—Differential compaction map of overlying mudstone beds in Mitsuke Tuff region.

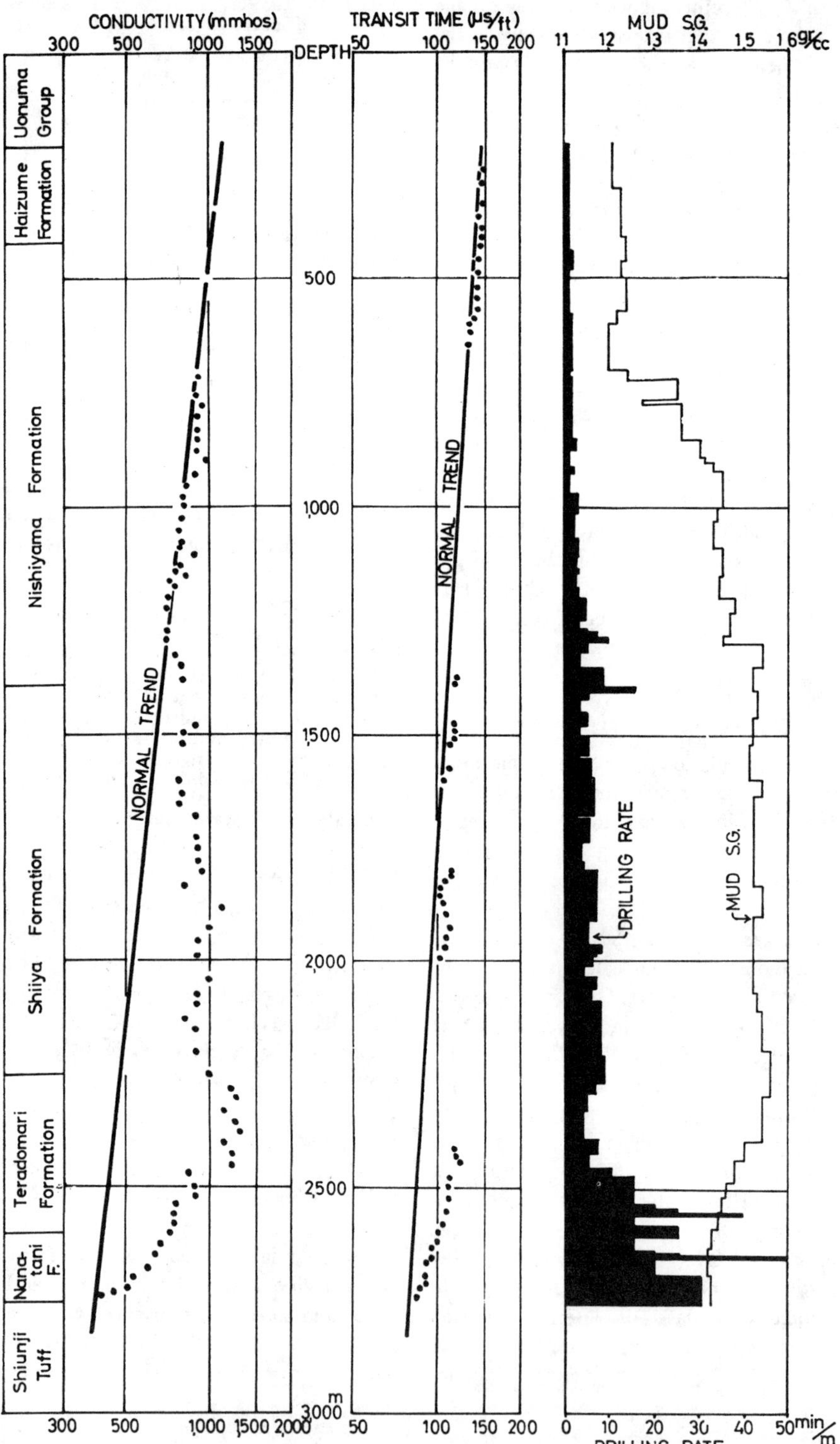

Fig. 10.—Mudstone conductivity, transit time, drilling rate, and specific gravity of drilling mud of Shiunji SK-21.

with the abnormally high mudstone porosity,[4] and the drilling rate generally increases in that section. As has been noted, if escape of fluid from mudstone is inhibited, compaction may be slight and high pressure–high porosity may result. Thus, the high porosity of these mudstone beds may be the result of inhibited water expulsion from the mudstone during subsidence. This suggests that in some periods of the geologic past the water in such mudstone beds would scarcely have moved.

According to Fons and Holt (1966), montmorillonitic shale generally is more resistant than illite or kaolinite shale to internal fluid migration and compaction.

The fluid pressures in such low-permeability mudstone would also become relatively higher than the pressures above and below the mudstone beds. It therefore seems reasonable that these mudstone beds would act as barriers to water movement, and compaction currents could move both downward and upward from the low-permeability zone.

A reliable mudstone porosity value is needed to evaluate the amount of compaction current, but no porosity data were taken in the Shiunji SK-21 well except a few spot sonic-log data. The mudstone porosity values are calculated from the mudstone resistivity levels measured by the induction log.

Archie (1942) defined the formation resistivity factor as:

$$Ro = F \times Rw \quad (5)$$

where F is formation resistivity factor, Ro is electrical resistivity of rock saturated 100 percent with water, and Rw is electrical resistivity of water.

For a pure mudstone, the resistivity of mudstone measured by the induction log approximates Ro, because mudstone can be assumed to be saturated with water. Though the resistivity of the water saturating the mudstone is unknown, it may be equal or related in magnitude to the resistivity of the water saturating nearby sandstone. Archie found that the formation resistivity factor, F, is related to porosity:

[4] If the resistivities of matrix and formation water are the same, the higher porosity mudstone contains more (salty) water and resistivity lower (conductivity is higher) than the lower porosity mudstone.

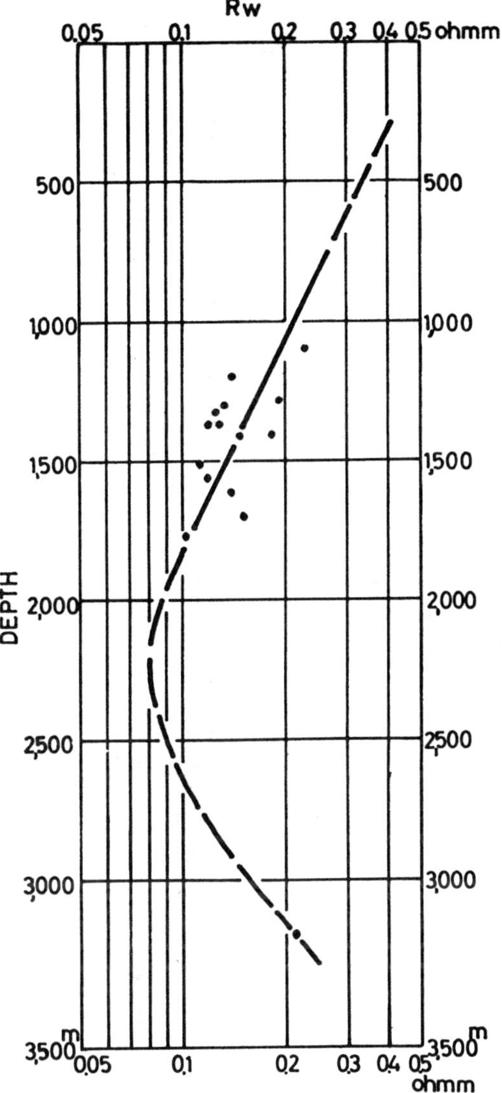

FIG. 11.—Relation between formation water resistivity, Rw, and depth at Shiunji gas field.

$$F = \frac{1}{\Phi^m} \quad (6)$$

where m is the cementation factor. If Rw and m are known for each depth, Φ can be calculated from Ro, or mudstone resistivity. Rw is derived from water analysis data by the method proposed by Dunlap and Hawthorne (1951). The relation between Rw and depth at Shiunji is shown in Figure 11.

Because gamma-gamma logs have been recorded

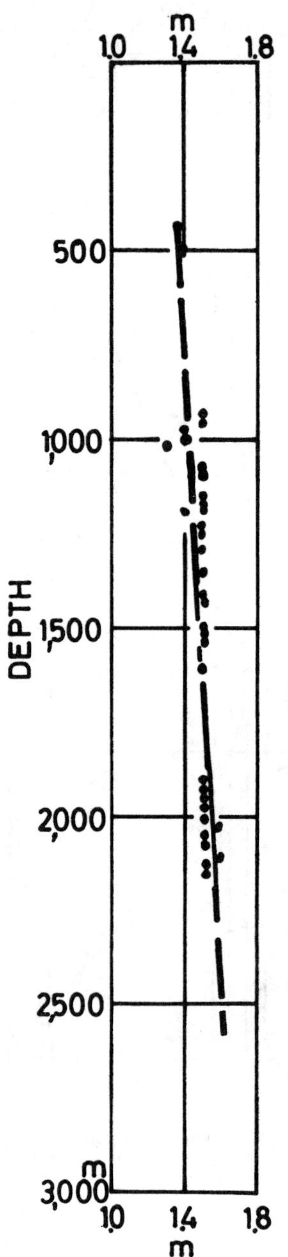

FIG. 12.—Relation between cementation factor, *m*, and depth of Shiunji gas field.

in Shiunji SK-7 and SK-10, near Shiunji SK-21, values of mudstone porosity, Φ, can be calculated easily from the gamma-ray counts per second.

The cementation factor, *m*, is determined from Φ, *Ro*, and *Rw*. Figure 12 shows the relation between *m* and depth as derived from SK-7 and SK-10.

The mudstone porosity values at Shiunji SK-21 are estimated from the induction log, based on these *m* and *Rw* values, and indicated in Figure 13. The normal porosity trend which corresponds to hydrostatic pressure in the subsurface also is estimated from the normal trend of conductivity shown in Figure 10, based on the aforementioned method (Fig. 13).

To calculate the volume of water expelled from source rocks from a given time to the present, it is necessary to know the original thickness of the source rock.

Assuming that the mudstone grain is the same before and after burial, the following relation obtains:

$$D(1 - \overline{\Phi}) = D'(1 - \overline{\Phi}') \qquad (7)$$

where D and D' are the thicknesses of mudstone before and after burial, respectively, and $\overline{\Phi}$ and $\overline{\Phi}'$ are the average porosity values before and after burial, respectively. The reconstruction of the formations at the end of deposition of the Shiiya Formation (the beginning of deposition of the Nishiyama Formation) is determined by the method described in equation 7 and shown in Figure 14. It is impossible to determine the vertical porosity distribution at the end of the Shiiyan Stage, but the writer assumes it conformed to the normal porosity trend of mudstone expressed in Figure 13, implying that at that time water could be expelled rather easily from the mudstone and a hydrostatic condition prevailed. The present porosity distribution, after burial, also is plotted in Figure 14. Because the Teradomari mudstone just above 1,340 m (Fig. 14) seems to have acted as a barrier to water movement, the water squeezed from the mudstone below 1,340 m would have moved downward and the water expelled above 1,340 m would have gone upward to the sea.

The volume of expelled water, W, can be calculated by the equation:

$$W = \overline{\Delta\Phi} \times V \qquad (8)$$

where $\overline{\Delta\Phi}$ is the mean mudstone porosity decrease during compaction, or $\overline{\Phi} - \overline{\Phi}'$, and V is the volume of mudstone before burial. The volume of water squeezed out downward, *Wd*, is determined, where the average porosity decrease of mudstone from 1,340 to 1,640 m, $\overline{\Delta\Phi_d}$, is 5 percent, by the

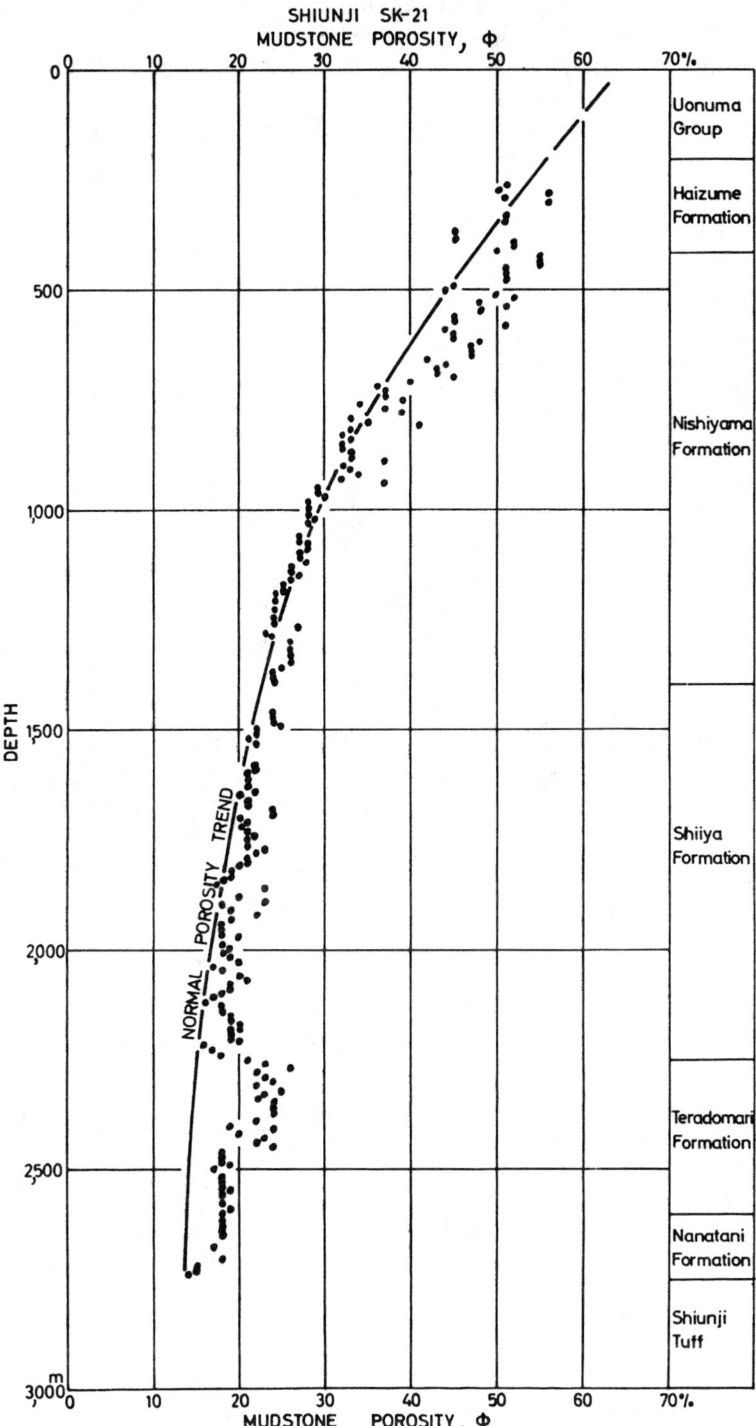

Fig. 13.—Mudstone porosity values of Shiunji SK-21.

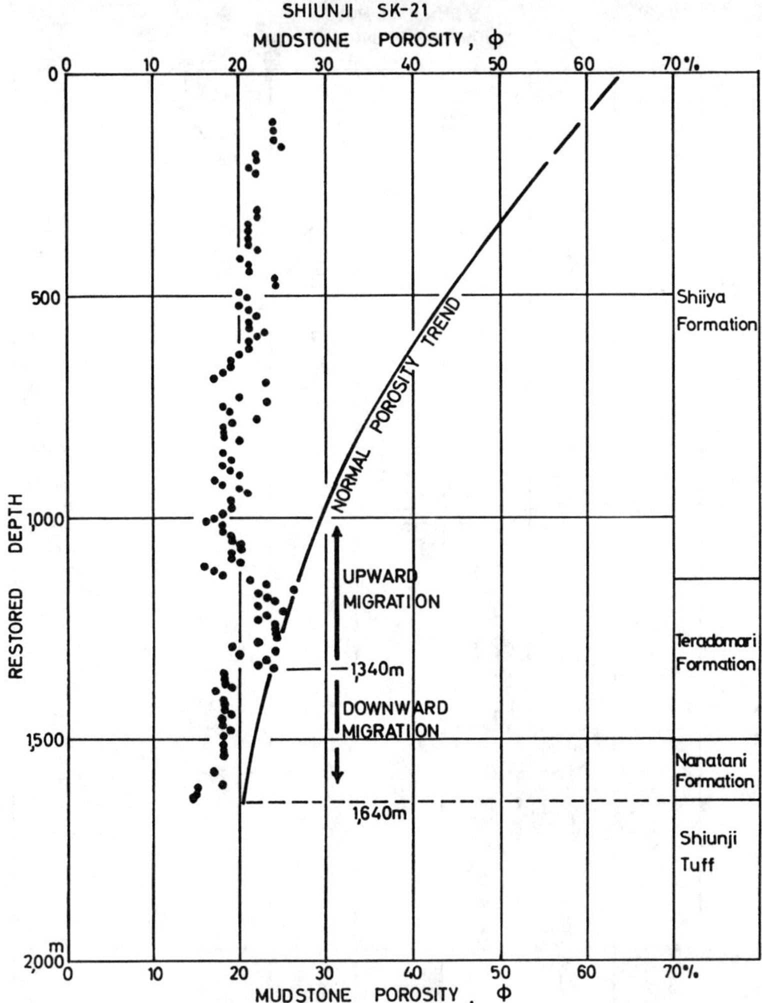

Fig. 14.—Reconstruction of formations at end of Shiiyan Stage and mudstone-porosity plot of Shiunji SK-21.

equation:

$$Wd = \overline{\Delta\Phi_d} \times V = 0.05 \times (1{,}640 - 1{,}340) \times 1 = 15m^3.$$

The volume of water expulsion per 1-m² unit of mudstone column after the beginning of the Nishiyaman Stage is 15 m³ downward. The volume of downward water expulsion accompanied by organic material seems to have been important in forming the hydrocarbon accumulation in the Shiunji Tuff.

Because the reconstruction of formations is very complicated and time consuming, the writer proposes one simple and convenient way of calculating a volume of downward compaction current.

For a volume of mudstone, equation 7 can be expressed:

$$V(1 - \overline{\Phi}) = V'(1 - \overline{\Phi}') \qquad (9)$$

where V and V' are the volumes of mudstone before and after burial, respectively. First, the present (after burial) vertical porosity distribution of mudstone in each well, as shown in Figure 15, must be established. Then, by assumption of the normal porosity trend, the average porosity below the barrier before burial, $\overline{\Phi}$, should be known.

In addition, if an average porosity below the barrier after burial, $\overline{\Phi}'$, and the present (after burial) volume of mudstone below the barrier,

V', is known, the mudstone volume before burial, V, can be estimated:

$$V = V' \frac{(1 - \overline{\Phi}')}{(1 - \overline{\Phi})}. \tag{10}$$

Therefore, a volume of downward compaction current, Wd, can be calculated:

$$Wd = \overline{\Delta\Phi V} = (\overline{\Phi} - \overline{\Phi}')V' \frac{1 - \overline{\Phi}'}{1 - \overline{\Phi}}. \tag{11}$$

This method is applied to Nagaoka Plain.

The volcanic reservoirs in Nagaoka Plain differ from the basement rock types at Shiunji in that there is source mudstone below the reservoirs. However, it is believed that hydrocarbons produced from the overlying mudstone could have been entrapped in the reservoirs more effectively because the overlying mudstone is also important as cap rock. It is doubted that the cap rock overlying the reservoir was present at the time the compaction current expelled fluids from the underlying mudstone. The overlying source mudstone thus seems to be more important than the underlying mudstone in the accumulation of hydrocarbons. Because of the much greater significance of the overlying mudstone in Nagaoka Plain, and the scarcity of data on the underlying mudstone, the writer has studied only the overlying rock.

The vertical mudstone porosity distributions of the main wells in the Nagaoka agglomerate region are determined from the sonic-log data, based on Figure 6 and illustrated in Figures 16–18. These sections suggest that the compaction current may have moved into the volcanic reservoirs from the Shiiya mudstone at Fujikawa and from the Nishiyama mudstone at Kumoide, Sekihara, and Katagai, indicating that these are the major source beds.

The clear porosity differences between the mudstone barriers and the underlying mudstone beds are observed at the Fujikawa, Kumoide, and Seki-

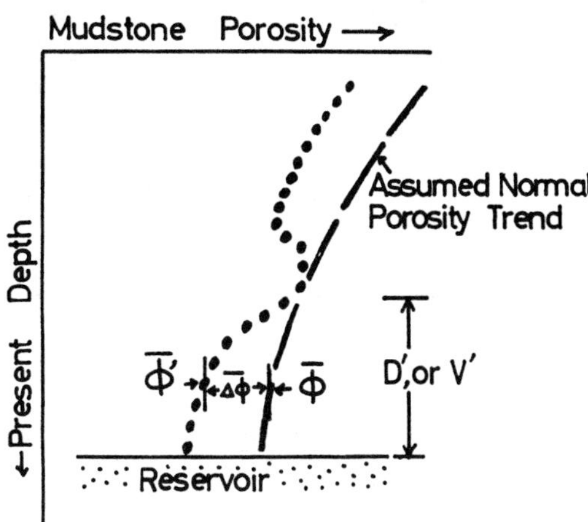

FIG. 15.—Diagram showing way of calculating volume of compaction current.

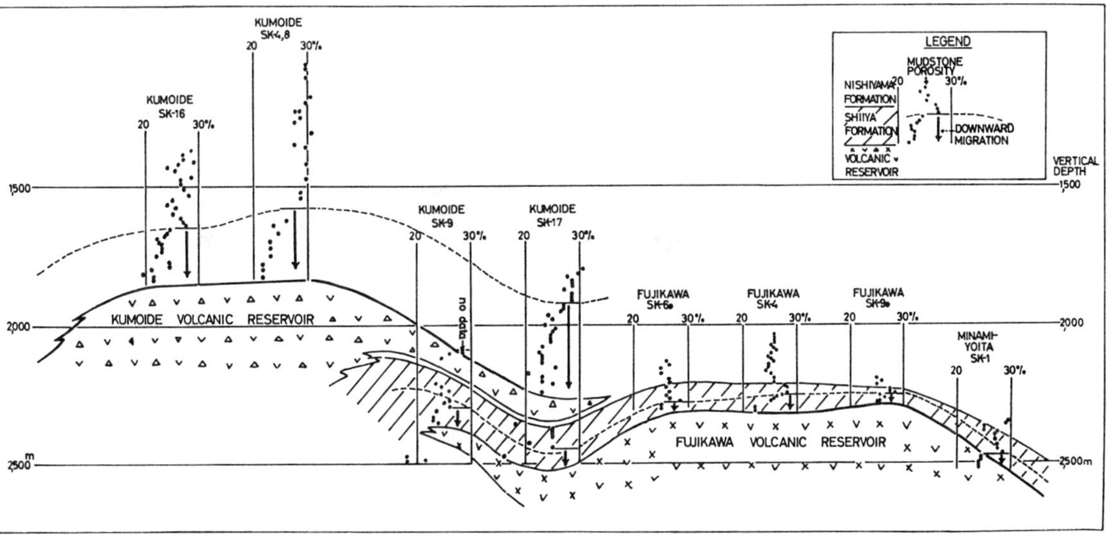

FIG. 16.—Vertical porosity distributions of mudstone in Fujikawa-Kumoide trend.

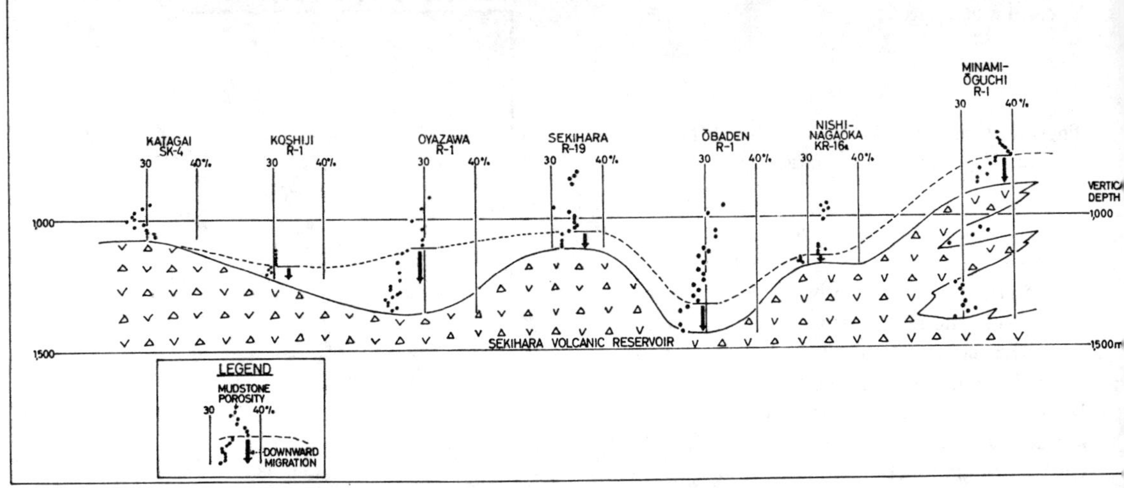

FIG. 17.—Vertical porosity distributions of mudstone at Nishi-Nagaoka, Sekihara, and Katagai oil and gas fields.

hara gas fields, but at Yūkyūzan and Minami-Nagaoka there seems to have been only slight downward migration.

Table V shows the volumes of downward compaction currents per unit mudstone column of base area 1 m² at the main wells in the Nagaoka agglomerate region. The current is largest at Kumoide and Oyazawa, and also may be important in the formation of hydrocarbon accumulations at Kumoide, Sekihara, and Katagai. Figure 19 illustrates the variation in present overlying mudstone thicknesses which caused the downward compaction current, or downward migration. In order to evaluate the total volumes of the

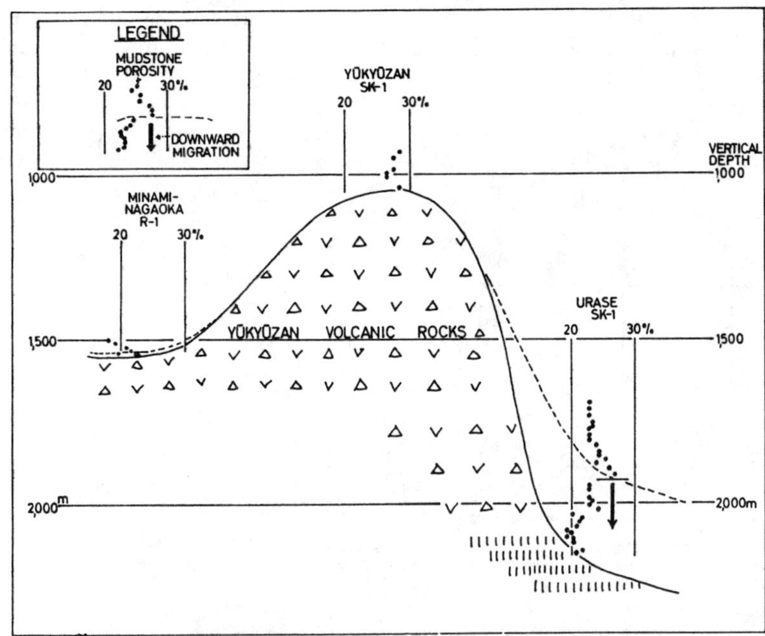

FIG. 18.—Vertical porosity distributions of mudstone at Yūkyūzan and Minami-Nagaoka.

Table V. Volumes of Downward Compaction Currents in Nagaoka Agglomerate Region

well	average value of porosity below the mudstone barrier		$\bar{\Phi}-\bar{\Phi}'$	$\frac{1-\bar{\Phi}'}{1-\bar{\Phi}}$	mudstone thickness below the barrier		volume of downward compaction current per $1m^2$ Wd(m^3)	block	representative well	$\bar{\Phi}-\bar{\Phi}'$	$\frac{1-\bar{\Phi}'}{1-\bar{\Phi}}$	mudstone volume below the barrier		total volume of downward compaction current total Wd (m^3)	overlying mudstone
	before burial $\bar{\Phi}$	after burial $\bar{\Phi}'$			after burial D'(m)	before burial D(m)						after burial V'(m^3)	before burial V(m^3)		
Minami-Yoita SK-1	0.28	0.24	0.04	1.06	0.5×10^2	0.53×10^2	2.1	A	Minami-Yoita SK-1	0.04	1.06	4.4×10^9	4.7×10^9	19×10^7	Shiiya
Fujikawa SK-4	0.28	0.24	0.04	1.06	0.7×10^2	0.74×10^2	3.0	B	Fujikawa SK-4	0.04	1.06	2.1×10^9	2.2×10^9	8.9×10^7	Shiiya
Kumoide SK-17	0.25	0.19	0.06	1.08	0.5×10^2	0.54×10^2	3.2	C	Kumoide SK-17	0.06	1.08	2.0×10^9	2.2×10^9	13×10^7	Shiiya
Kumoide SK 17	0.28	0.24	0.04	1.06	3.3×10^2	3.5×10^2	14.0	C	Kumoide SK-17	0.04	1.06	4.6×10^9	4.9×10^9	20×10^7	Nishiyama
Kumoide SK-4,8	0.28	0.23	0.05	1.07	2.7×10^2	2.9×10^2	14.5	D	Kumoide SK-4,8	0.05	1.07	7.7×10^9	8.2×10^9	41×10^7	Nishiyama
Minami-Ōguchi R-1	0.39	0.35	0.04	1.07	1.3×10^2	1.4×10^2	5.6	E	Minami-Ōguchi R-1	0.04	1.07	5.9×10^9	6.3×10^9	25×10^7	Nishiyama
Nishi-Nagaoka KR-16-1	0.32	0.28	0.04	1.06	0.3×10^2	0.32×10^2	1.3	F	Nishi-Nagaoka KR-16-1	0.04	1.06	4.2×10^9	4.5×10^9	18×10^7	Haizume
Ōbaden R-1	0.28	0.25	0.03	1.04	1.1×10^2	1.15×10^2	3.5	G	Ōbaden R-1	0.03	1.04	2.6×10^9	2.7×10^9	8.1×10^7	Nishiyama
Minami-Nagaoka R-1	0.21	0.20	0.01	1.01	0.1×10^2	0.10×10^2	0.1	H	Minami-Nagaoka R-1	0.01	1.01	0.55×10^9	0.55×10^9	5.5×10^7	Haizume
Sekihara R-19	0.34	0.32	0.02	1.03	0.6×10^2	0.62×10^2	1.2	I	Sekihara R-19	0.02	1.03	2.1×10^9	2.2×10^9	4.3×10^7	Nishiyama
Oyazawa R-1	0.29	0.24	0.05	1.07	2.5×10^2	2.7×10^2	13.5	J	Oyazawa R-1	0.05	1.07	2.2×10^9	2.4×10^9	12×10^7	Nishiyama
Koshiji R-1	0.30	0.29	0.01	1.01	0.5×10^2	0.50×10^2	0.5	K	Koshiji R-1	0.01	1.01	3.4×10^9	3.4×10^9	3.4×10^7	Nishiyama
Urase SK-1	0.26	0.22	0.04	1.05	2.3×10^2	2.4×10^2	9.6	L	Urase SK-1	0.04	1.05	7.8×10^9	8.2×10^9	33×10^7	Nishiyama

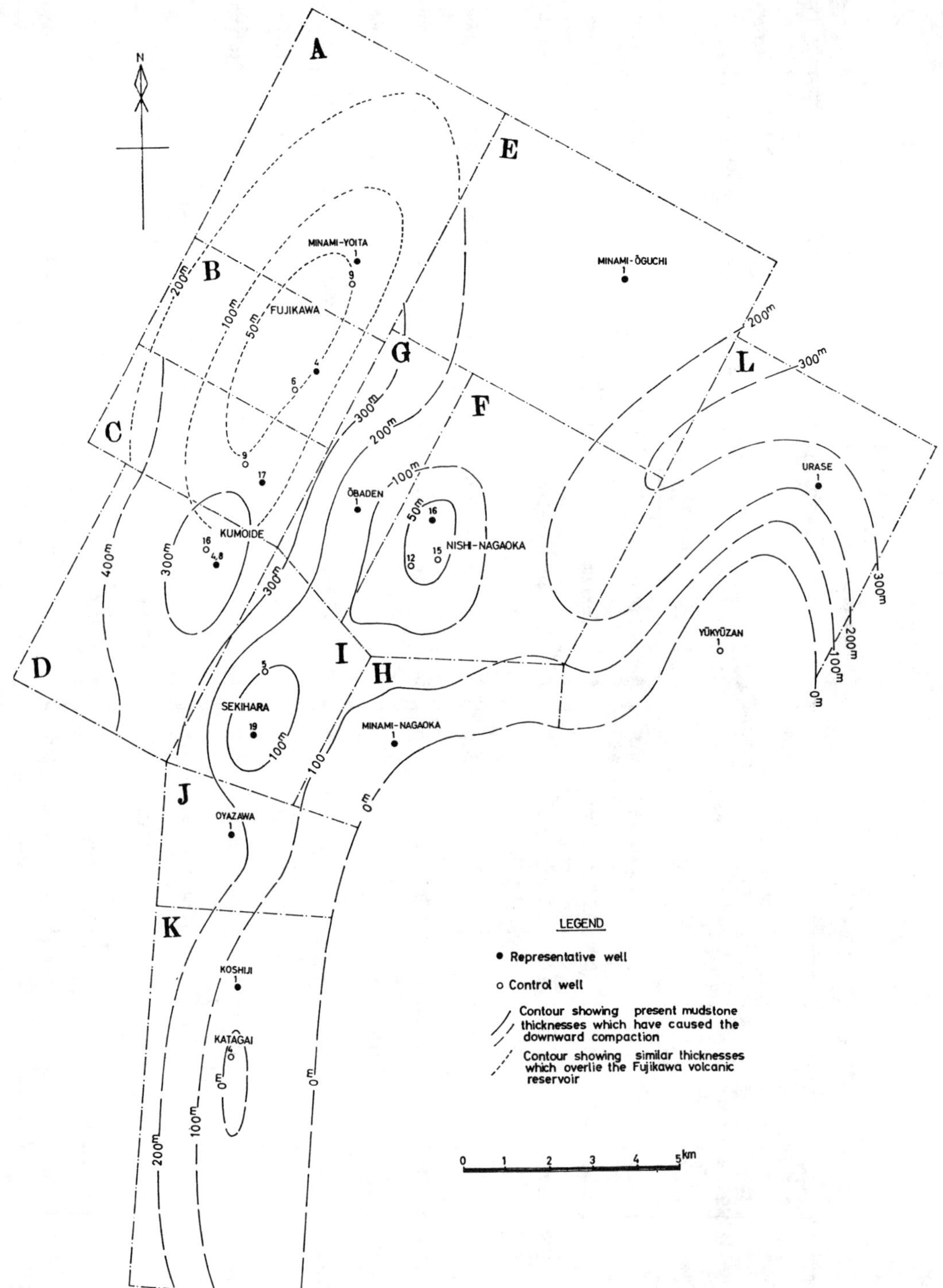

FIG. 19.—Map showing present mudstone thicknesses which have caused downward compaction currents in Nagaoka agglomerate region.

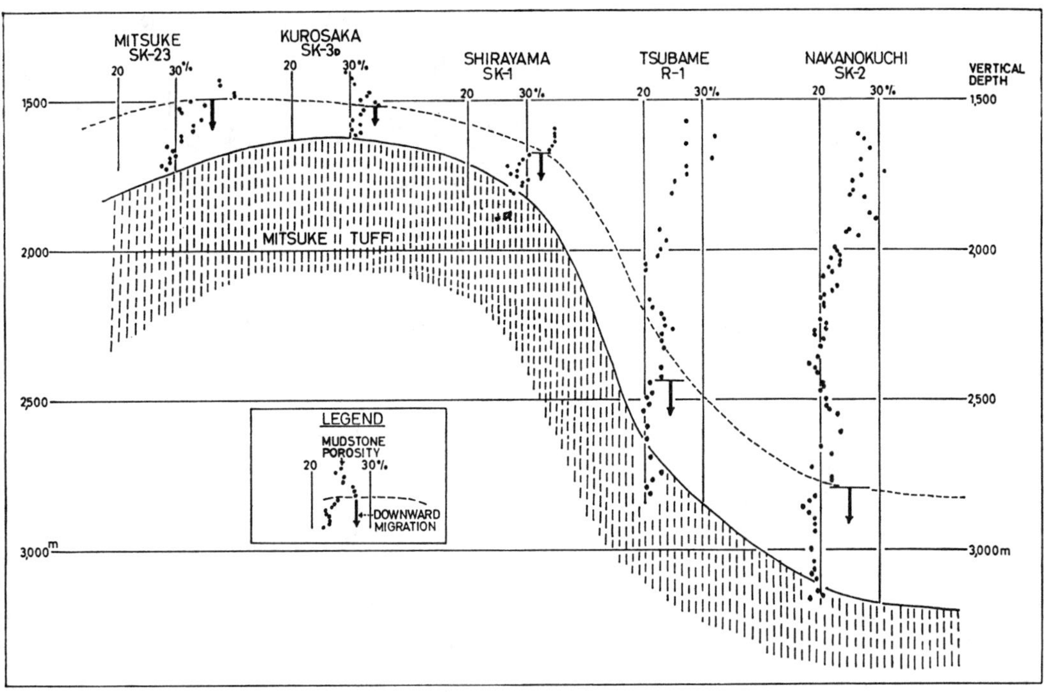

Fig. 20.—Vertical porosity distributions of mudstone in Mitsuke Tuff region.

downward current, as shown in Figure 19, the area is divided into many blocks, based on the well locations, the locations of two synclines (between ABCD and EGI, and between EFH and L), and the distribution of the volcanic reservoirs. Table V also shows the total volumes of the downward compaction current of these blocks. The volumes of the current from the overlying mudstone beds seem to have depended on the permeability of the reservoirs as well as the character of the mudstone.

Figure 20 indicates the vertical porosity distribution of the overlying mudstone beds in the Mitsuke Tuff region. There is a clear porosity difference between the mudstone barrier and the underlying mudstone at the Mitsuke oil field. However, there is no clear contrast in the mudstone porosity values at the Tsubame and Nakanokuchi.

Table VI shows the volume of the downward compaction current per 1-m² unit mudstone column at the main wells in the Mitsuke Tuff region; the thickness of the overlying mudstone which caused the downward compaction current is shown, divided into five blocks, in Figure 21. The total downward compaction current of each block also is listed in Table VI, which shows that the volume of the current is greatest at Mitsuke. This suggests that the tuff beds would have been the most permeable at Mitsuke during the compaction of the overlying source mudstone, and the downward compaction current from the mudstone thus might have been expelled most easily there.

Reservoir Fluids in Nagaoka Plain

Gas

Table VII shows the composition of gas produced from the main wells in Nagaoka Plain. Gases of Kumoide, Fujikawa, and Mitsuke contain more heavy hydrocarbons than those of the other fields. The reservoirs in the stratigraphically lower strata generally contain more heavy hydrocarbons, because the amount of organic extract is the richest in the Nanatani-lower Teradomari mudstone beds and decreases upward in the Tertiary strata. To explain the differences in methane content between Mitsuke and Higashi-Sanjō, or between Kumoide and Sekihara, Kujiraoka (1965) offered the following hypothesis:

> Oil and gas, which were originated from the Teradomari and Nanatani muddy sediments, might have been entrapped in the Higashi-Sanjō gas field as in

TABLE VI. VOLUMES OF DOWNWARD COMPACT

well		average value of porosity below the mudstone barrier		$\Phi - \Phi'$	$\frac{1-\Phi'}{1-\Phi}$	mudstone thickness below the barrier		volur of do ward comp tion curre per Wd
		before burial Φ	after burial Φ'			after burial D'(m)	before burial D(m)	
Nakanokuchi	SK-2	0.21	0.19	0.02	1.03	3.8×10^2	3.9×10^2	7.8
Tsubame	R-1	0.23	0.21	0.02	1.03	4.0×10^2	4.1×10^2	8.
Shirayama	SK-1	0.33	0.29	0.04	1.06	1.5×10^2	1.6×10^2	6.
Kurosaka	SK-3D	0.34	0.31	0.03	1.05	1.0×10^2	1.05×10^2	3.
Mitsuke	SK23	0.39	0.31	0.08	1.13	2.4×10^2	2.7×10^2	21.

the case of the Mitsuke oil field, but the beds overlying the reservoir of this field were almost eroded out through the upheaval at the end of the Siiyan stage, and so the pool must have been destroyed. It is presumed that the methane-rich gas in Higashi-Sanjō lately originated from the muddy rocks of the lower Nishiyama, and then migrated into the same reservoir during the time from the late Nishiyaman to the Haizumean stages. While, in such gas fields as Sekihara and Katagai, gas was entrapped in later stage than in Kumoide, and only gases from comparatively younger source rocks took part in forming the pools.

The writer could not determine which mudstone was the source of hydrocarbons at Higashi-Sanjō because of the lack of vertical mudstone porosity data. But it can be assumed that the hydrocarbon sources for Higashi-Sanjō and Mitsuke were different, and the pool at Higashi-Sanjō might have been destroyed and refilled because the overlying Teradomari mudstone is very thin there (Fig. 3).

OIL

Figure 22 illustrates the relations between oil-condensate gravity and depth in the Nagaoka agglomerate reservoir. According to Holmquest (1966), certain lithologic types of source bed possess distinct hydrocarbon characteristics; gray shale may produce oils with lower gravity (higher API gravity) at a given depth than black and dark shale. This implies that source beds deposited in open-marine waters produce lower gravity (higher API gravity) oil than source beds deposited under stagnant-marine conditions.

Figure 22 suggests that the Kumoide condensate and the Sekihara-Katagai oils are different in source from the Fujikawa condensate. This harmonizes with the apparent source beds in those fields, as shown in Figures 16 and 17. Consequently, oil and condensate in Kumoide, Sekihara, and Katagai may have been derived from the Nishiyama or gray mudstone, and the Fujikawa condensate may have been generated from the Shiiya or dark-gray mudstone. In addition, the crude oil produced from Kumoide SK-13 may be of the same origin as Fujikawa condensate. The origin of the Nishi-Nagaoka oil is not identified in Figure 22.

Figure 23 similarly shows the relation between oil gravity and depth at Mitsuke, Kurosaka, and Higashi-Sanjō. Figure 23 may suggest a difference in source beds between Mitsuke and Higashi-Sanjō which coincides with Kujiraoka's (1965) explanation.

WATER

Regional changes in the water of a rock unit are seen easily in isoconcentration maps, which show the changing concentration of water in a continuous reservoir. Figure 24 indicates the concentration of sodium chloride in the waters which were obtained from the upper part of the Nagaoka agglomerate reservoir. The general northwestward increase in NaCl concentration shown in Figure 24 suggests that diffusion occurred southeastward. However, some anomalies are ob-

RRENTS IN MITSUKE TUFF REGION

ock	representative well		$\Phi - \Phi'$	$\frac{1-\Phi'}{1-\Phi}$	mudstone volume below the barrier: after burial V' (m^3)	mudstone volume below the barrier: before burial V (m^3)	total volume of downward compaction current: total Wd (m^3)	overlying mudstone
A	Nakanokuchi	SK-2	0.02	1.03	16×10^9	16.5×10^9	33×10^7	Shiiya
B	Tsubame	R-1	0.02	1.03	21×10^9	22×10^9	43×10^7	Teradomari
C	Shirayama	SK-1	0.04	1.06	11×10^9	12×10^9	47×10^7	Nanatani
D	Kurosaka	SK-3b	0.03	1.05	6×10^9	6.5×10^9	19×10^7	Teradomari
E	Mitsuke	SK23	0.08	1.13	13×10^9	15×10^9	120×10^7	Teradomari

served at the Kumoide, Sekihara, and Katagai gas fields, where the structurally higher wells produced less concentrated water than the lower wells. It is not clear whether these anomalies are due to density stratification of water in the subsurface or to the water distillate which is produced with gas from gas wells.

Figure 25 shows the isoconcentration of NaCl in the Mitsuke Tuff région. Water-analysis data from Mitsuke, Kurosaka, and Higashi-Sanjō are based on production samples, and therefore are very reliable. The other data are unreliable for illustrative purposes.

Abnormal Pressure in Nagaoka Plain

Abnormal pressures are observed in the volcanic reservoirs of the Fujikawa and Kumoide gas fields in the western Nagaoka Plain and in the reservoirs of the Mitsuke oil field in the eastern part. Hydrostatic pressures are measured in the reservoirs of such fields as Katagai, Nishi-Nagaoka, and Higashi-Sanjō. Table VIII shows the initial bottom-hole pressures measured at several wells in Nagaoka Plain.

Abnormal pressures in hydrocarbon accumulations can be caused by density differences of hydrocarbon and water. The excess pressure from hydrostatic pressure in this case is:

$$P = (\rho_w - \rho_h)gh$$

where P is excess pressure in a hydrocarbon accumulation, ρ_w and ρ_h are, respectively, the densities of the water and of hydrocarbons in the reservoir, g is the acceleration of gravity, and h is the height above the free-water level. Because the gas-water contact in the volcanic reservoir of the Fujikawa gas field is estimated to be 2,360 m below sea level, the height of the gas column up to the depth where the bottom-hole pressure was measured is about 70 m. If it is assumed, from the analysis of water and gas, that ρ_w and ρ_h are 1.02 and 0.24, respectively, the excess pressure above the hydrostatic pressure is 5.5 kg/cm^2. Similarly, the excess pressure in the Kumoide volcanic reservoir is 17.2 kg/cm^2, where $h = 209$ m, $\rho_w = 1.02$, $\rho_h = 0.20$. However, because the pressures measured in these fields are much higher than the calculated values, the abnormal pressures cannot be explained fully by these calculations.

The writer prepared a piezometric surface map (Fig. 26) of the volcanic reservoirs on the basis of the pressure data measured at the wells in Nagaoka Plain. A piezometric surface is defined as an imaginary surface that everywhere coincides with the static level of the water in the aquifer, or the surface to which the water from a given aquifer will rise under its full head.

If the water in the reservoir is static, the piezometric surface is level. If the water is moving, the piezometric surface is inclined in the direction of the movement.

The piezometric surface of the Nagaoka ag-

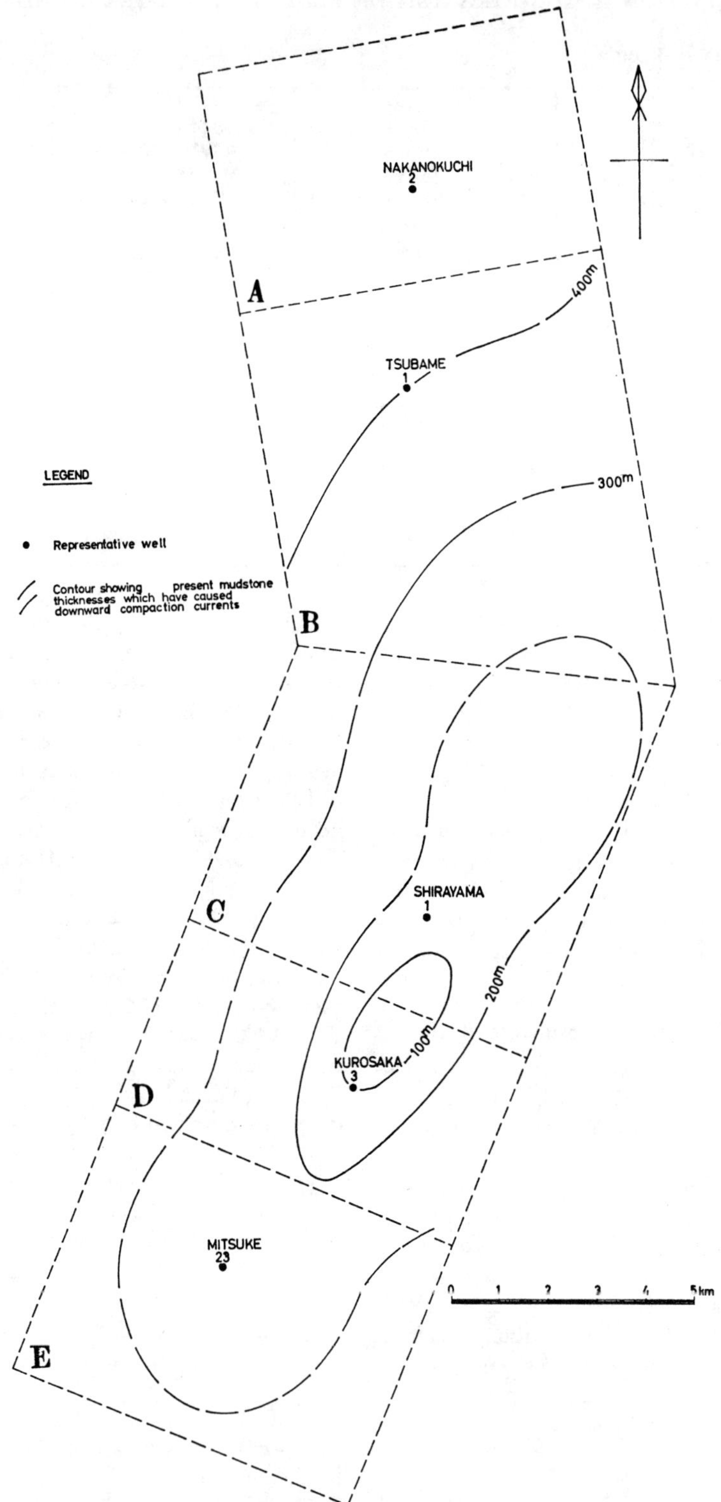

FIG. 21.—Map showing present mudstone thicknesses which have caused downward compaction currents in Mitsuke Tuff region.

TABLE VII. COMPOSITION OF GAS PRODUCED FROM MAIN WELLS IN NAGAOKA PLAIN
(Redrawn from Kujiraoka, 1965)

well	formation	depth of completion	specific gravity of gas	chemical composition of gas						
				hydrocarbon (%)						
				CH_4	C_2H_6	C_3H_8	C_4H_{10}	C_5H_{12} & higher	CO_2	N_2
Nishi-Nagaoka KR9-1	Nishiyama	1,169.76 / 1,200.15	0.563	98.5	1.4	trace	-	-	0.1	-
Katagai SK-1	Nishiyama	1,011.00 / 1,094.00	0.5593	99.26	0.30	0.14	0.05	0.02	trace	0.23
Sekihara R-3	Nishiyama	1,127.5 / 1,130.5	0.559	99.2	0.7	trace	-	-	0.1	-
Kumoide SK-5	Nishiyama	1,725.1 / 1,761.0	0.6334	89.27	6.74	2.23	1.05	0.57	0.11	0.03
Fujikawa SK-1b	Shiiya	2304.80 2326.26 / 2316.03 2333.36	0.6248	90.42	6.30	2.09	0.86	0.33	trace	-
Mitsuke SK-8	Nanatani	1,691.00 / 1,710.00	0.6444	88.53	6.23	2.37	1.15	0.73	trace	0.99
Higashi-Sanjō SK-1	Nanatani	1,215.93 / 1,265.50	0.5732	95.75	0.79	0.02	trace	-	0.07	3.37

glomerate reservoir is at a height of 1,300 m above sea level at Fujikawa, and is inclined eastward at a slope of 100 m per 300 m. Does this indicate that there is water movement in the reservoir?

Hubbert (1953) found the following relation.

$$\tan\theta = \frac{\Delta z}{\Delta x} = \frac{\rho_w}{\rho_w - \rho_h}\frac{\Delta h}{\Delta x}$$

where $\Delta z/\Delta x$ is the rate of inclination of oil-water or gas-water contact and $\Delta h/\Delta x$ is the rate of inclination of the piezometric surface.

If one assumes from the analysis $\rho_w = 1.02$, $\rho_h = 0.2$, and $\Delta h/\Delta x = 1/3$ in the Nagaoka aglomerate reservoir, the angle of inclination of the gas-water contact can be shown to be

$$\tan\theta = \frac{1}{1.02 - 0.2} \times \frac{1}{3} = 0.407$$

$$\theta = 22°.$$

If this calculation is reasonable, a 22° inclination of gas-water contact should be present in the reservoir. However, no inclinations are observed at any fields in the area.

If it is assumed that the water in the reservoir is moving eastward, are there any intakes in the west where new water enters? There is no outcrop of the volcanic reservoir in the western hilly lands. Therefore, there may not be a sufficient supply of water from the west.

Finally, there is no land on the western side of the plain more than 1,300 m in altitude.

The piezometric surface of the Mitsuke Tuff (Fig. 26) is inclined northwestward and the maximum height is about 900 m at the Mitsuke oil field. No water movement has been observed in the Mitsuke Tuff, for reasons similar to those described for the Nagaoka agglomerate region.

Consequently, permeability barriers seem to be present in the volcanic reservoirs between Fujikawa and Kumoide, Kumoide and Sekihara, and between Mitsuke and Higashi-Sanjō, and the pressures are very different. Because of these barriers the pools maintain high pressure levels, and can be considered gas fields completely isolated from each other.

ABNORMAL PRESSURE HYPOTHESES

Reservoir pressures in the subsurface usually approximate hydrostatic pressure, which equals weight of the water column from the reservoir to the surface. A pressure that materially exceeds the weight of an equivalent column of the reservoir water is "abnormal," and a pressure that is materially less is "subnormal" (Levorsen, 1954). Reservoir pressure will not exceed the weight of the overlying rock column, or the petrostatic pressure.

Dickinson (1951) reported an instance of abnormal pressure in the Gulf Coast equal to 0.87 times petrostatic, and Kok and Thomeer (1955) cited an example of 0.90 times petrostatic. More data on abnormal pressures were complied by Rubey and Hubbert (1959). Dickinson (1951) also reported the first recognized association of

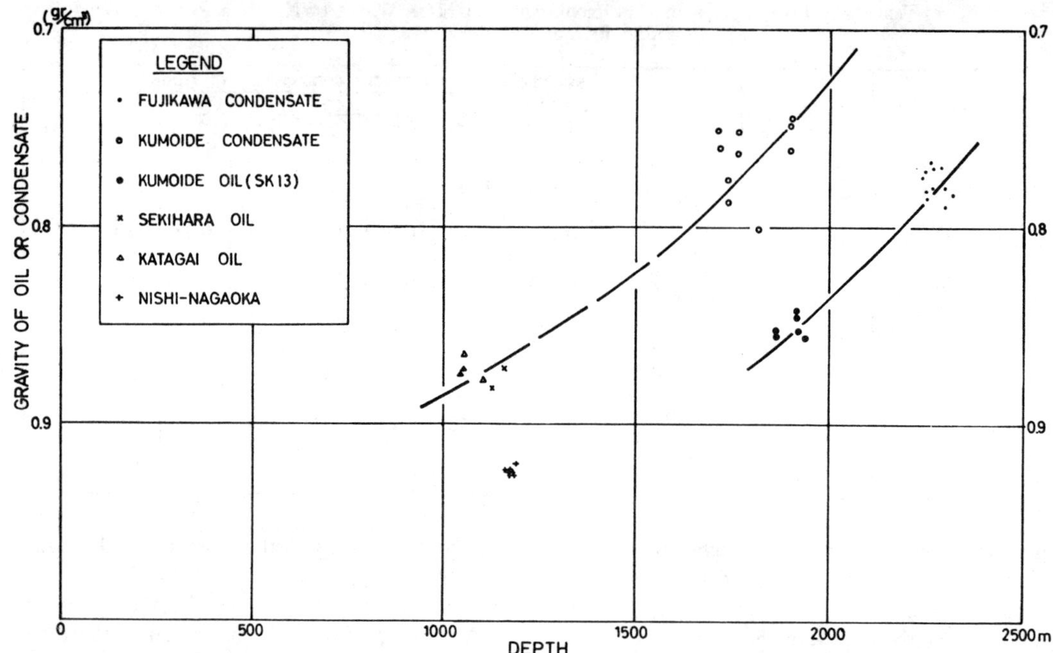

Fig. 22.—Relation between oil-condensate gravity and depth in Nagaoka agglomerate region.

abnormal pressures with the relative proportion of sand and shale in a geologic column. Abnormal pressures can be influenced also by the mean formation permeability, elapsed time since deposition, rate of deposition, and the amount of overburden (Hottmann and Johnson, 1965).

Thomeer and Bottema (1961) indicated two types of deposit suitable for the occurrence of abnormal pressure.

1. Rock salt or similar deposits of very great lateral extent, such as are present in the North German Zechstein basin. Rock salt is completely impervious to

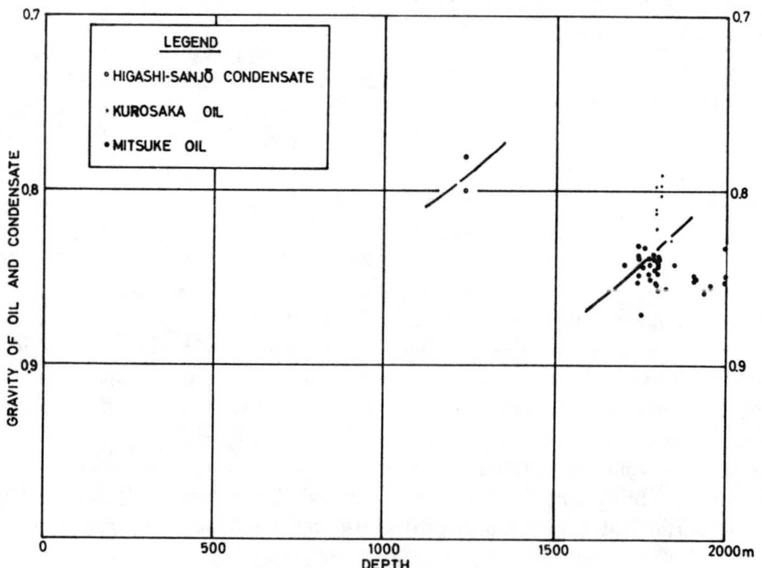

Fig. 23.—Relation between oil-condensate gravity and depth in Mitsuke Tuff region.

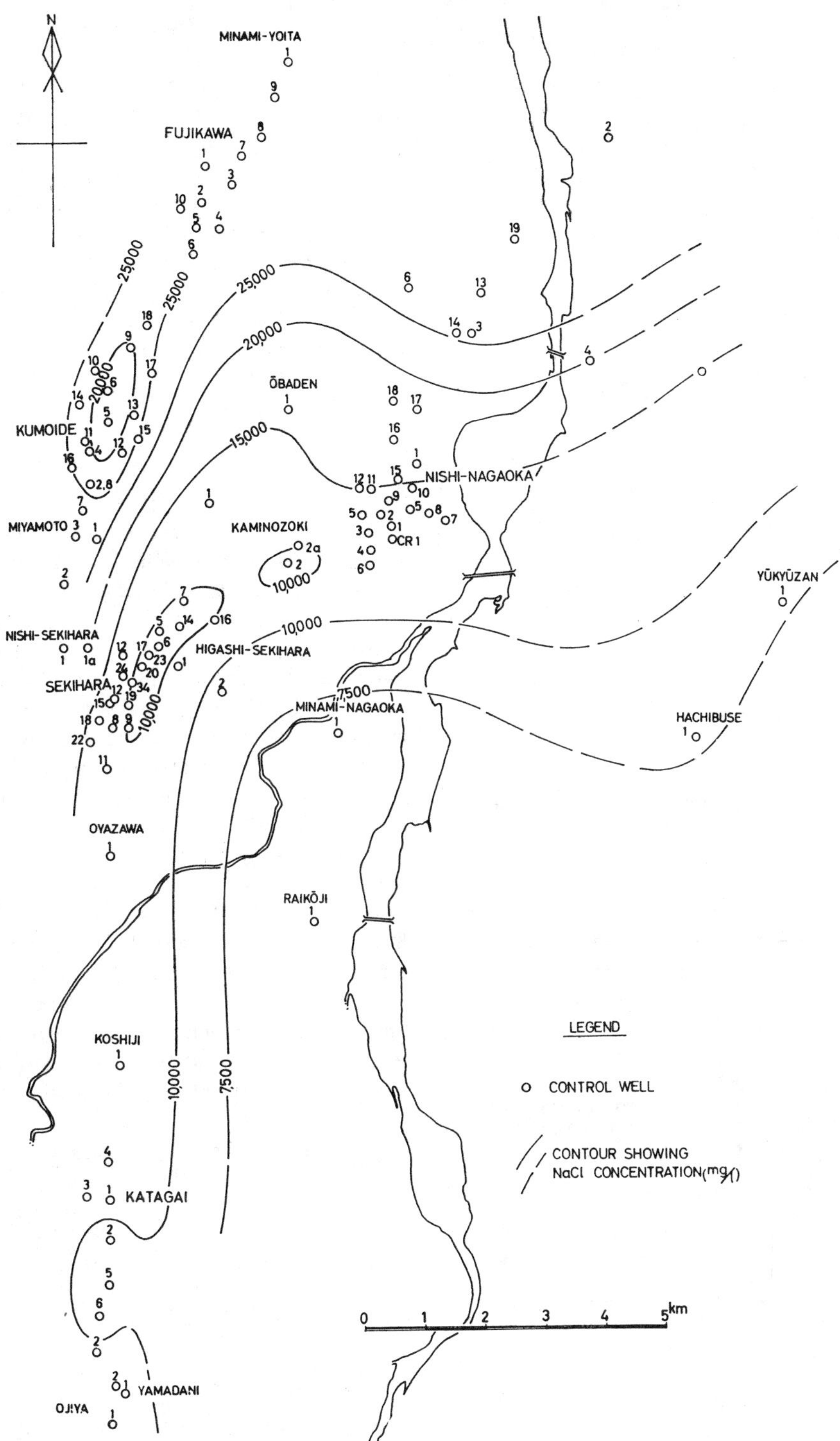

FIG. 24.—Isoconcentration map (of NaCl) of Nagaoka agglomerate reservoir.

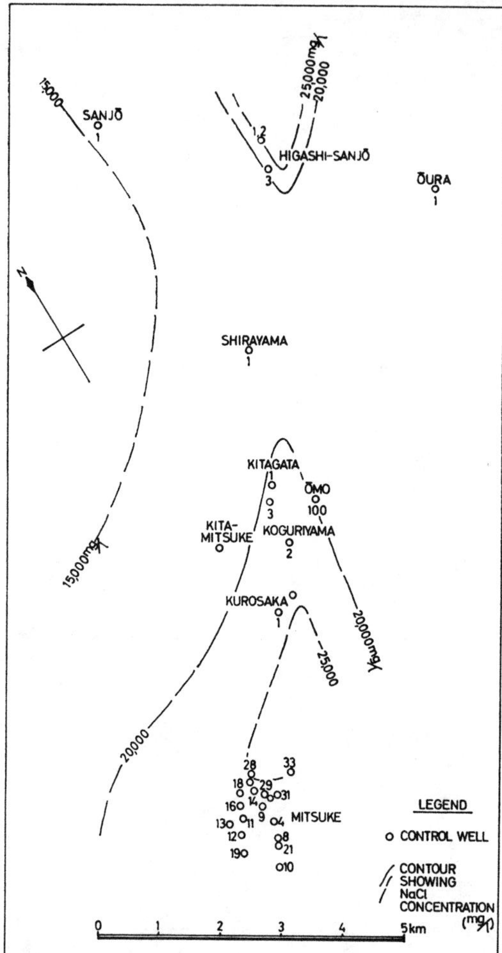

FIG. 25.—Isoconcentration map (of NaCl) of Mitsuke Tuff reservoir.

fluid flow and even under high tectonic stress will not break, but rather be transformed pseudo-plastically (recrystallization effect). If a way of escape for compressed fluid is likewise excluded by the horizontal extent of the deposit, all fluids in all porous formations below this salt deposit should be found under a reservoir pressure equalling the full value of the rock pressure. This situation is permanent as long as the salt deposits are not removed by erosion or otherwise.

2. Clay or shale deposits of large extent and thickness. Clay and shales are not strictly impermeable but their permeability is generally very low. As a result, the escape of water out of the sediments through an overlying shaly column is retarded.

Consequently, conditions favorable for abnormal reservoir pressures below shale or mudstone may be found in some young sedimentary basins, where the thick shale or mudstone bodies have been deposited relatively rapidly across large areas.

Levorsen (1954) pointed out as other possible causes of pressure increases secondary precipitation or cementation, earthquakes or tectonic movements, temperature changes, *etc.*

The cause of subnormal pressure is less well understood than the cause of abnormal pressures, but possibly the expansion of elastic reservoir rocks as confining and compacting pressures were reduced by erosion has increased the volume of the pore space in the reservoirs and thus decreased the hydrostatic pressure in the reservoirs (Russell, 1951).

CAUSES OF ABNORMAL PRESSURES IN NAGAOKA PLAIN

The reservoir pressures in the Fujikawa and Kumoide gas fields and the Mitsuke oil field are abnormally high, and the reservoir pressures in Nishi-Nagaoka, Katagai, and Higashi-Sanjō approximate hydrostatic pressure. The initial bottom-hole pressures in these fields (Table VIII) are plotted against depth in Figure 27, which indicates that the reservoir pressures increase with depth and the excess pressures above hydrostatic pressure also generally increase with depth. This suggests that the overburden may influence the reservoir-pressure increase.

In addition, overlying mudstone beds of the Nishiyama-Shiiya Formations at Fujikawa and Kumoide are much thicker where abnormal pressures are measured than at Nishi-Nagaoka and Katagai where hydrostatic pressures are observed. Similarly, the Teradomari mudstone overlying the reservoir is thicker at Mitsuke, where pressures are abnormal, than at Higashi-Sanjō, where pres-

TABLE VIII. BOTTOM-HOLE PRESSURES MEASURED IN MAIN WELLS IN NAGAOKA PLAIN

well		reservoir	depth(m)	bottom hole pressure(kg/cm²)
Fujikawa	SK-1b	agglomerate	2,310.42	367.58
Kumoide	SK-5	agglomerate	1,745.00	251.20
	1a	sandstone	1,394.79	157.07
Sekihara	R-3	agglomerate	1;129.00	117.50
	SK-1	sandstone	817.50	74.50
Katagai	SK-1	agglomerate	1,052.50	99.76
Nishi-Nagaoka	KR-5₂	agglomerate	1,172.18	117.72
	R-3	sandstone	1,144.50	106.10
Mitsuke	SK-3	sandstone	1,201.27	125.53
	4	lava	1,744.90	264.29
Kurosaka	SK-1	lava	1,816.00	256.62
Higashi-Sanjō	SK-1	tuff	1,248.82	132.33

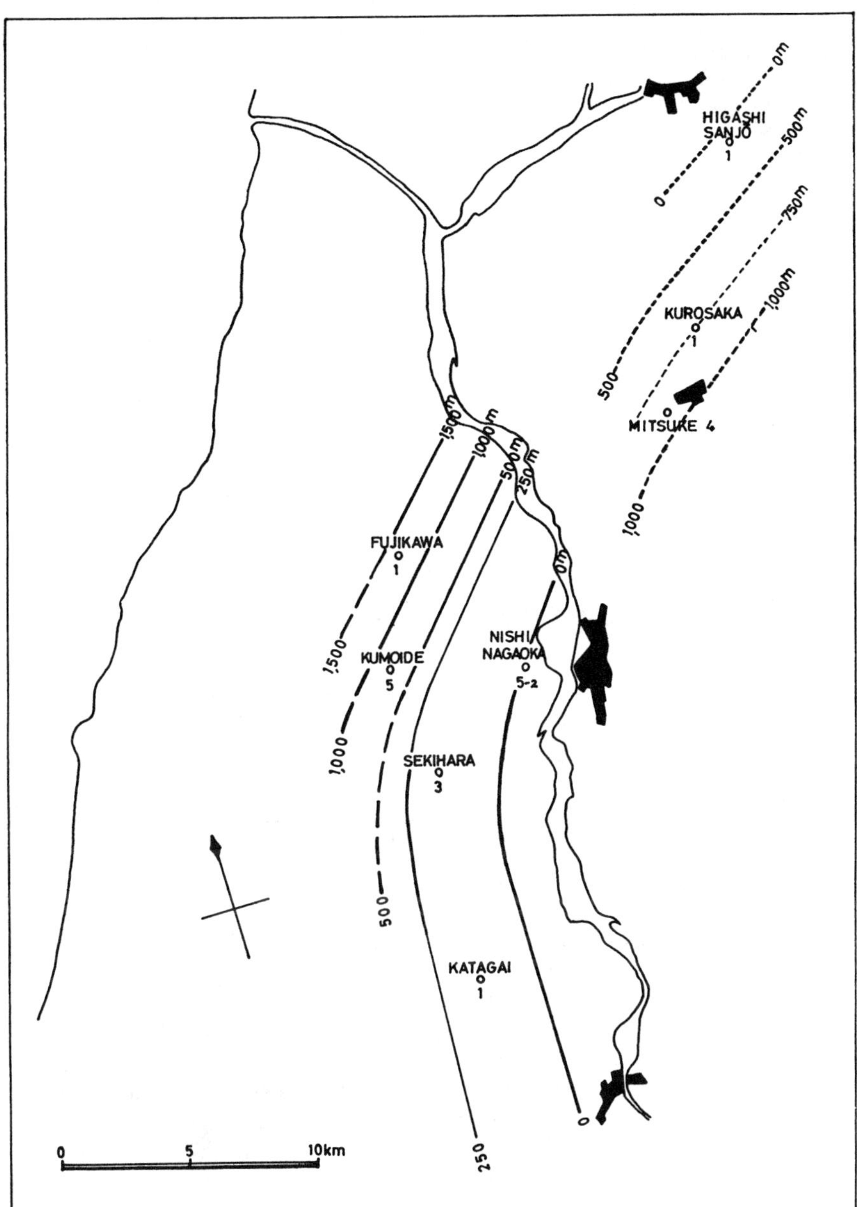

FIG. 26.—Piezometric surface map of volcanic reservoirs in Nagaoka Plain.

sure is approximately hydrostatic. Thus, the overlying mudstone beds seem to have been important in the generation of the abnormal pressures in the district.

Levorsen's supplementary explanations of abnormal pressure are only partly applicable to Nagaoka Plain.

Secondary precipitation or cementation cannot be considered as the cause of the abnormal pressure in the study area. The volcanic reservoir rocks and adjacent mudstone beds are scarcely metamorphosed. Moreover, there seems to be no difference between Fujikawa-Kumoide and Katagai-Nishi-Nagaoka, or Mitsuke and Higashi-Sanjō.

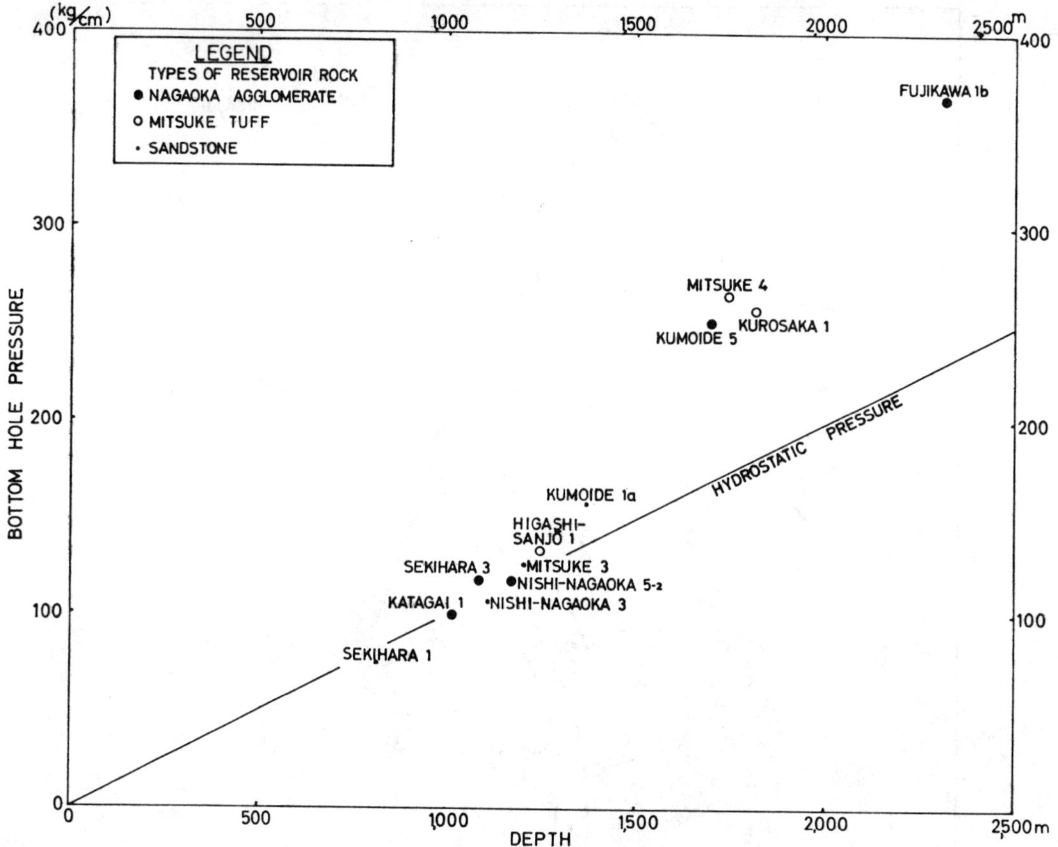

FIG. 27.—Relation between bottom-hole pressure and depth in Nagaoka Plain.

The abnormal pressure cannot be explained completely by earthquakes or tectonic movements. The relation between reservoir pressure and tectonic movement should be studied in the future, because of the thrusts west of the Fujikawa-Kumoide trend and east of the Higashi-Sanjō-Mitsuke trend, bounding the adjacent hilly lands. The structures beneath the Nagaoka Plain, however, are rather gentle. The excess pressure at Fujikawa is much greater than that at Kumoide, although the distances from these fields to the thrust and the structural situations of these fields are very similar.

Because temperature increases with subsidence at a linear rate of approximately 4.5°C/100 m in the study area, Figure 27, which shows the relation of depth and bottom-hole pressure, may also be read for temperature.

Apparently, the abnormal reservoir pressures in Nagaoka Plain may be ascribed to the thick mudstone seal, which has retarded the escape of fluid, and the overburden pressure.

Conclusion

Water may move in a carrier bed from a place where more water is expelled to a place where less water is squeezed out. The directions of compaction currents in the volcanic reservoirs are estimated from the horizontal porosity distributions of the overlying mudstone beds in the Nagaoka Plain. The present locations of the hydrocarbon accumulations can be explained easily by these mudstone porosity distributions, or the differential compaction in the district.

In the Nagoaka Plain these porosity distributions correspond closely to the present distribution of structures. In the Kumoide-Fujikawa trend, however, the overlying mudstone porosity values increase northward, although the present structure plunges north. This discordance between porosity distribution and present structure can be explained by the variation of the overlying mudstone thickness and the abnormally high pressure. Because the overlying Haizume and Nishiyama Formations thicken markedly southward, these formations seem to have subsided more deeply and have been more compacted in the south than in the north in the early stages of compaction, and compaction currents may have moved northward in the volcanic reservoirs. The higher abnormal pressure then was generated in the northern part and the compaction of the mudstone beds might have been greatly interrupted.

Mudstone compaction may be due to water expulsion, because water is almost incompressible. If escape of water is inhibited, compaction may not be great, and high porosity–high fluid pressure will result. Because such a low-permeability mudstone is thought to be a seal to water movement, and fluid pressure seems to become relatively higher than the pressures above and below the mudstone, compaction currents could move downward from the low-permeability zone and move upward above the zone.

There are clear porosity differences between the mudstone barrier and the underlying mudstone at Fujikawa, Kumoide, Sekihara, and Mitsuke. These porosity differences, which indicate qualitatively the volumes of the downward compaction currents, seem to be related to the permeability of the underlying reservoirs, as well as properties of the mudstone. The volumes of the downward compaction currents are estimated from the vertical porosity distributions and the porosity decrease of the mudstone beds below this low-permeability zone in Nagaoka Plain. These volumes appear to be related to the present hydrocarbon reserves in the district. It also can be assumed from the porosity distributions that the main source beds are the Shiiya mudstone at Fujikawa, the Nishiyama mudstone at Kumoide, Sekihara, and Katagai, and the Teradomari mudstone at Mitsuke.

Hydrostatic pressures are measured in the reservoirs of Katagai, Nishi-Nagaoka, and Higashi-Sanjō, whereas abnormal pressures are observed in the volcanic reservoirs of the Fujikawa and Kumoide gas fields, and the Mitsuke oil field. These abnormal pressures suggest the presence of permeability barriers in the volcanic rocks. The depth of the reservoirs and the thicknesses of the sealing mudstone beds are greater at Fujikawa, Kumoide, and Mitsuke than at Katagai, Nishi-Nagaoka, and Higashi-Sanjō. Consequently, abnormal pressures in Nagaoka Plain may be ascribed to the thick mudstone seals and overburden pressures.

Selected References

Abaie, I., *et al.*, 1963, History and development of the Alborz and Sarajeh fields of Central Iran: 6th World Petroleum Cong. Proc., sec. 1, p. 697–713.

Archie, G. E., 1942, The electrical resistivity log as an aid in determining some reservoir characteristics: Am. Inst. Min. Metall. Eng. Trans., v. 146, p. 54–62.

——— 1950, Introduction to petrophysics of reservoir rocks: Am. Assoc. Petroleum Geologists Bull., v. 34, no. 5, p. 943–961.

Athy, L. F., 1930, Density, porosity, and compaction of sedimentary rocks: Am. Assoc. Petroleum Geologists Bull., v. 14, no. 1, p. 1–24.

Bille, J. L., 1964, Gravity for the geologist: World Oil, v. 159, no. 5, p. 147–150, no. 6, p. 115–118.

Chave, K. E., 1960, Evidence of history of sea water from chemistry of deeper subsurface waters of ancient basins: Am. Assoc. Petroleum Geologists Bull., v. 44, no. 3, p. 357–370.

Dickinson, G., 1951, Geological aspects of abnormal reservoir pressures in the Gulf Coast region of Louisiana, U.S.A.: 3d World Petroleum Cong. Proc., sec. 1, p. 1–16; 1953, Am. Assoc. Petroleum Geologists Bull., v. 37, no. 2, p. 410–432.

Dunlap, H. F., and R. R. Hawthorne, 1951, The calculations of water resistivities from chemical analysis: Am. Inst. Min. Metall. Eng. Trans., v. 192, p. 373–375.

Engelhardt, W. von, and K. H. Gaida, 1963, Concentration changes of pore solutions during the compaction of clay sediments: Jour. Sed. Petrology, v. 33, p. 919–930.

Ferguson, L., 1963, Estimation of the compaction factor of a shale from distorted brachiopod shells: Jour. Sed. Petrology, v. 33, p. 796–798.

———1964, A comparison of two techniques for measuring shale compaction: Jour. Sed. Petrology, v. 34, p. 694–695.

Fons, L., and O. Holt, 1966, Formation log pressure data can improve drilling: World Oil, v. 163, no. 4, p. 70–74.

Foster, J. B., *et al.*, 1966, Estimation of formation pressures from electrical surveys—offshore Louisiana: Jour. Petroleum Technology, v. 18, no. 2, p. 165–171.

Frederick, W. S., 1966, Hydrodynamics aid exploration and development: World Oil, v. 163, no. 4, p. 66–69.

Gussow, W. C., 1954, Differential entrapment of oil and gas: a fundamental principle: Am. Assoc. Petroleum Geologists Bull., v. 38, no. 5, p. 816–853.

Hedberg, H. D., 1936, Gravitational compaction of clays and shales: Am. Jour. Science, v. 31, no. 4, p. 241–287.

Holmquest, H. J., 1966, Stratigraphic analysis of source-bed occurrences and reservoir oil gravities: Am. Assoc. Petroleum Geologists Bull., v. 50, no. 7, p. 1478–1486.

Hosoi, H., 1963, Akita Ken, Yamagata Ken ni okeru

Sekiyu no Dai lji Ido (First migration of petroleum in Akita and Yamagata Prefectures): Japanese Assoc. Mineralogists, Petrologists and Econ. Geologists Jour., v. 49, no. 2, p. 43–55, no. 3, p. 101–114.

Hottmann, C. E., and R. K. Johnson, 1965, Estimation of formation pressures from log-derived shale properties: Jour. Petroleum Technology, v. 17, no. 6, p. 717–722.

Hubbert, M. K., 1953, Entrapment of petroleum under hydrodynamic conditions: Am. Assoc. Petroleum Geologists Bull., v. 37, no. 8, p. 1954–2026.

——— and W. W. Rubey, 1959, Role of fluid pressure in mechanics of overthrust faulting, I: Geol. Soc. America Bull., v. 70, no. 2, p. 115–166.

Hunt, J. M., 1953, Composition of crude oil and its relation to stratigraphy in Wyoming: Am. Assoc. Petroleum Geologists Bull., v. 37, no. 8, p. 1837–1872.

Ishiwada, Y., 1964, Kambara Sōjo Shisui ni Tsuite (Kambara stratigraphic well): Chishitsu News, no. 115, p. 1–9.

Kok, P. C., and J. H. M. A. Thomeer, 1955, Abnormal pressures in oil and gas reservoirs: Geologie en Mijnbouw, Aug.

Kujiraoka, A., 1962, Arayasō no Imisuru Mono (Meaning of the Araya Facies): Japanese Assoc. Petroleum Technologists Jour., v. 27, no. 6, p. 519–556.

——— 1965, Volcanic activity and its influence on the migration and accumulation of oil and gas in Nagaoka Plain, Japan: Contribution from the Government of Japan to ECAFE 3d Petroleum Symposium, Tokyo, Japan.

Landes, K. K., 1951, Petroleum geology: New York, John Wiley & Sons, 660 p.

——— *et al.*, 1960, Petroleum resources in basement rocks: Am. Assoc. Petroleum Geologists Bull., v. 44, no. 10, p. 1682–1691.

Levorsen, A. I., 1954, Geology of petroleum: San Francisco, W. H. Freeman Co., 703 p.

MacGregor, J. R., 1965, Quantitative determination of reservoir pressures from conductivity log: Am. Assoc. Petroleum Geologists Bull., v. 49, no. 9, p. 1502–1511.

Magara, K., 1965, Compaction factor no Motomekata (Determination of compaction factor): Japanese Assoc. Petroleum Technologists Jour., v. 30, no. 3, p. 146–148.

——— 1966a, Nagaoka Heiya no Kazangan Yusō ni tsuite (Volcanic reservoir rocks and reservoir pressure in Nagaoka District, Niigata Prefecture): Japanese Assoc. Petroleum Technologists Jour., v. 31, no. 1, p. 22–29.

——— 1966b, Sonic Log no Tankō eno Ōyō (Application of sonic log for exploration): Japanese Assoc. Petroleum Technologists Jour., v. 31, no. 2, p. 51–58.

——— 1966c, Kensō Data ni yoru Ijōkoatsusō no Hantei (Estimation of abnormal pressures from log data): Japanese Assoc. Petroleum Technologists Jour., v. 31, no. 2, p. 67–70.

——— 1966d, Kensō Data ni yoru Yusōatsu no Suitei-Shiunji Gas Den ni okeru Kentō (Estimation of reservoir pressures from electrical logging—Shiunji gas field, Niigata Prefecture): Japanese Assoc. Petroleum Technologists Jour., v. 31, no. 6, p. 266–273.

———, 1967, Shiunji Gas Den ni okeru Kahōidō no Kanōsei ni tsuite (Possibility of downward migration at the Shiunji gas field): Japanese Assoc. Petroleum Technologists Jour., v. 32, no. 1, p. 1–8.

Martin, R., 1964, Buried hills hold key to new Mississippian pay in Canada: Oil and Gas Jour., v. 62, no. 42, p. 158–162.

Matsuzak, D. R., 1965, Adjustment of electric log values for preparation of subsurface facies maps: Am. Assoc. Petroleum Geologists Bull., v. 49, no. 2, p. 126–138.

Matsuzawa, A., 1961, 1962, Taisekigan no Mitsudo to Chishitsu Kōzō (On the relationship between the density of sedimentary rock and the subsurface geology): Butsuri Tanko (Geophys. Exploration), v. 14, p. 195, v. 15, p. 1.

Mitsuchi, T., 1960, Sekiyu no Idō ni tsuite no Kōsatsu (Considerations upon the migration of petroleum, with special reference to some of the Japanese oil fields): Japanese Assoc. Petroleum Technologists Jour., v. 25, p. 200–237.

Miyazaki, H., 1966, Gravitational compaction of the Neogene muddy sediments in Akita oil fields, northern Japan: Jour. Geosciences, Osaka City Univ., v. 9, art. 1, p. 1–23.

Mochizuki, H., 1962, Niigata Ken Ka Shin Daisankei no Sekiyu Chishitsugaku Teki Kōsatsu (Petroleum geology of Neogene formations of Niigata Prefecture, with special reference to the accumulation of oil and the structural development in the Chūetsu District): Japanese Assoc. Petroleum Technologists Jour., v. 27, no. 6, p. 557-585.

Moore, E. J., *et al.*, 1966, Determining formation water resistivity from chemical analysis: Jour. Petroleum Technology, v. 18, no. 3, p. 373–376.

Ōta, I., and T. Oki, 1965, Nakajō-Shiunji Gas Den (Nakajō-Shiunji gas fields): Japanese Petroleum Inst. Jour., v. 8, no. 2, p. 111–114.

Philipp, W., *et al.*, 1964, History of migration in the Gifhorn trough (NW-Germany): 6th Petroleum Cong. Proc., sec. 1, paper 19, p. 457–481.

Pirson, S. J., 1958, Oil reservoir engineering: New York, McGraw-Hill, 735 p.

——— 1964, Projective well log interpretation: World Oil, v. 159, no. 2, no. 4, no. 6.

Roach, J. W., 1965, How to apply fluid mechanics to petroleum exploration: World Oil, v. 160, no. 4, no. 5.

Rogers, L. C., 1966, How Shell controls Gulf Coast pressures: Oil and Gas Jour., v. 64, no. 20, p. 264–266.

Rubey, W. W., and M. K. Hubbert, 1959, Role of fluid pressure in mechanics of overthrust faulting, II: Geol. Soc. America Bull., v. 70, p. 167–206.

Russell, W. L., 1951, Principles of petroleum geology: New York, McGraw-Hill Book Company, 508 p.

——— 1961, Reservoir water resistivities and possible hydrodynamic flow in Denver basin: Am. Assoc. Petroleum Geologists Bull., v. 45, no. 12, p. 1925–1940.

Schlumberger, Log interpretation charts, Schlumberger Well Surveying Corp., Houston, Texas.

Schwade, I. T., S. A. Carlson, and J. B. O'Flynn, 1958, Geologic environment of Cuyama Valley oil field, California, *in* Habitat of Oil, Am. Assoc. Petroleum Geologists, p. 78–98.

Shelton, J. W., 1962, Shale compaction in a section of Cretaceous Dakota Sandstone, northwestern North Dakota: Jour. Sed. Petrology, v. 32, p. 873–877.

Snarskiy, A. N., 1964, Relationship between primary

migration and compaction of rocks: Petroleum Geology, v. 5, no. 7.

Takeuchi, Y., 1962, Niigata Ken Chūbu ni okeru Yuden no Seisei ni Kansuru Kenkyū-Toku ni Ōmo, Mitsuke Yuden ni tsuite (A study on the growth of oil fields in the middle parts of Niigata Prefecture —especially about Ōmo and Mitsuke oil fields): Japanese Assoc. Petroleum Technologists Jour., v. 27, no. 6, p. 587–613.

Talash, A. W., and P. B. Crawford, 1965, An improved method for calculating water resistivities from chemical analysis: Jour. Petroleum Technology, v. 17, no. 12, p. 1396–1398.

Terzaghi, K., and R. B. Peck, 1948, Soil mechanics in engineering practice: New York, John Wiley & Sons.

Thomeer, J. H. M. A., and J. A. Bottema, 1961, Increasing occurrence of abnormally high reservoir pressures in boreholes, and drilling problems resulting therefrom: Am. Assoc. Petroleum Geologists Bull., v. 45, no. 10, p. 1721–1730.

Tixier, M. P., 1961, Interpretation in very shaly sand: Houston, Texas, Schlumberger Well Surveying Corp.

Wallace, W. E., 1965, Logs measure abnormal subsurface pressures: Oil and Gas Jour., v. 63, no. 27, p. 102.

Wyllie, M. R. J., *et al.*, 1956, Elastic wave velocities in heterogeneous and porous media: Geophysics, v. 21, no. 1, p. 41–70.

——— *et al.*, 1958, An experimental investigation of factors affecting elastic wave velocities in porous media: Geophysics, v. 23, no. 3, p. 459–493.

Reprinted from:
BULLETIN OF THE AMERICAN ASSOCIATION OF PETROLEUM GEOLOGISTS
VOL. 53, NO. 1 (JANUARY, 1969), PP. 73–93, 17 FIGS., 1 TABLE

Diagenesis of Gulf Coast Clayey Sediments and Its Possible Relation to Petroleum Migration[1]

JOHN F. BURST[2]
Philadelphia, Pennsylvania 19102

Abstract The writer has established reference intervals in the subsurface on the basis of apparent systematic interlayer water loss of swelling clay minerals. The intervals are used in much the same manner as the familiar indicators for metamorphism, but are present at sufficiently shallow depths to be evident within oil-bearing strata of the Gulf Coast. The resulting conclusion is that clay-mineral diagenesis indicators may prove to be important petroleum-evaluation markers as well as fundamental properties of sedimentary basins.

Sedimentary basins are viewed as combinations of gases, liquids, and semisolids distributed through a solid matrix. During geologic development the interstitial components segregate by migration and produce various commercially exploitable concentrations. Water, the principal fluid component of the sedimentary section, is thought to migrate in three separate stages. Initially, pore water and excessive (more than two) clay-water interlayers are removed by the action of overburden pressure. This initial water flow (which is essentially completed after the first few thousand feet of burial) reduces the water content of the sediment to about 30 percent, most of which is in the semisolid interlayer form. A second stage of dehydration is thought to occur when the heat absorbed by the buried sediment becomes sufficiently great to mobilize the next-to-last water interlayer in an $M(H_2O)_x + \Delta H_r = I + XH_2O$ fashion. The final stage of sediment dehydration which extracts the last remaining water monolayer from clay lattices is apparently very slow, even by geologic standards, requiring tens or possibly hundreds of millions of years depending upon the geothermal and burial history of the sediment.

Petroleum hydrocarbons which are distributed throughout the matrices of potential source beds in normal frequencies of 300-3,000 ppm are thought to be too sparse to initiate continuous fluid flow. In normal marine sediments, however, the water associated with clay minerals is present to a considerable depth in the order of 200,000 ppm, and therefore it is reasoned that this phase forms the connection between petroleum source and reservoir beds.

The first and last dehydration stages are probably unimportant in Gulf Coast oil migration, inasmuch as they occur, respectively, at levels too shallow and too deep to intersect the interval of maximum liquid petroleum availability. The amount of water in movement during the second stage, at a level which *does* intersect this interval, is 10-15 percent of the compacted bulk volume and represents a significant fluid displacement capable of redistributing any mobile subsurface component. A measure of the degree to which the second-stage interlayer water has been discharged into the system can be noted on X-ray diffractograms. The movement appears to occur in a relatively restricted, depth-dependent temperature zone in which the average dehydration temperature of the points measured is 221°F. With the use of an empirically derived P/T curve and a geothermal-gradient map, a set of regional subsurface dehydration contours can be constructed. A plot of 5,368 liquid petroleum production depths referenced to this dehydration "surface" shows an almost perfect Gaussian distribution. It seems significant that, although the dehydration depths range from 4,000 to 16,000 ft, hydrocarbon production depths are distributed in a statistically consistent relation to the calculated clay-dehydration contour surface.

INTRODUCTION

General Concept

An evolutionary process operates in sedimentary sections between the time when terrigenous wastes are introduced into a sedimentation basin and the time when they are recovered from deep burial as indurated rocks. Changes in rock character occur principally in response to the pressure and thermal effects of an increasing overburden mantle. As a time function, the process must certainly be continuous; however, in normal petrologic description, certain degrees of change have been categorized in terms such as "unconsolidated," "lithified," and "metamorphosed." The ordering sequence of these descriptions generally parallels an energy gradient, as may be inferred from a simple burial-depth plot. Actually the categories are integrals of several energy components which effect changes in the physical appearance of the rock specimen.

This classification of gross textural type can be refined on a mineralogic basis by use of zonal designations based on shallowest appearances of *in situ* diagenetic minerals or mineral groups, such as glauconite, illite, zeolites, or minerals of the greenschist facies. From this classification, a further refinement now is being considered wherein isoenergetic surfaces are established, based on the physico-chemical condi-

[1] Manuscript received, July 24, 1967; accepted, February 3, 1968.

[2] Director, Research and Development, General Refractories Company.

The writer is indebted to G. M. Griffin, University of South Florida (Tampa), and D. B. Shaw, Milchem Corp., who contributed generously in time and thought to the development of this paper, and to Shell Development Company, which released it for publication.

tions at which certain mineral groups reequilibrate in the subsurface. Most of the equilibrations are accomplished by phase changes involving short-range migrations of atoms and bond-energy adjustments. Specifically, this paper concerns illitization, during which a major change in bulk density, as well as crystal-lattice dimensions, takes place as a result of the expulsion of the intracrystalline water. It is thought that the intracrystalline water is converted to capillary water and removed from the vicinity of the "mother structure" by migration through formational porosity. The fluid release associated with this transition may alter significantly the distribution of all the mobile constituents of subsurface rocks, including petroleum.

The work reported here shows that clay dehydration and the consequent fluid flow begin within rather narrow limits of minimum energy input. The study further suggests that this minimum or threshold energy constitutes a point in the diagenetic continuum at which the bonding and hydrostatic pressure energies holding water within the clay lattices no longer are able to withstand the energy forces related to geothermal gradient which are working to expel the water. This energy threshold is thought to constitute an equilibration plateau in the rock-alteration process which can be of primary significance in the classification of buried sediments and the disposition of subsurface fluids.

Observed Diagenesis Indicators

Clay-mineral changes.—Evidence of systematic diagenetic conversions in clay minerals were observed several years ago in an extensive survey of clay-mineral relations in the subsurface Wilcox Formation (Burst, 1959). Montmorillonite, a common swelling clay constituent of Wilcox outcrop material, is less evident in samples from below 3,000 ft than in those from above, and normally is not found in an unmixed[3] state below the 9,000–10,000-ft depth level. Below 14,000 ft, swelling capabilities of the clay minerals are virtually eliminated (Burst, 1959). Similar findings in the Upper Cretaceous of North Africa were noted by Dunoyer de Segonzac (1964).

The progressive lattice mixing with depth in Wilcox sediments is illustrated in Figure 1 by the variation in clay-lattice measurements recorded after treatment with ethylene glycol. A fully expandable montmorillonite lattice will absorb ethylene glycol to the extent that its "c"-axis unit-cell dimension measures approximately 17.0Å. The illite-type lattice does not absorb ethylene glycol and thus records the same 10.0Å measurement after treatment as before. Intermediate values of the "c"-axis measurement indicate random mixings of the two lattice types in proportion to the degree of mixing.

This phenomenon can be observed in studies of a single well by a simple plot of the 001 measurement of the glycol-treated three-layer lattice *versus* burial depth. In a presentation such as Figure 2, in which the expanded "c"-axis lattice parameters are plotted against recovery depth, it can be seen that, for several thousand feet below the surface, Gulf Coast sediments contain fully swelling, three-layer clay, as indicated by the repeated 17Å measurements recorded for shallow-recovery specimens. Normally in such a set of measurements, the 17Å component disappears at a certain depth and does not reappear in any deeper sample. Generally this type of change has been attributed to variations in sediment source. However, after a careful study of several dozen wells in many sedimentary basins, the writer has concluded that this systematic elimination of swelling clay is a valid reflection of diagenetic development in that it is unidirectional. The alteration is consistently from swelling to nonswelling lattices with depth. It is obvious, therefore, that the shallowest depths at which fully swelling montmorillonite is no longer in evidence can be contoured as a level of uniform diagenesis which may have possibilities as a subsurface drilling marker. Sediments represented by the 17Å measurements are from many different facies which contain different amounts of montmorillonite. The measurement indicates the swelling capability of the montmorillonite component regardless of its concentration.

A more sophisticated and possibly more definitive estimate of Gulf Coast clay dehydration can be made by using MacEwan's (1956,1958) and MacEwan and Ruiz's (1959) system for analysis of the 002/003[4] mixed-layer doublet

[3] The term "mixing" is applied to clay-mineral complexes in which two or more different mineral lattices are interspersed randomly to the extent that X-ray diffraction from the combination produces an averaged cell dimension which is respresentative of neither individual component. In this instance the mixing refers to layers of nonswelling clay lattices interspersed within layers of swelling lattice types.

[4] According to Brown (1961): "No indices can be assigned to the peaks of an interstratified mixture, but for two-component mixtures, they are conveniently labelled by means of the peaks of the pure phases towards which they migrate: *e.g.* 002/003 is a peak which moves continuously from the 002 position of one component (which we take to be the one of lower spacing) to the 003 position of the other."

WELL AND DEPTH	X-RAY PATTERN AFTER GLYCOL TREATMENT
ATLANTIC GOODPINE WINN PH., LA. 2,109	
SILBERMAN-AYCOCK RAPIDES PH., LA. 6,061	
J. B. EDWARDS #1 RAPIDES PH., LA. 10,010	
SHELL-LUMA DARBONNE ALLEN PH., LA. 12,515	

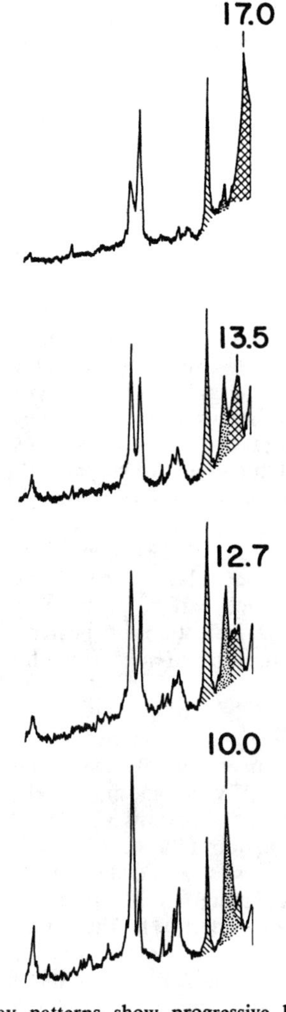

FIG. 1.—Lattice mixing in Wilcox (Eocene) sediments. X-ray patterns show progressive loss of expandable material with depth.

to estimate the percentage of dehydrated lattices in a mixed-layer component. Random interlayering of the 002 illite reflection with the 003 montmorillonite reflection is expressed by intermediate peak location and, in the case of progressive dehydration of expandable lattices, by movement of the montmorillonite 003 peak toward the smaller dimension normal for the 002 illite line. The end-point positions of the 002 illite reflection and the 003 montmorillonite (glycol expanded) reflection are recorded in the X-ray spectrum at approximately 5.00Å and 5.67Å, respectively. The latter relation is illustrated by curves A and B in Figure 3.

Figure 3C shows the type of pattern to be expected in this part of the spectrum from a physical mixture of montmorillonite and illite, each phase essentially devoid of interlayering. Figure 3D illustrates the type of pattern which results from a mixture of illite (large peak) and interstratified illite/montmorillonite (the small peak area denoted by the distribution

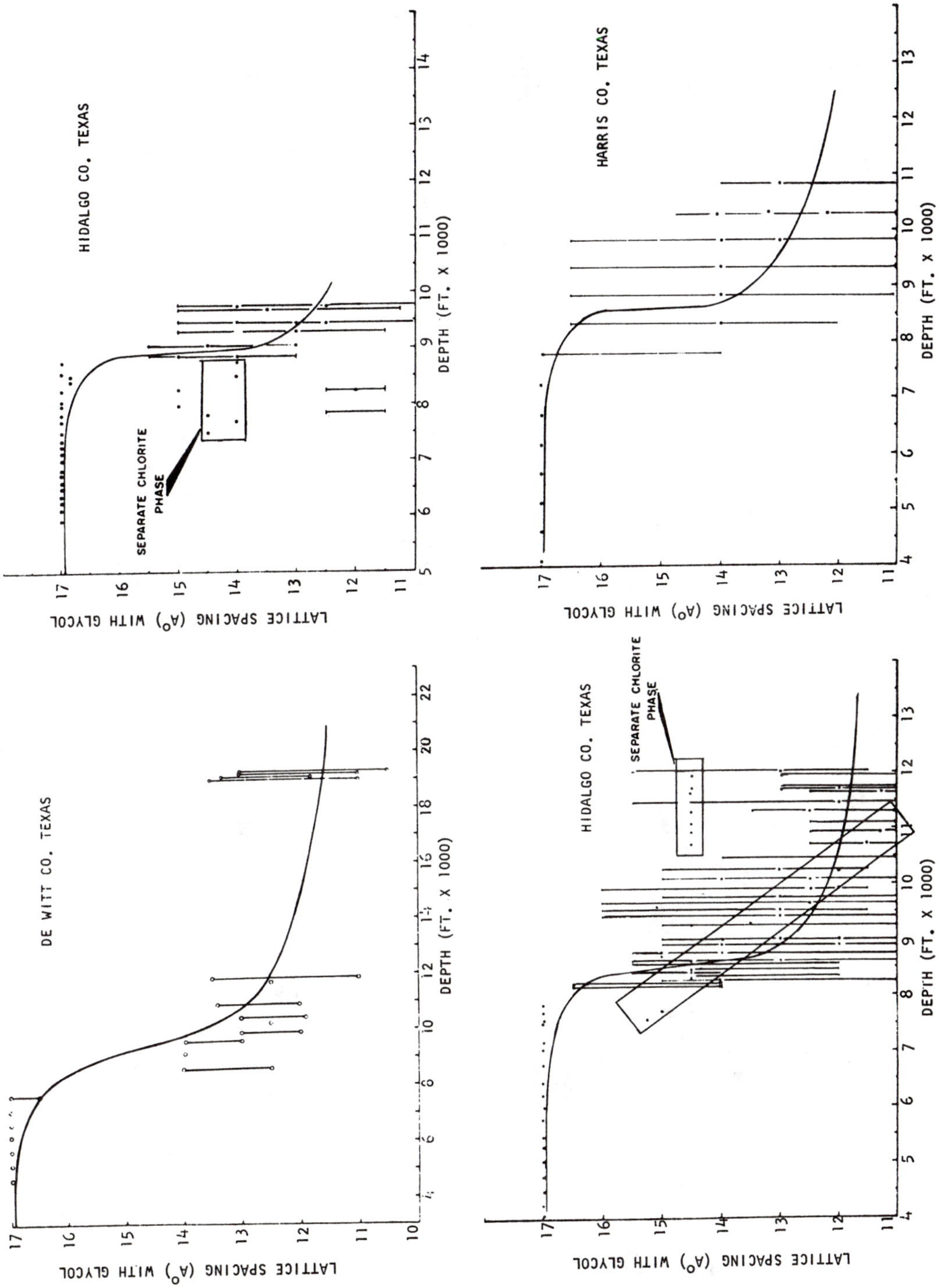

FIG. 2.—Clay-lattice variation with depth in Gulf Coast.

curve). In the latter example (Fig. 3D) the montmorillonite component has become interlayered and its peak has migrated toward the illite end of the spectral section, where it is referred to as the 002/003 peak. G. M. Griffin (personal commun.) has devised a graphic method for converting MacEwan's calculated interlayer 002/003 peak positions directly into

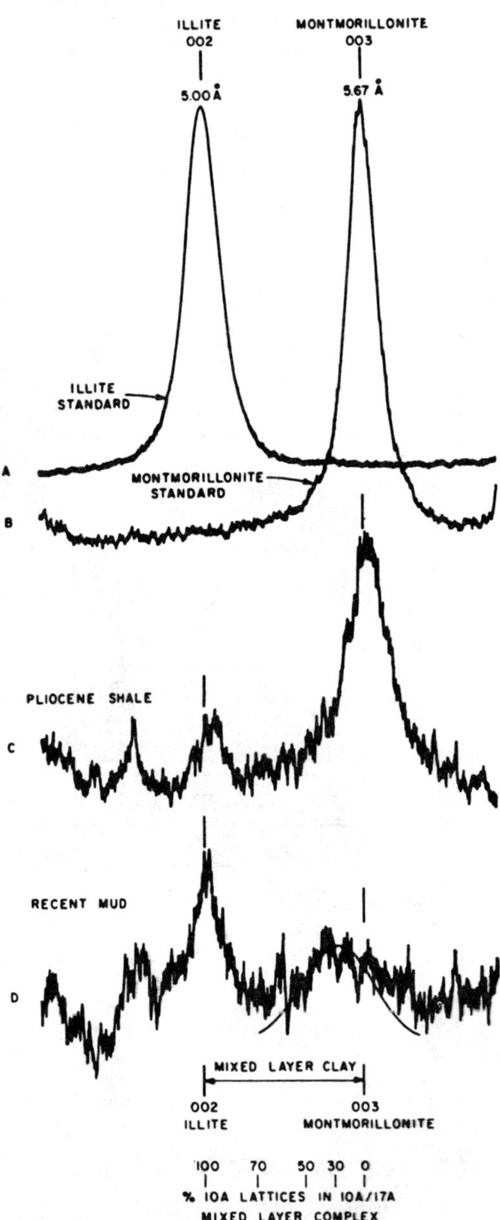

FIG. 3.—Montmorillonite *003* peak movement in response to lattice interlayering.

percent interlayer composition. His system is represented by the percentage scale at the bottom of Fig. 3 and shows that about 25–30 percent of the montmorillonite lattices in the material represented by Fig. 3D has already been dehydrated. The advantage of this approach is that the gradual subsurface dehydration of montmorillonite lattices, which is reflected only periodically by changes in 001 peaks, can be quantified continuously by measurement of the 002/003 doublet. A comparison of the types of data produced by the measurement methods is shown in Figure 4, where the 001 peaks (upper curves) in the low 2Θ spectral region are compared with the 002/003 peak (lower curves) character of the same sample.

The 002/003 method was used to characterize the clay sediments in a 15,000-ft Chambers County, Texas, wildcat. The well was sidewall cored at 100- and 200-ft intervals below surface casing for the purpose of evaluating the theory of systematic Gulf Coast clay-sediment dehydration (Figs. 1, 2).

A plot of the percent nonexpandable lattices (based on the 002/003 peak position) *versus* sample depth produces what appears to be the normal pattern of Gulf Coast clay-sediment dehydration (Fig. 5). From 2,600 ft to about 8,500 ft, a rather consistent base line of about 15–20 percent dehydrated lattices is established which indicates that the interlayer water content of shaly sediments in the shallow-burial range of the well is relatively constant. In addition, it was established that recent clay from the Mississippi River, which is thought to be mineralogically similar to clay introduced from the ancient source of Tertiary sediments in the Gulf Coast region, also contains about 20 percent dehydrated lattices. The conclusion follows, therefore, that there has been essentially no interlayer dewatering in the well above the 8,500-ft depth level.

In the depth interval from about 8,500 to 12,500 ft, 002/003 reflections become relatively diffuse, but peaking is sufficient to reflect a systematic increase in nonexpandable layers from 20 to 65 percent. From 12,500 ft to total depth, no further reduction in expandable layers is apparent.

Significant dehydration, therefore, appears to be confined to a 4,000±-ft interval beginning at about 8,500 ft which corresponds to a temperature interval of 65°F between 210 and 275°F on the basis of published data for the true-equilibrium gradient in the area (Moses, 1961). The conversion from depth references

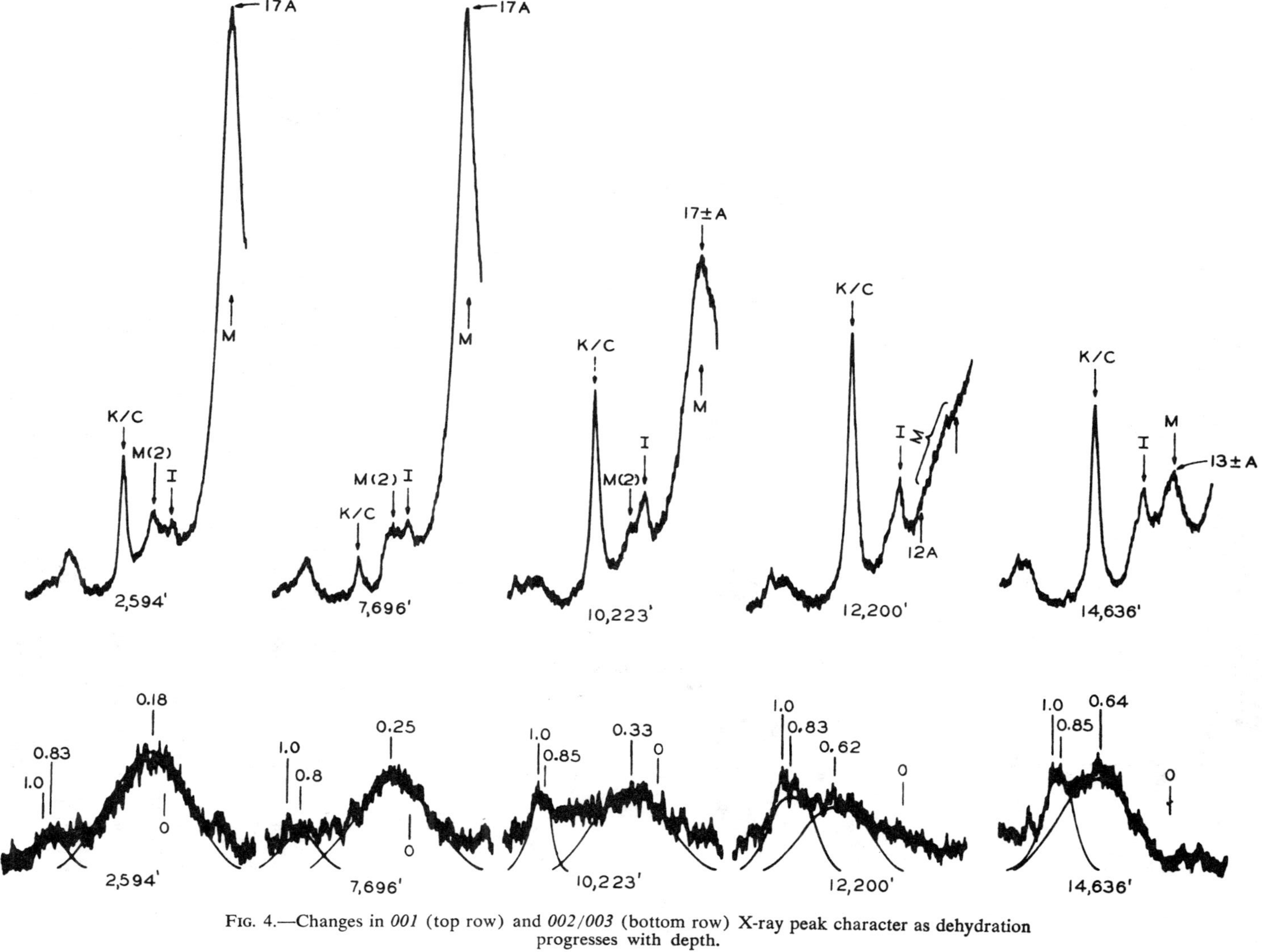

FIG. 4.—Changes in *001* (top row) and *002/003* (bottom row) X-ray peak character as dehydration progresses with depth.

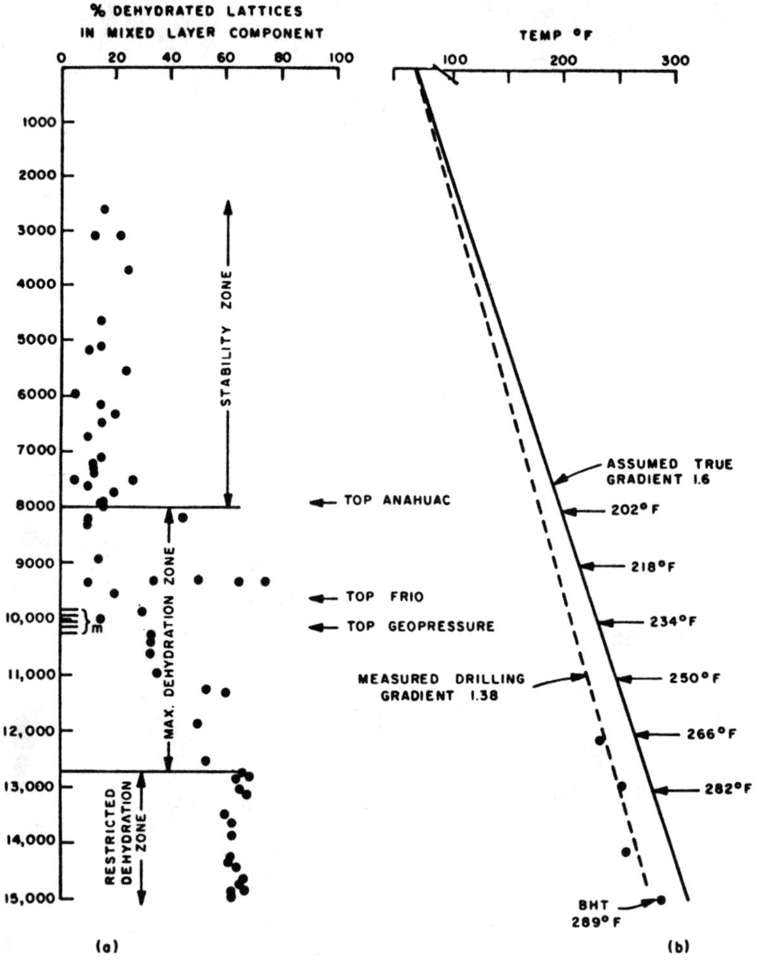

Fig. 5.—Stepwise dehydration of clay in Gulf Coast well, Chambers County, Texas.

to temperature references is important, because it has been found that clay diagenesis is essentially a temperature-controlled process, the effects of which can be interpreted more usefully in terms of burial temperature rather than burial depth.

Clay-Mineral Dehydration and Subsurface Fluid Redistribution

Factors Influencing Clay Dehydration

For many years the thought has persisted that clay compacted (*i.e.*, discharged its water content) in response to sediment overburden pressure. Van Olphen (1963) has demonstrated, however, that the pressure necessary for removal of interlayer water between clay platelets at 25°C is considerably in excess of that normally present in sedimentary columns. According to his calculations, pressures of 65,000–70,000 psi are necessary to remove the last water layer and 30,000 psi is needed to remove the second-to-last water interlayer. Therefore, it must be concluded that, at relatively low temperatures, liquid discharge from clay sediments in response to loading must be confined mostly to pore-water effluent.

Van Olphen (1963) also concluded that complete water removal rarely occurs at the overburden pressures normally present in sedimentary sections. However, his conclusions are based on adsorption/desorption isotherms rather than geologic observations. Many clay lattices in Paleozoic and older (pre-Miocene) Tertiary sediments appear to have been dehydrated completely under normal overburden pressures. Also, partial dehydration at levels where overburden pressure is much less than those predicted from the Van Olphen calculations has been observed in relatively young sediments. Obviously, a force other than overburden pressure must be effective.

Because of the paucity of fully swelling clay material in Paleozoic rocks, its restricted presence in older Tertiary rocks, and its abundance in younger Tertiary and Quaternary sediments, the thought also has persisted that clay hydration state is a function of geologic age. From his work with clay sediments of several geologic ages and his general review of the literature on clay diagenesis, Powers (1959) concluded that mixed layering in the montmorillonite-illite series is not a function of geologic age *per se*, but probably is related to a unique depth below the surface, "... depending on the permeability, porosity, chemistry and probably the temperature of the sediment from the time of its deposition to the time of its greatest depth of burial." He was unable to vary his depth estimates in terms of the modifiers he had recognized, possibly because he was searching for the key within the general field of ionic substitution in the clay lattices. In a more recent article, Powers (1967) again keys clay diagenesis to specific depths as a combined effect of ion exchange and overburden pressure.

Clay compaction as a result of interlayer water discharge is postulated to be an essentially temperature-dependent phase change in which a silicate lattice hydrate (montmorillonite) begins to dehydrate at a critical temperature in the 200–230°F range, regardless of burial depth. Within the geologic framework of normal sediments, therefore, temperature is proposed as a more effective dehydration agent than either pressure (depth of burial) or age. The phase transition can be described by the equation

$$M(H_2O)_x = I + XH_2O + \Delta H_r$$

where

M = a hydrated clay lattice,
$(H_2O)_x$ = the water held within the hydrated lattice,
I = a dehydrated lattice,
XH_2O = capillary (pore) water, and
ΔH_r = the heat of reaction (H is negative, consequently the reaction is endothermic).

Interlayer water dehydration should not begin in sediments until they have reached the critical subsurface conditions required by this phase transition.

Dehydration Sequence

Observations made during the course of this study suggest that water is extracted from subsurface clays in three separate stages, as illustrated in Figure 6. Initially (stage 1), pore water and excessive (more than two) water interlayers are removed by the action of overburden pressure. In volume, this represents by far the largest water removal in the process of sediment dehydration. Shortly after settling on the basin floor, marine clay sediments of the Mississippi delta type contain approximately 70–80 percent water by volume. The initial dehydration (commonly referred to as "compaction"), which should be essentially completed in the first few thousand feet of burial, reduces the water content to approximately 30 percent (about 20–25 percent interlayer water and 5–10 percent residual pore water).

In stage two, pressure is relatively ineffective as a dehydrating agency because of the increased density of the interlayer water packet, and the sediment remains in a state of quasi-equilibrium as it continues to absorb heat generated deeper in the geologic section. When the heat accumulation is sufficient to mobilize the interlayer water, one of the two remaining interlayers (statistically averaged) is discharged into the bulk system. The amount of water in movement should constitute 10–15 percent of the compacted bulk volume. This movement is the most significant fluid displacement subsequent to the initial pore-water drainage, and is capable of redistributing mobile subsurface components. In stage three, the final water increment, which approximates capillary water density, gradually is forced out of the clay lattice and the pores as sediment temperature increases. The rate of this last stage of dehydration is apparently slow even by geologic standards, requiring tens or possibly hundreds of millions of years for completion, depending on the geothermal history of the location and burial history of the rock. Figure 7, in which claystone porosity is plotted against geologic time for the pertinent compaction interval, illustrates the rate at which the third dehydration stage takes place.

The volumetric relation of the various solid and liquid components of the sediment considered for this illustration was compiled from data on the Mississippi delta type of clay mud which eventually might be converted into subsurface shale. The solid part of the original sediment is estimated on the basis of laboratory

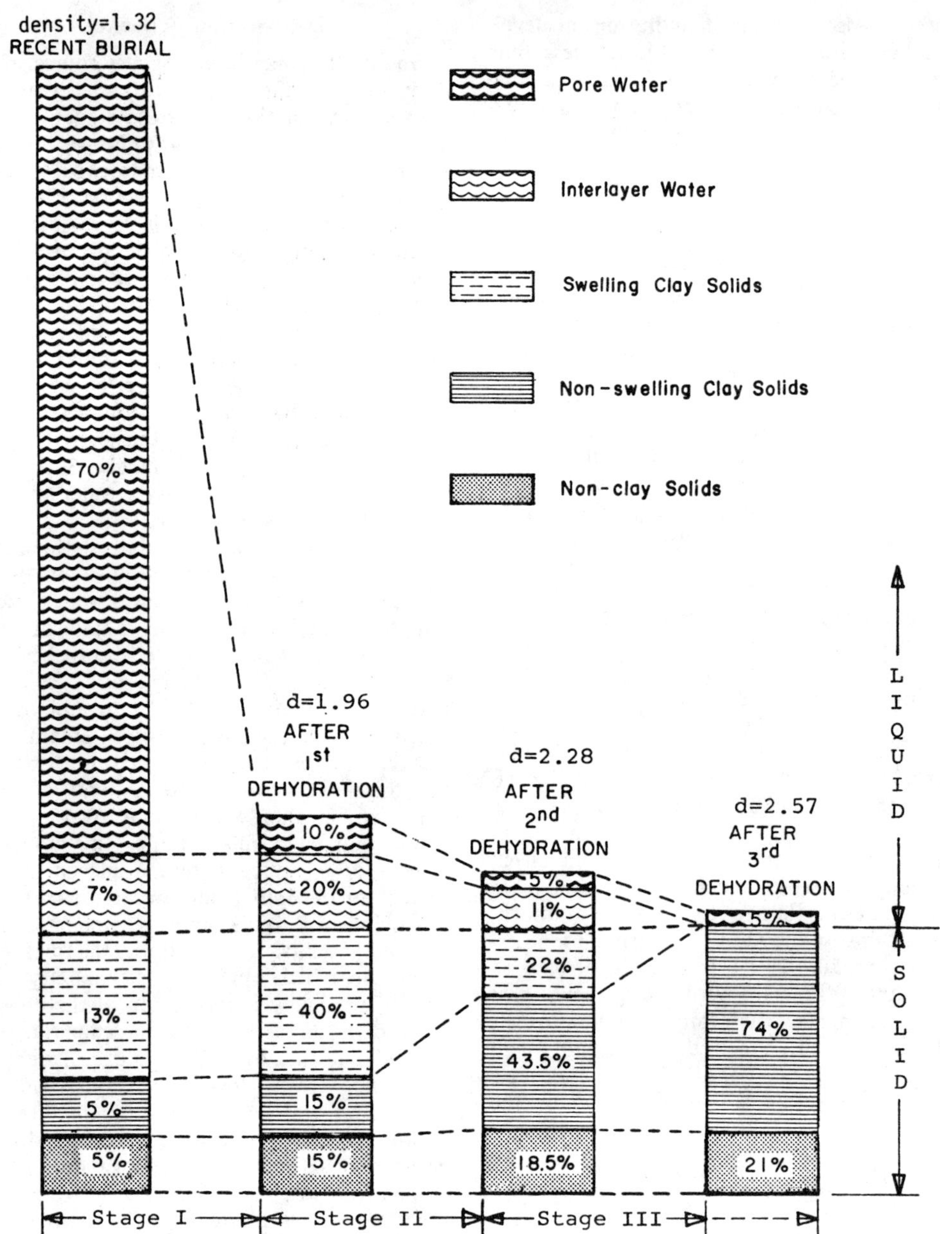

FIG. 6.—Marine shale bulk composition during dehydration.

measurements to be 80 percent clay, of which 80 percent is of the swelling type.

Interlayer Water Density Inversions

An interesting calculation pertinent to this work indicates that the density of water entrapped between layers of vermiculitic swelling clay changes during subsurface dehydration and actually traverses the unity value at least twice. A three-stage dehydration sequence based on these calculations is presented in Table I. Adaptation of these calculations to the

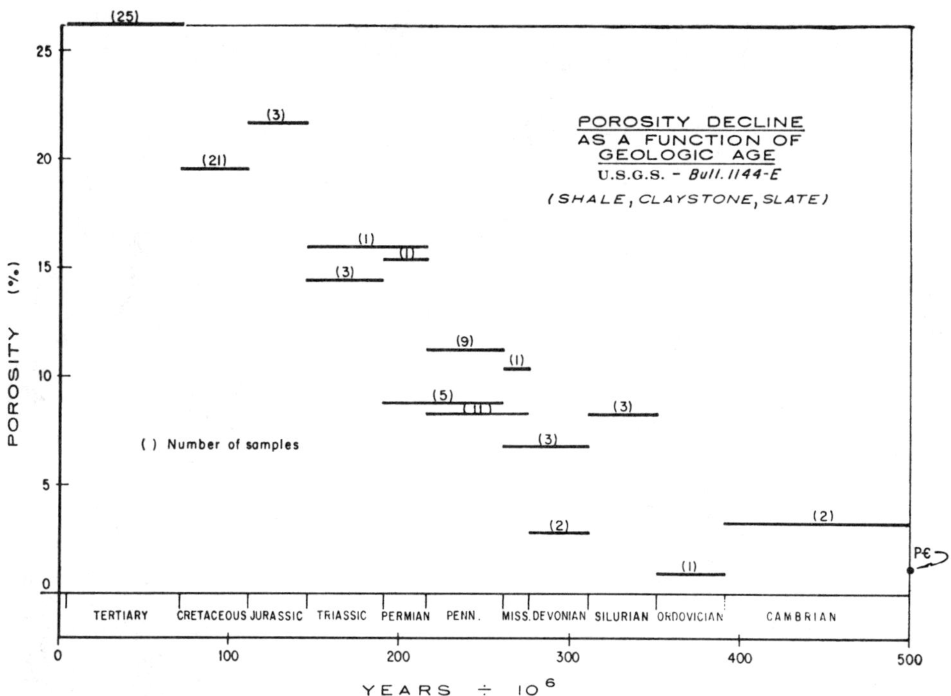

FIG. 7.—Porosity decline as a function of geologic age. Modified from Manger (1963).

Tertiary sediments of the Gulf Coast could have far-reaching consequences in the interpretation of fluid pressures and distribution.

First-stage dehydration.—At the depositional interface in the sedimentary basin and in the shallow-burial part of the subsurface sedimentary column, swelling clay material holds several layers of water molecules between the flat, essentially two-dimensional clay lattices (Fig. 8a). The two water layers that lie adjacent to clay lattices are affected markedly by the surface geometry of the lattices and are distributed across the lattice face in a configuration which produces a water density somewhat less than one. Between these two semibound layers are additional water layers with considerable freedom of movement and probably in free interchange with outside pore water. Their density is assumed, therefore, to be the same as that of pore water (1.00 for fresh water, ~ 1.04 for seawater of normal salinity). Volumetrically, the entire interlayer water block occupies more

Table I. Calculated Interlayer Water Density

		Hexagonal Water Net No Overpacking		*Overpacked due to Exchange Site Effects*	
(1) Number Water Layers	*(2) Total Thickness Interlayer Water (A)*	*(3) No. H_2O Molecules*	*(4) Density*	*(5) No. H_2O Moles. with Exch. Site Effect*	*(6) Overpacked Density*
1	3	4	0.85	4.25	0.94
2	5	8	1.02	9.	1.15
3	8	12	0.96	13.	1.04**
4	10.5	16	0.975	17.	1.04**
		Probable limit of crystalline swelling			
5	13.0	20	0.985	21.	1.035
6	15.5	24	0.99	25.	1.03
7	18.	28	0.995	29.	1.03
8	20.5	32	1.00*	33.	1.03

* Fresh water density.
** Seawater density; 35,000 ppm salinity

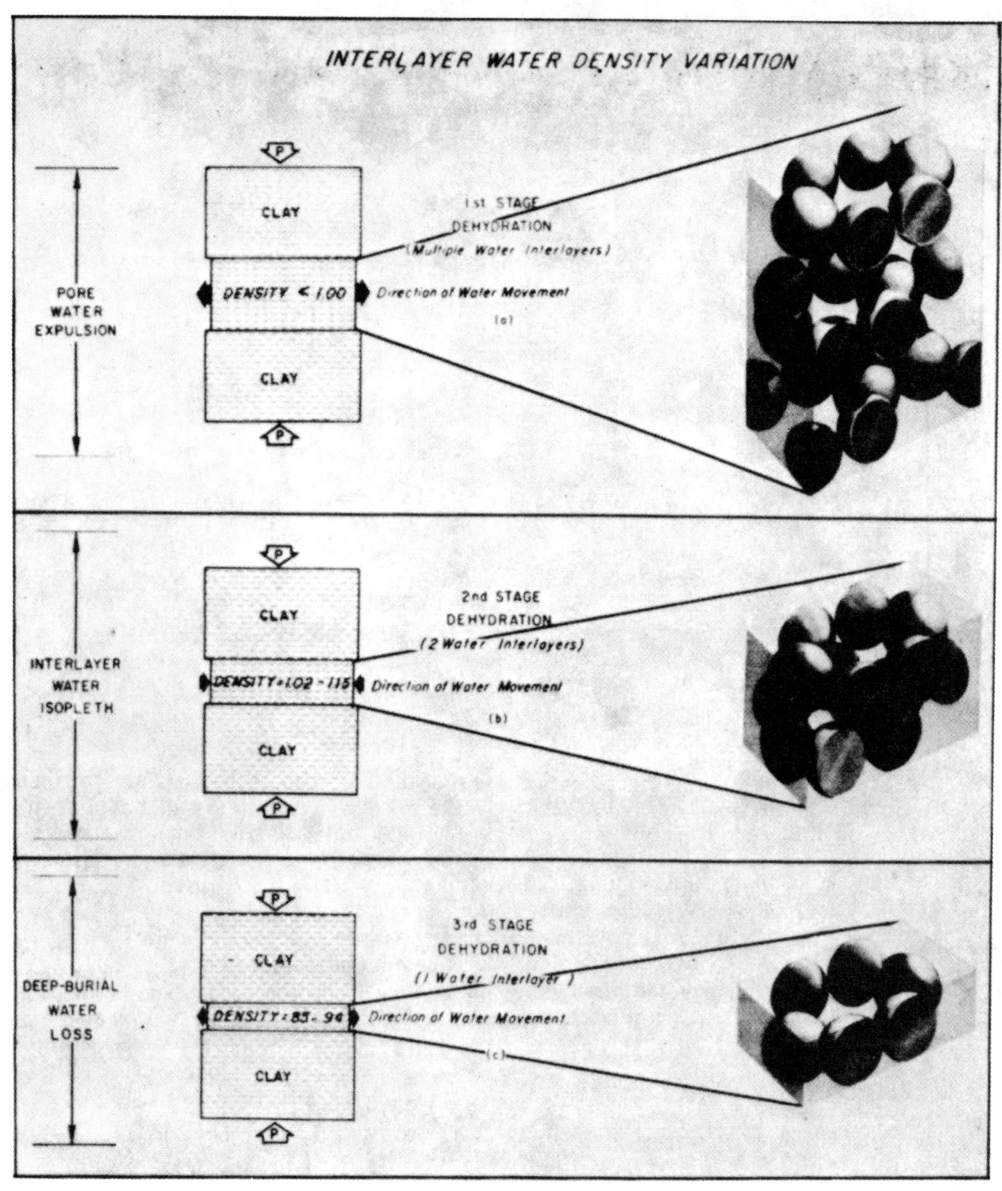

FIG. 8.—Interlayer water density variation.

space inside the clay lattice than it would in interparticle pores.

Overburden pressure initiates dewatering by removing excess (more than two) water layers from the clay. The actual mechanism of this dehydration is not clear, but two explanations are feasible, either of which leads to the same result. If the excess water layers actually do exchange freely with pore water and act independently of the two lattice-bound layers, the rate of initial water loss is controlled by the magnitude of the hydraulic pressure gradient between the clay particle and the water-table surface. If, however, the excess water layers must be considered part of a unit which also includes the relatively immobile lattice-bound water, the density of the unit is less than one and overburden-pressure effects, in accordance with the Clausius-Clapeyron relation, would reduce the systemic energy by transferring the interlayer water to a more dense configuration outside the lattice. In either case, the dewatering of the clay lattice continues until the two lattice-oriented layers remain (Fig. 8b).

Second-stage dehydration.—At this stage of dewatering a significant change is recorded in the interlayer water phase. From X-ray analysis, it is known that the two bound water layers occupy a vertical distance of about 5Å and that a unit of this water has the same horizonal dimensions as a unit cell of clay, 5.2 × 9.0 Å. The unit water block, therefore, is 5 × 5.2 × 9 Å and contains eight molecules of water arranged as in the three-dimensional illustration (Fig. 8b). The density of this liquid equals the number of molecules considered, multiplied by the molecular weight of one molecule, further multiplied by the reciprocal of Avogadro's number, all divided by the volume which the molecules occupy in space.

$$\text{Density} = \frac{\text{number molecules} \times \text{mole. wt.} \times \dfrac{1}{\text{Avogadro's number}}}{(\text{cell length} \times \text{width} \times \text{height})}$$

or interlayer water density;

$$\text{two-layer case} = \frac{8 \times 18 \times 1.66}{9 \times 5.2 \times 5} = 1.02.$$

Actual measurements on vermiculite, which has the type of structure being considered (Bradley and Serratosa, 1960), show that, because of the attractive forces generated by exchange sites within clay-layer lattices, the average two-layer water pack is infilled or overpacked by an average of one water molecule per unit, in which case the density is

$$D = \frac{9 \times 18 \times 1.66}{9 \times 5.2 \times 5} = 1.15.$$

The density of interlayer water (theoretically as high as 1.15) is significantly higher than the density of pore water (1.00–1.04); thus, a unit of water in the interlayer position occupies less space than it would within a pore. The pressure effect of overburden actually resists the transfer of water from the lattice to the pores and, in all practical aspects, interlayer dewatering ceases. The sediment is in a state of quasi-equilibrium and a steadily increasing pressure is applied from overburden buildup.

The quasi-equilibrium is disrupted by increasing temperature, which eventually overwhelms the retarding pressure forces and releases one of the water layers to the bulk system. The result is a major rearrangement of subsurface fluids, for the fluid released in the second stage is approximately 10–15 percent of the bulk volume (Fig. 6).

Third-stage dehydration.—X-ray measurements show that one layer of water occupies a vertical distance of approximately 3 Å between the clay lattices. This measurement is slightly larger than half of the two-layer distance because of a closest approach restriction, which is apparently the result of a Van der Walls contact between the water molecules and the clay surface. The density of the single water layer is calculated therefore as

$$D = \frac{4 \times 18 \times 1.66}{9 \times 5.2 \times 3} = 0.85.$$

Actual measurements (Bradley and Serratosa, 1960) show an overpacking caused by exchange-site forces which average ¼ molecule of water per unit. Density is actually then

$$D = \frac{4.25 \times 18 \times 1.66}{9 \times 5.4 \times 3} = 0.94.$$

Calculations were not included in the foregoing description of multilayer dehydration (first stage) because data available on those systems are much less reliable than data for the one- and two-layer systems. Calculations were made, however, and tabulated in Table 1, column 2, for as many as eight water interlayers on the basis of the formulas:

one-layer case,

2.5 Å + 0.5 Å closest approach restriction;

two-layer case,

2(2.5 Å) + 0.5 Å closest approach restriction
− 0.5 Å overlap allowance;

n-layer case, where $n > 2$,

n(2.5 A) + 0.5 Å closest approach restriction.

It was not necessary to consider overlap in the multilayer case because the constant interchange of fluid between the layers and the pore water maintains a 2.5 Å spacing for molecular passage.

Density values calculated by using these spacings and the number of water molecules normally in each layer are shown in Table I, column 4. A density of 1.0 is assumed for all water in excess of two layers. These data show a gradual density *decrease* from 1.0 for eight layers of water to a calculated value of 0.96 for three interlayers. The density gradient inverts to 1.02 at the two-interlayer dehydration level and reinverts to 0.85 at one interlayer.

Recalculation of the molecular population and interlayer water density values to include the extra water overpacked because of exchange-site attraction produces the data tabulated in columns 5 and 6. A gradual *increase* in density occurs during dehydration to a three-interlayer water level, and a sharp increase to 1.15 occurs if the two-layer overpacked and overlapped configuration is assumed. Further dehydration to one interlayer reduces the density to 0.94. Whenever the interlayer water density exceeds the density of the pore water, loss of interlayer water ceases until temperature is raised enough to initiate second-stage dehydration.

Interlayer Water Reactions

An analogy can be made between the temperature behavior of the second-stage, interlayer-water, capillary-water transition and the common ice-water transition. In each system, the molecules of one phase are in rigid confinement in comparison to their state in the other phase, and in each system the molecular mobility can be increased easily by the addition of heat.

Molecules of water are oriented within the swelling clay lattice in a pattern much like their orientation in ice. The water molecules are keyed to the clay surface, at least partly, by charges from the hexagonally positioned clay-lattice components and would describe, in plan view, hexagonal patterns similar to those present in ice (and observed in snowflakes), although considerably more densely packed. In each instance (ice and interlayer water) the molecules are in an arrested vibration of equilibrium which is broken when sufficient heat is applied to the system.

When the molecular velocity becomes sufficient to allow the molecules to escape from their oriented pattern and move freely about each other, ice is changed to water and, according to the analogy, interlayer water is free to move out of the clay lattice and into adjacent capillaries. By analogy, therefore, clay dehydration is pictured as a semisolid-liquid phase transition.

The effects of pressure and temperature on this particular phase transition are opposed to each other because the density of interlayer water is thought to be greater than that of capillary water. The thermodynamic system involved at this stage of compaction actually requires volume expansion rather than contraction, as is suggested by the bulk-volume changes observed in compacting sediments. The phase transition consists of conversion of the more dense, highly oriented, semisolid interlayer water into a less dense liquid phase. According to the Clausius-Clapeyron equation, this conversion prescribes a negative role for pressure; an increase in pressure will raise the dehydration temperature by maintaining the interlayer water in its semisolid state. Although there are no precise, direct measurements for interlayer water density, calculations from structural data indicate that it can be somewhat greater than unity. An indirect measurement, apparently corroborating this calculation, was made recently by M. H. Waxman (personal commun.), who used a dilatometer procedure in which the total volume (corrected for thermal expansion) of a vermiculite-water system was shown to increase about 12 percent as interlayer water was released to the pore system.

This observation helps explain why the extraction of interlayer water from expanded clay lattices by the application of high pressures at room temperatures is not easily accomplished. Further, it modifies the usual theories which prescribe increased overburden pressure as the agency responsible for shale compaction. In fact, increased overburden pressure in normal sedimentary rocks probably has little positive effect on, and may actually retard, the part of the compaction which results from expulsion of the second-to-last water interlayer.

From the Clausius-Clapeyron equation it also can be demonstrated that the effect of temperature is much greater than that of pressure. A pressure increase of several thousand atmospheres is approximately equivalent to a 10 percent rise in temperature. Thus, it appears that pressure changes with depth in the sedimentary section are too small to have thermodynamic significance in clay-water systems, and that the effect of pressure in sediments buried deeper than about 10,000 ft can be considered constant. Sedimentary overburden may function more effectively in some clay-compaction processes as a heat accumulator than as a pressure component.

Three-Stage Water-Escape Curve

Powers (1959, 1967) published a sketch (Fig. 9a) illustrating a two-stage subsurface dehydration system for montmorillonitic sediments. This curve has been adjusted and rein-

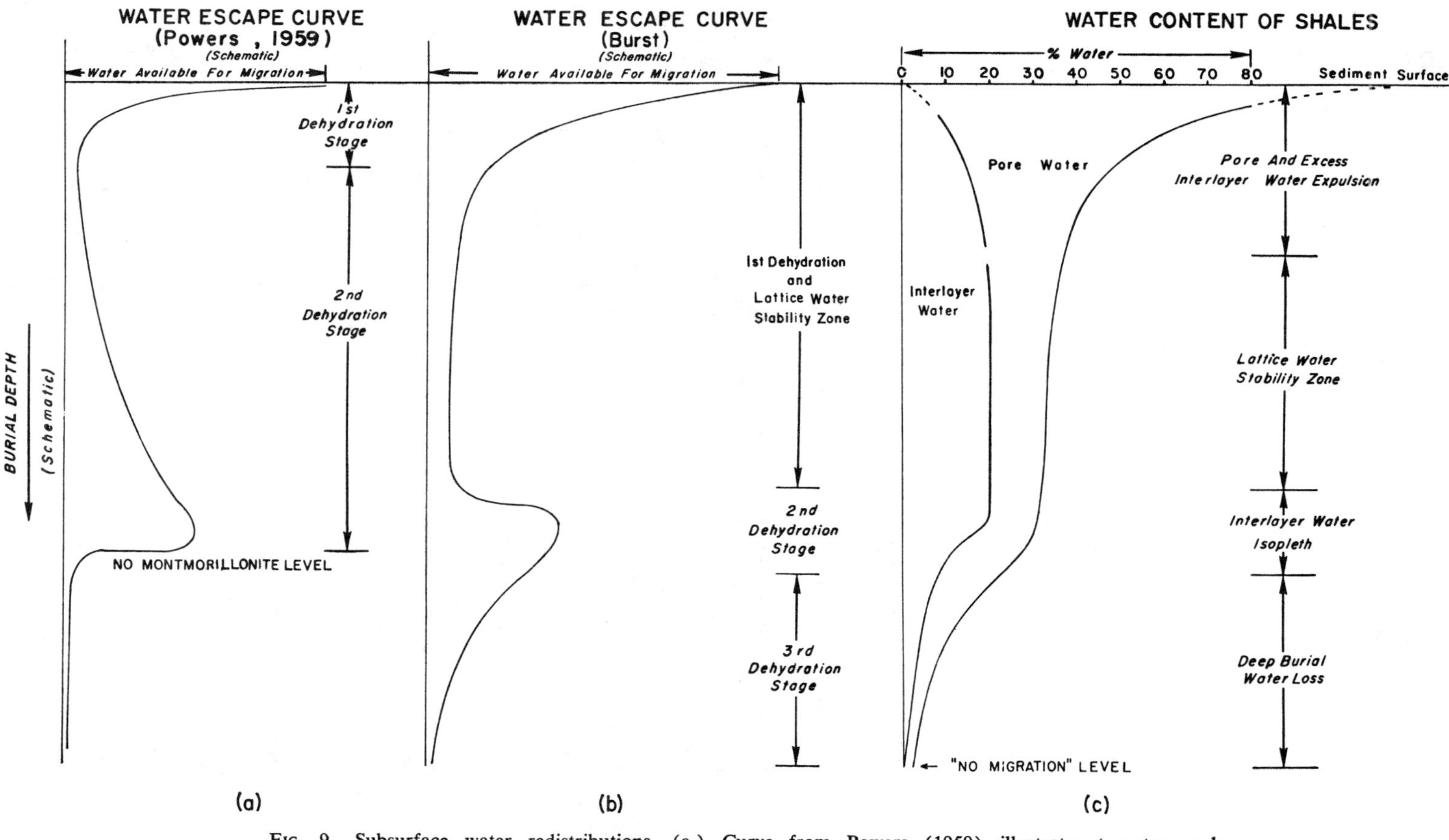

Fig. 9.—Subsurface water redistributions. (a.) Curve from Powers (1959) illustrates two-stage subsurface dehydration system for montmorillonitic sediments. (b.) Curve from Powers adjusted and reinterpreted as three-stage system. (c.) Water-content of curve reflects percentage at any stage of dehydration.

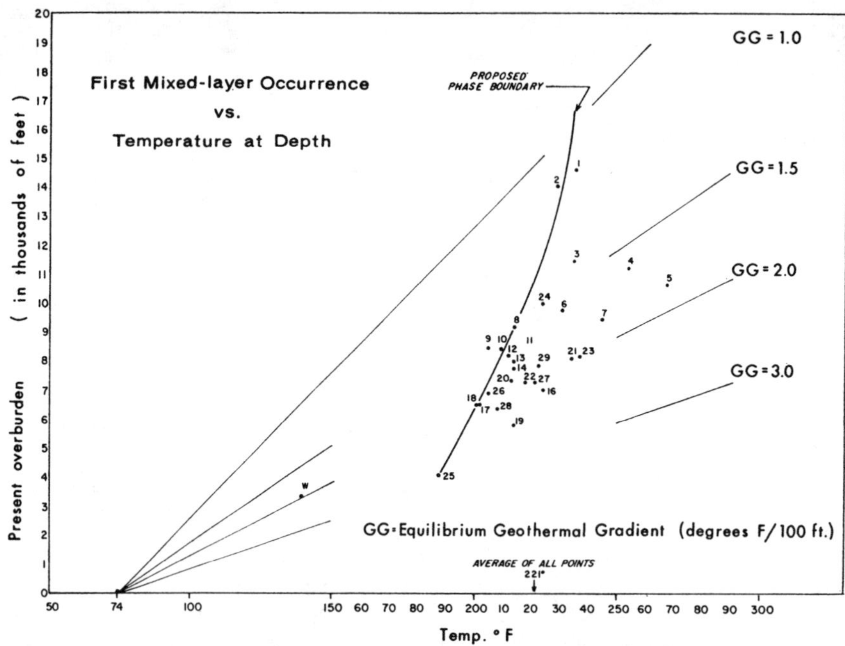

FIG. 10.—First mixed-layer occurrence *vs.* temperature at depth.

terpreted (Fig. 9b) as a three-stage system in accordance with the dehydration sequence heretofore described, to which some quantification (Fig. 9c) can be adapted. The additions to the Powers curve are (1) the interjection of an extensive zone of significantly reduced dewatering and (2) the consequent change in slope of the part of the curve representing second-stage dehydration. According to the current interpretation, the second stage of dehydration approaches a maximum abruptly and declines asymptotically through the third stage to a minimum at depth.

The water-content curve (Fig. 9c) reflects the amount of water present in a shale at any stage of dehydration and is based upon a sediment that contains 80 percent clay, 80 percent of which is montmorillonite. A similar clay was partitioned previously.

Two implications of possible significance are shown in Figure 9c. First, the total water content of shale decreases continuously with depth, thereby continuously restricting the possible avenues for fluid distribution. At a certain depth, illustrated schematically at the base of Figure 9c, the water channels become disconnected and permeability declines to the point that fluid migration is no longer possible. Second, the second hydration stage is completed in a relatively short time and initiates a considerable fluid discharge. The first and last dehydration stages are probably unimportant in Gulf Coast oil migration, inasmuch as they occur at levels that are too shallow and too deep, respectively, to intersect the interval of maximum liquid petroleum availability. However, the amount of water in movement during the second stage, at a level which does intersect this interval, is 10–15 percent of the compacted bulk volume and represents a significant fluid displacement capable of redistributing any mobile subsurface component.

Clay Dehydration Versus Petroleum Occurrence

The shallowest discernible subsurface clay-lattice mixing was plotted on a graph (Fig. 10) of recovery depth *versus* calculated *in situ* temperatures for what is thought to be a statistically sufficient number of trials. A greater degree of confidence was assigned to the most shallow dehydration under the assumption that, in some places, true shallowest occurrences may have been overlooked because of sparse sampling or inconclusive X-ray data. A best-fit drawn through these shallowest dehydration points is designated as the phase boundary in the aforementioned system $M(H_2O)_x = I + xH_2O + \Delta H_r$. The slope of this boundary demonstrates that, regardless of overburden build-

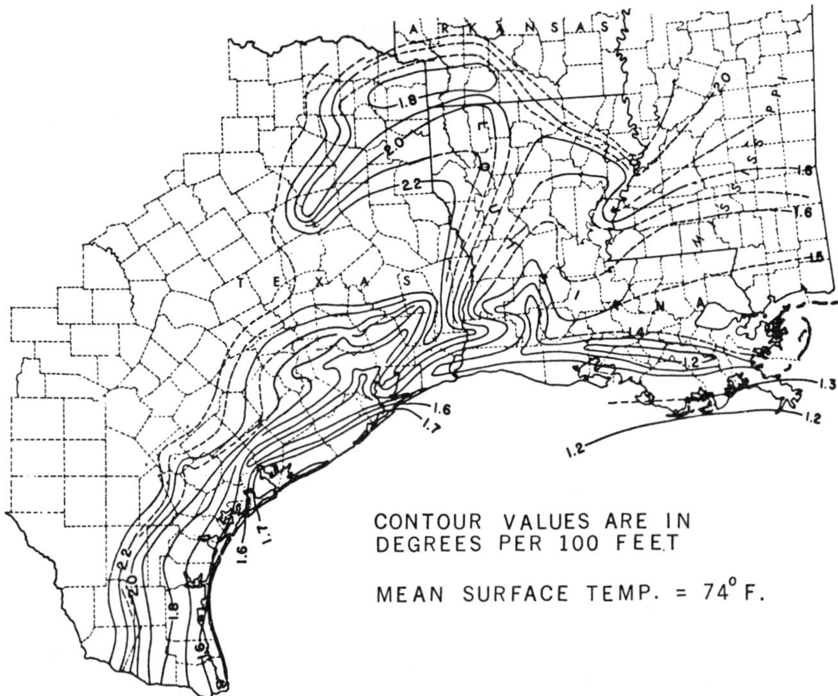

FIG. 11.—Geothermal-gradient contours (after Moses, 1961). Contour values in degree per 100 ft.

up, only an associated temperature increase can force the system to dehydrate.

On a contour map of true geothermal gradients, Moses (1961) recorded variations in the Gulf Coast area ranging from 1.2 to 2.2°F per 100-ft burial depth. Interpolating from Moses' map (Fig. 11) and using the phase boundary in Figure 10, D. B. Shaw (personal commun.) estimated the clay-dehydration depths and temperatures in more than 2,000 fields listed by the National Oil Scouts and Landmen's Association (1958). He found that, although the estimated burial depths for clay dehydration range from 4,200 to 15,900 ft, the temperature of conversion can be limited to the phase-boundary range 182–232°F.

Subsea depths of petroleum reservoirs in the 2,000 Gulf Coast oil fields are distributed in accordance with the bimodal plot in Figure 12, which includes data on 5,368 production levels. The character of the curve is largely a reflection of the average drilling depth in Gulf Coast oil exploration and offers little predictability.

Theoretical second-stage dehydration depths at each of the 5,368 locations were calculated by using the phase-boundary curve (Fig. 10) and the geothermal-gradient map (Fig. 11). The difference between dehydration depth and production depth is shown as a distribution function in Figure 13. Class limits were chosen at 1,000-ft intervals above and below the reference dehydration zero point. It seems significant that, although the dehydration depths range from 4,000 to 16,000 ft, hydrocarbon production depths are normally distributed about a mean 1,500 ft above the theoretical second-stage clay-mineral dehydration level. This distribution can be explained by a model in which hydrocarbon movement was initiated at least partly by water expulsion within the second-stage dehydration level and transported upward to the nearest trap. The reference to upward migration is made on a statistical basis, in the realization that local migration can be in any direction. The ultimate fluid discharge from the sediments must be at the surface; therefore, in a plot of 5,000+ instances an upward migration tendency should be shown. The large number of points also minimizes obvious violations of the hypothesis, such as rapid redistribution along newly formed fractures.

The number of data points included in the frequency plot is necessarily limited by the number and depth of wells drilled. Although

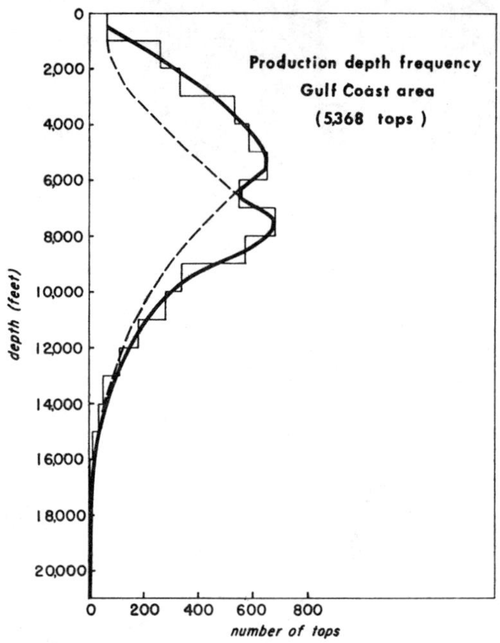

FIG. 12.—Production-depth frequency in Gulf Coast area using data from 5,368 production levels.

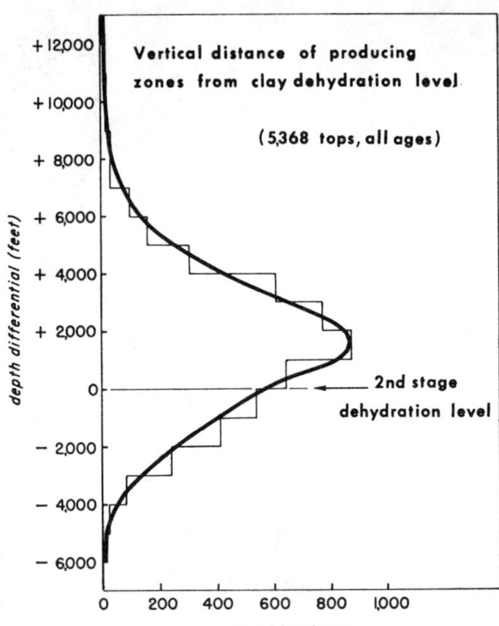

FIG. 13.—Vertical distance of productive zones from clay-dehydration level, Gulf Coast area, using data from 5,368 production levels.

more data are available from above the second-stage dehydration level than from below it, the character of the frequency distribution does not seem to be affected. It must be agreed that the shallow parts of the Gulf Coast Tertiary basin have been sampled sufficiently for statistically reliable analysis. Therefore, the discovery of numerous deep (17,000–23,000 ft) production zones would be the only possible agency for frequency alteration. Because the data are several years old (1958) and subsequent drilling has failed to produce an excep-

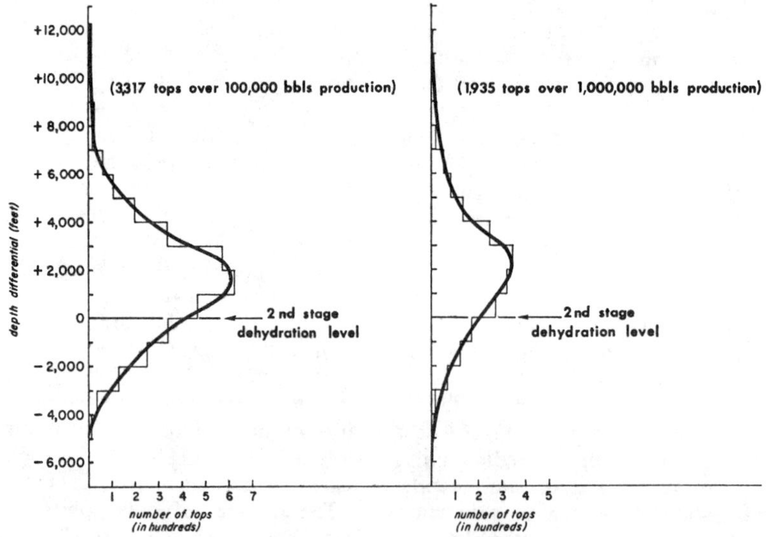

FIG. 14.—Vertical distance of producing zones from clay-dehydration level, Gulf Coast area.

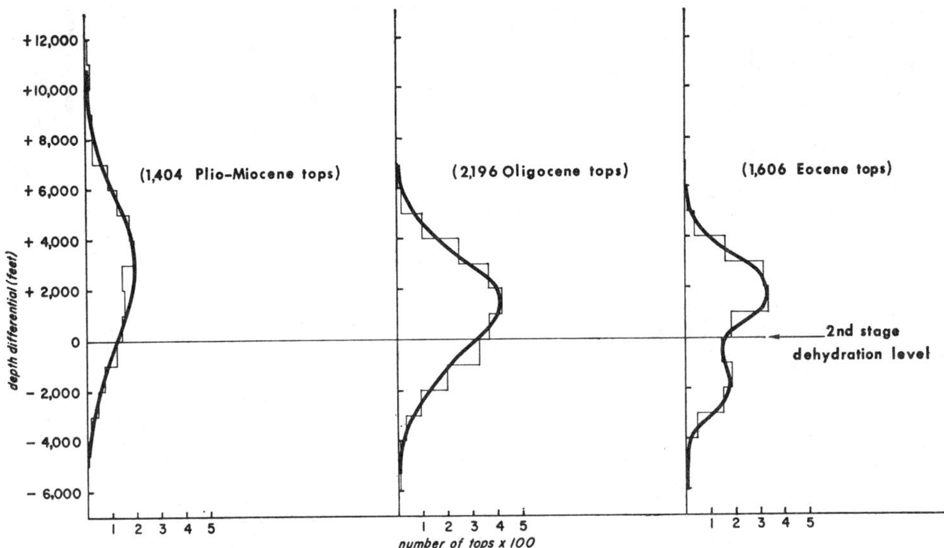

Fig. 15.—Vertical distance of production from clay-dehydration level.

tional deep-discovery pattern, the validity of the curve, as well as its usefulness in production predictability, must be considered.

A breakdown of the production *versus* dehydration data shows that the suggested conclusions are independent of field size (Fig. 14) and geologic age (Fig. 15) of the producing formation for Gulf Coast Tertiary sediments, and it appears that considerable predictability of subsurface petroleum occurrence can be obtained from careful geothermal-gradient and clay-dehydration work.

Conclusions and Projections

Subsurface Fluid Distribution Model

The dewatering of clay-rich sediments, long a subject of investigation by petroleum geologists, has been presented as a three-stage process which consists of expulsion of (1) pore water and excess (more than two) water interlayers, (2) second-to-last water monolayer, and (3) last water monolayer. The first and third stages are thought to be kinetic processes and the second a thermodynamic process. Most pore and excess interlayer water is removed from the sediment relatively early in its compaction history. It is believed to flow as a result of hydraulic-gradient differentials and commonly is referred to as being "squeezed" out of the sediment by overburden pressure. After this initial dewatering, the water-escape rate is reduced considerably, and during a subsequent period of overloading only minor pore-water loss continues.

A second major dewatering which appears to be thermally activated occurs deeper in the section (in the interval between 3,000 and 15,000 ft). It results in a discharge of interlayer water molecules as their vibrations attain sufficient velocity to break the bonding forces holding them to the mineral lattice faces. Plots of burial depth and temperature conditions under which this dehydration has actually occurred suggest that the water discharge may be repressed by overburden loading and therefore the release depends essentially on thermal buildup. The suggestion is supported by theoretical calculations on the distribution of water molecules within the interlayer space, which show that the density of interlayer water changes during dehydration and may become greater than unity at one stage of the dehydration sequence. From Clausius-Clapeyron considerations, it is concluded that this second stage of dehydration is a temperature-dependent water transition, in which pressure is an inhibitor and to which the familiar expression "squeezing out" should not be applied. An empirical pressure-temperature curve has been constructed to illustrate that additional accumulation of heat is necessary to perform the transition at greater burial depth.

A third stage of dehydration which progresses very slowly finally eliminates the last interlayer of water. At this point no further com-

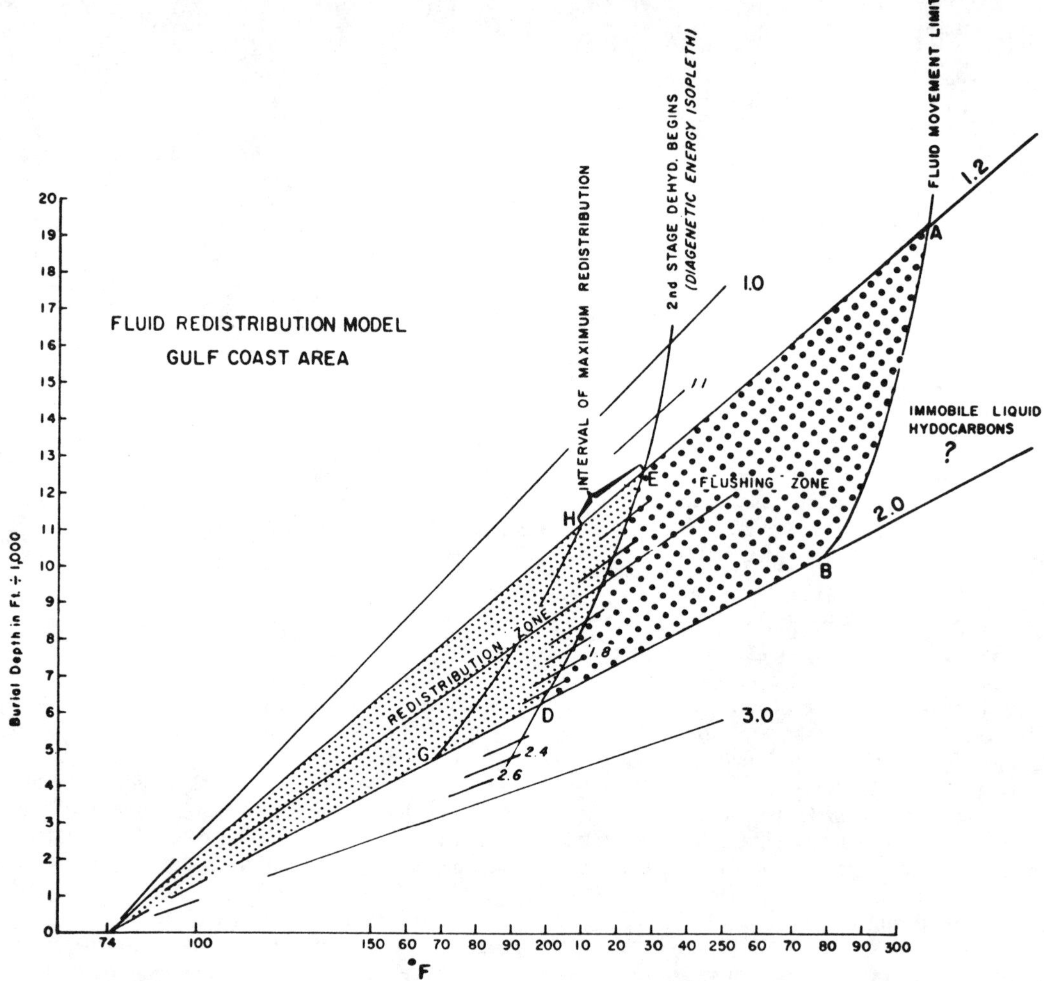

FIG. 16.—Fluid-redistribution model, Gulf Coast area.

paction or fluid migration resulting from dewatering of clay minerals can be considered possible. Third-stage dehydration is slow because the continuous permeability of the water-egress system deteriorates rapidly after the total shale-water content falls below 10 percent.

Clay-Mineral Dehydration Model

The three-stage dehydration sequence corresponds to the proposed petroleum-distribution system, illustrated in Figure 16, which consists of (1) a burial level deep in the subsurface at which liquid hydrocarbons are immobile and at which no appreciable liquid movement of any type can take place, (2) an intermediate interval wherein hydrocarbons are being flushed by clay-generated formation water, and (3) a relatively shallow interval wherein hydrocarbons are being redistributed and fluids in general are being transported to the surface. The relation between this model and actual petroleum distribution can be judged from a comparison of the fluid-distribution model and the petroleum-distribution curve shown in Figure 13. The second-stage dehydration level is indicated on both figures.

Fluid redistribution within the second (intermediate) stage dehydration level is of particular interest. Large-scale water release and the ensuing movement in any bed that contains available petroleum hydrocarbons may initiate petroleum migration. Below the second-stage

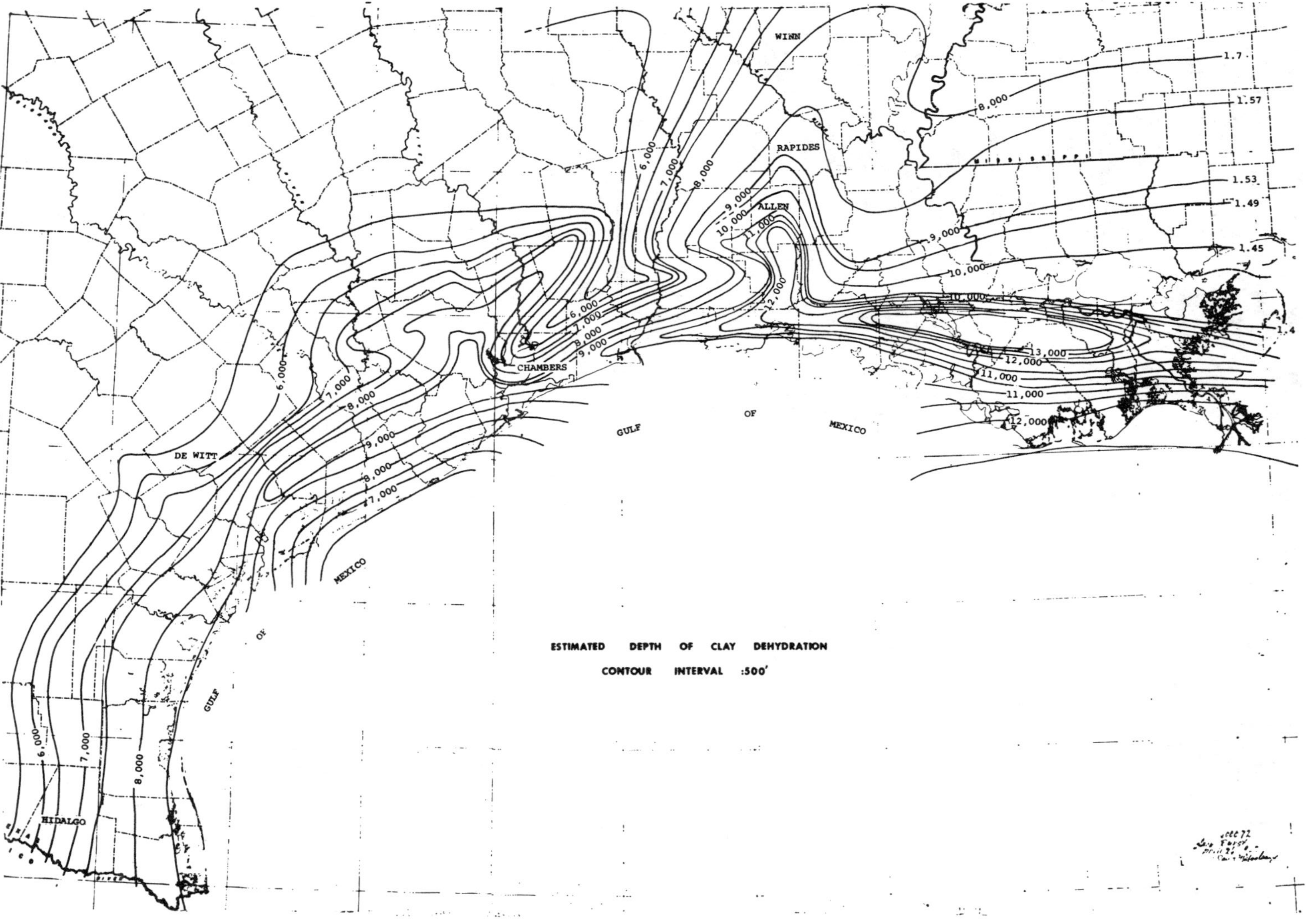

FIG. 17.—Estimated depth of clay dehydration, Gulf Coast area.

dehydration level, oil traps are statistically fewer because of flushing, spilling, breaching, and possible conversion to the gaseous state. Above, they are filled statistically in relation to their distance from the major source of interlayer-water effluent. The net result is an essentially normal distribution of petroleum-containing intervals. This pattern apparently has been established for the Gulf Coast, as the several thousand data points indicate, and it will be difficult to alter. Provided that the basin has not already been drilled to its full potential in the optimum interval, most future discoveries of liquid hydrocarbons probably will be from a 2,000-ft temperature-dependent interval extending a thousand ft above and below what has been designated as the top of the second-stage interlayer-water dehydration level. A proposed structure contour of this calculated level is shown in Figure 17.

On the basis of the thesis developed herein, the estimated dehydration structure surface is remarkably similar to the geothermal-gradient map shown in Figure 11. It is modified by converting geothermal gradients to dehydration depths *via* the proposed phase boundary constructed in Figure 10. Significant features are a "trough" in southeast Louisiana which indicates the possibility of deeper petroleum deposits in that area, and a shallow reentrant just west of the Texas-Louisiana border which delineates an essentially oil-barren locality.

References Cited

Bradley, W. F., and J. M. Serratosa, 1960, A discussion of the water content of vermiculite, *in* A. Swineford, ed., Clays and clay minerals: 7th Natl. Clays and Clay Mineral Conf. Proc., Pergamon Press, 369 p.

Brown, G., ed., 1961, The X-ray identification and crystal structures of clay minerals: London, Mineralogical Soc. (Clay Minerals Grp.), 418 p.

Burst, J. F., 1959, Postdiagenetic clay mineral environmental relationships in the Gulf Coast Eocene, *in* A. Swineford, ed., Clays and clay minerals: 6th Natl. Clays and Clay Mineral Conf. Proc., Pergamon Press, 411 p.

Dunoyer de Segonzac, G., 1964, Les Argiles du Crétace Superior dans le bassin de Douala (Cameroun): Problems de diagenese: Alsace-Lorraine Service Carte Géol. Bull., t. 17, fasc. 4, p. 287–310.

MacEwan, D. M. C., 1956, Fourier transform methods I, a direct method of analyzing interstratified mixtures: Kolloidzeitschrift, v. 149, p. 96–108.

——— 1958, Fourier transform methods II, Calculation of diffraction effects for different types of interstratification: Kolloidzeitschrift, v. 156, p. 61–67.

———and A. A. Ruiz, 1959, Fourier transform methods III, curves for some practically important cases: Kolloidzeitschrift, v. 162, p. 93–100.

Manger, G. E., 1963, Porosity and bulk density of sedimentary rocks: U.S. Geol. Survey Bull. 1144–E, p. 34–40 and Table 3.

Moses, P. L., 1961, Geothermal gradients: World Oil, v. 152, no. 6, p. 79–82.

National Oil Scouts and Landmen's Association, 1958, Oil and gas field development in the United States and Canada, Yearbook, Review of 1957: Austin, Texas, Natl. Oil Scouts and Landmen's Assoc., v. 28, 1136 p.

Powers, M. C., 1959, Adjustment of clays to chemical change and the concept of the equivalence level, *in* A. Swineford, ed., Clays and clay minerals: 6th Natl. Clays and Clay Mineral Conf. Proc., Pergamon Press, 411 p.

——— 1967, Fluid release mechanisms in compacting marine mudrocks and their importance in oil exploration: Am. Assoc. Petroleum Geologists Bull., v. 51, p. 1240–1253.

Van Olphen, H., 1963, Compaction of clay sediments in the range of molecular particle distances, *in* Clays and clay minerals: 11th Natl. Clays and Clay Mineral Conf. Proc., New York, Macmillan Co., p. 178–187.

Reprinted from:
BULLETIN OF THE AMERICAN ASSOCIATION OF PETROLEUM GEOLOGISTS
VOL. 53, NO. 1 (JANUARY, 1969), PP. 171–173

Indication of Proximity of High-Pressure Fluid Reservoir, Louisiana and Texas Gulf Coast[1]

JAMES M. FORGOTSON, SR.[2]
Shreveport, Louisiana 71101

Abstract The writer suggests that a method is available to Gulf Coast operators for detecting the proximity of high-pressure fluid reservoirs. Use of this method may avoid costly blowouts or saltwater flows. Mud logging companies have reported that the presence of high background gas and high trip gas, together with lower than normal shale density, may signify the proximity of a high-pressure fluid reservoir. The writer, however, found that in many areas low-density shales are not accompanied by other background gas or trip gas. Even the presence of low-density shale and low resistivity on an electric log does not indicate the proximity of such a high-pressure fluid reservoir. What the writer has found is that a minimum increase of 200 percent in the shale penetration rate indicates that a high-pressure fluid reservoir is nearby.

This note is written to help operators of drilling wells in the Gulf Coast of Louisiana and Texas avoid expensive and dangerous high-pressure blowouts and/or uncontrollable saltwater flows.

Formulas have been devised for use with electrical logs to determine the pore pressure of shales by using depth and resistivity as variables. Graphs drawn from these formulas indicate the expected pressure of fluid in a reservoir just under or horizontally adjacent to the shale. The pressure of the reservoir fluids being known, the weight of the mud to be used can be determined, and the minimum fracture point of the formation estimated. The combination of these parameters indicates to the prudent operator the point at which protective casing should be set.

An electric log of a well close to the proposed well usually will supply all the data necessary provided, however, that the formations in the proposed well are the same as, and at the same relative depth as, those in the well of which the electric log is available. The proposed well might not be close to such a well, or it might pass through a fault plane which had not been anticipated. The electric log analysis is a helpful tool, but it cannot be used as total protection to the operator. An electric log of the drilling well is not available as the well is being drilled; furthermore, it is not practical to stop the drilling process and run an electric log every time there is a change in the formation, unless there is a reason to suspect that the change is indicative of the proximity of a high-pressure fluid reservoir.

Research by mud logging companies and drilling technicians has resulted in the detection of the possible proximity of a high-pressure fluid reservoir by an analysis of the density of the shale being drilled.[3] It has been demonstrated that normal shale follows a normal density trend with depth, and that a change in the trend line of shale density is an indication of possible drilling trouble. The presence of high background and high trip gas, together with lower than normal density of the shale, is the signal of the possible proximity of a high-pressure fluid reservoir.

It is to be noted, however, that in many areas low-density shales are found without an increase in either background gas or trip gas, and hence do not indicate necessarily the possible proximity of a high-pressure fluid reservoir. The shale below the particular low-density shales has normal density, but generally the low-density shales are close to a fluid reservoir of normal pressure. The writer is not offering herein any explanation of the low-density shale but is merely stating that they are found in wells drilled in the Louisiana and Texas Gulf Coast (*e.g.*, wells drilled by Glasscock-Chapman, Inc., on the eastern side of the Iota field in Acadia Parish, Louisiana). An examination of shale density logs and electric logs indicates that these occurrences are widespread; furthermore, low-density shale usually does not have low resistivity.

Low-density shales which do have low resistivity indicated on the electric logs are considered to be indicative of the possible proximity of a high-pressure fluid reservoir. Generally, they have a faster drill penetration rate than normal density shales for any specified depth, and the increase in drilling rate in a shale body is indicative of the presence of low-density shales that are also of low resistivity on the electric log.

[1] Manuscript received, May 15, 1968; accepted, June 19, 1968.

[2] Independent operator.

[3] Any reservoir present within the shale body will be high pressure.

An increase in the rate of penetration in a shale body, whether or not there are (1) low-density shale, (2) increased gas background, (3) increased trip gas, or (4) low resistivity on electric log, is the indication of the proximity[4] of a high-pressure fluid reservoir.

An electric log of Pan American Petroleum No. 1 Louvierre, drilled in Sec. 95, T9S, R6E, St. Martin Parish, Louisiana, was run at a depth of 12,565 ft in a hole carrying 10.4 lb/gal mud weight. The electric log indicated that a sandstone at the bottom of the wellbore was flowing salt water. The mud weight was increased to 11.6 lb/gal, and the well ceased to flow. The well drilled to 13,031 ft, where a drilling break cut 11.6 lb/gal mud back to 11.0 lb/gal. The mud was equalized in approximately two hours at 12.4 lb/gal. A drilling break at 13,255 ft started to flow against 12.6 lb/gal mud weight. The flow was uncontrollable and, because of possible lost circulation by high mud weight, the drill stem was cemented in the hole, and the well was killed. There was no mud log run on the well, but shale density values were observed and recorded. There was no indication reported of low-density shale in the wellbore which would indicate the possible proximity of a high-pressure fluid reservoir. Analysis of the electric log run to 12,565 ft indicated no high pore pressure in the shale.

The writer recently completed drilling of the No. 1 M. C. Babin *et al.* approximately 3,500 ft east and south of the Louvierre well. This well, in Sec. 94, T9S, R6E, was drilled without incident to 12,824 ft, where the second electric log was run. Mud weights were used to control caving shale and any potential water flow by using the Louvierre well as a guide. However, at 13,190 ft low-density shale was encountered, but only one low-density sample was logged. An electric log was run[5] because of an increased rate of penetration. The electric log indicated no low-resistivity shale. Analysis of the log gave a shale fracture point well in excess of any mud weight that would be necessary and a controlling mud weight much lower than was being used.

Drilling was resumed and the rate of penetration increased, but the density of the shales went back to the normal trend line. There was no indication of any low-density shale or of an increase in either background gas or trip gas. Sandstone was found at 13,404 ft and the well was drilled to 13,408 ft before drilling could be stopped. The well was shut in for flow tests, but it did not flow; however, within minutes after an attempt was made to circulate the cuttings out from the bottom, the well was flowing. When the blowout preventers were closed, the pressures immediately jumped to 200 and 300 lb, respectively, on the drill pipe and the annulus. The mud weight was increased until a small amount of lost circulation was noted even though lost circulation material was added to the mud. At this point the drill stem was cemented in the hole, and the well was killed.

Detailed information is available through the compiled electric log and the mud log which are being released in Lafayette, Louisiana.

It is the writer's opinion that an increase in the rate of penetration of shales, in which the increase is a minimum of 200 percent of normal, is an indication of the proximity, either vertically or horizontally, of a high-pressure fluid reservoir. The situation described is explained by the juxtaposition across a fault of a high-pressure fluid reservoir on the upthrown side, probably in the *Nodosaria blanpiedi* interval, and the reservoirs and shale of normal pressure in the *Nonion struma* zone are on the downthrown side of the fault.

The great differential pressure across the fault is sufficient to allow the high-pressure fluids to invade normal-pressure fluid reservoirs, and to invade horizontally and frac[6] the shales which are in direct contact with the high-pressure reservoir. The fault also became a passageway for upward movement of the high-pressure fluids, which fractured with lessening intensity the formations above them on the downthrown side of the fault until the containing pressure of the formation and the diminishing pressure of the invading fluids equalized. Thus, the 12,542-ft sandstone in the Pan American Petroleum Corporation well flowed against the mud column, and as the well was drilled deeper the pressure of each succeeding fluid reservoir became greater until one was encountered which made the well uncontrollable. This reservoir probably has the same pressure as that of the high-pressure reservoir on the upthrown side of the fault.

Sklar Producing Company No. 1 St. Martin Bank and Trust Company well was drilled in Sec. 30, T9S, R5E, St. Martin Parish, Loui-

[4] A high pressure reservoir will be present, but it might be in contact horizontally with the shale.

[5] Driller's and mud logger's depth was 13,230 ft, but electric log logged to 13,220 ft.

[6] Fracture a formation, horizon, or bed by the introduction of a foreign fluid under sufficient pressure.

siana, on the southeast flank of the Anse La-Butte dome. A sandstone encountered between 7,978 and 8,124 ft in the *Nonion struma* zone, showed oil to a depth of approximately 8,012 ft. The well was deepened to 9,750 ft and bottomed in the *Nonion struma* zone. A few thin sandstone streaks were found. The shale resistivity was approximately 1 ohm to total depth, which is normal for normal-pressure formations at that general depth. The shale density also was normal for that formation at that depth. No increase in pressure was indicated.

The well was plugged back to 6,695 ft and sidetracked in a direction approximating N68°W. The sandstone that had been encountered in the straight hole was reached at 7,450 ft in the directional hole, and was approximately 110 ft horizontally from its location in the original hole. The bottom of the sandstone was approximately 7,592 ft. The entire sandstone was oil saturated.

It is the opinion of some that a fault had been crossed, but the writer had anticipated 72° dip and believes that the great change in structural position of the sandstone was due to an 85° dip.

The sidetracked hole was deepened to 8,301 ft. Practically the entire *Nonion struma* zone had been penetrated at this point, and from the paleontologist's report of the *Nodosaria blanpiedi* zone was anticipated within a few feet. The oil-bearing sandstone found in the sidetracked hole has the same bottom-hole pressure as the oil-water contact of the sandstone in the straight hole.

The formations penetrated below the oil-bearing sandstone are sandy shale and sandstone in thin streaks. Most of the sandstone beds are hard and nonporous, but all indicated either oil or gas by shows in the drilling fluid, fluorescence in the samples and in some samples, fluorescence in the drilling fluid.

The pressures normal for these zones in the straight hole are the same in the sidetracked hole, but the formations were penetrated more than 500 ft higher structurally; therefore, the same bottom-hole pressure in the sidetracked hole required much heavier mud than was required in the straight hole. Furthermore, the shale density in the sidetracked hole is extremely low, becoming 2.09 toward the bottom, and the resistivity of the shale on the electric log indicated 0.5 ohm or less.

The sidetracked hole reached older formations than the straight hole did, but at no place were the two holes more than 150 ft apart; therefore, the writer concluded that the high pressure in the sandstone and the low density and low resistivity of the shale were caused by the same bottom-hole pressure, but at a shallower depth, as the same strata less than 150 ft away horizontally. This indicates to the writer that the hydrostatic head on the formations is probably that of the oil/gas/water contact, and in this example, the difference in pressure is great enough to invade the shales at some point between the two wells.

The electric logs, dipmeter log, mud log, *etc.*, are available in Lafayette, Louisiana.

Reprinted from:
BULLETIN OF THE AMERICAN ASSOCIATION OF PETROLEUM GEOLOGISTS
VOL. 56, NO. 3 (MARCH, 1972), PP. 528–536, 1 TABLE

Pressure-Depth Relations in Appalachian Region[1]

WILLIAM L. RUSSELL[2]
College Station, Texas 77843

Abstract The pre-Pennsylvanian reservoirs of the Appalachian region are predominantly lenticular. Fluid pressures in many gas fields in these reservoirs are subnormal, but some are normal and a few are above normal. Each reservoir, across large areas, has a characteristic departure from normal of the mean values of the fluid pressures. Abnormally low fluid pressures tend to occur in lenticular reservoirs closely associated with shales in areas which have undergone erosion. A possible explanation is that erosion causes a reduction in the fluid pressure in the pore space of shales, and that this reduction is transmitted to the closely associated reservoir rocks. The pressure reduction in shales may be due to the increase in pore volume and absorption of water in clay minerals as the overburden pressure decreases, and to the absorption of water during mineral transformations that occur because of the decrease in temperature.

Introduction

The Appalachian region is particularly favorable for the study of pressure-depth ratios in lenticular reservoirs. Abnormally low, normal, and abnormally high fluid pressures occur there in reservoirs of diverse lithology associated with varied degrees of deformation. Sufficient data are now available to indicate the relation of the fluid pressures to stratigraphy, lithology, and intensity of deformation, and to suggest possible causes of some of these relations. Research of this type is especially needed because little is known about the fluid pressures of lenticular reservoirs in areas where considerable erosion has taken place.

Definitions

The pressure-depth ratios given in this paper are the original fluid pressures in psi divided by the depths in feet (pressure-depth ratios in psi/ft may be converted to kg/sq cm/m by multiplying by 0.2307). Piezometric elevations or elevations of piezometric surfaces corresponding to the fluid pressures are the elevations of the tops of columns of water which balance the pressures. In this paper fluid pressures are considered normal if they balance columns of water extending to the overlying surface, or to the outcrops of rocks stratigraphically equivalent to the reservoirs. In the Appalachian region there are considerable variations in the elevations of the outcrops of the reservoir rocks, in the elevations of the ground surface over the reservoirs, and in the density of the subsurface waters. Because of these conditions precise distinctions beween normal and abnormal fluid pressures are difficult to make. Pressure-depth ratios in the Appalachian region are in this paper considered subnormal if below 0.4 and above normal if above 0.6.

Corrections and Errors

Nearly all wells in the Appalachian region have been drilled by cable tools or by rotaries using air as the drilling fluid. Fluid pressures of gas wells drilled by these methods are commonly measured at the surface. Fluid pressures at the tops of the producing zones were determined from the pressures at the surface by applying corrections obtained from charts by Crawford *et al.* (1954). These corrections have been applied to the reported pressures in the Appalachian region and in the Arkoma basin, unless they were recorded as bottomhole pressures. The corrections increase the values of the pressures measured at the surface by from 5 to 31 percent. Water or oil in the borehole above the top of the reservoir lowers the value of the pressure measured at the surface. Accordingly, pressure data were not used if this condition was reported. The shut-in time before the pressures were measured varies from less than 20 to more than 200 hours. Only the final pressures are available. Each pressure measurement used in compiling Table 1 is the original fluid pressure of a gas pool. Errors of instrumental origin may give pressures that are either too high or too low. On the other hand, values which are always too low are produced by lack of sufficient shut-in time to reach pressure equilibrium, and by drainage of reservoir fluids towards older wells in the same continuous reservoir.

[1] Manuscript received, March 3, 1971; accepted, June 21, 1971.

[2] Department of Geology, Texas A&M University. The writer thanks O. L. Haught, William S. Lytle, and Larry D. Woodfork for data on fluid pressures in the Appalachian region, and Jocelyn Tommerup for information on statistical problems and for making calculations.

Table 1. Pressure-Depth Ratios of Some Lenticular Reservoirs*

Reservoir, Age, and Location	Pressure-Depth Ratios		
	Mean	Standard Error of Mean	Number of Determinations
Lower Oligocene to upper Miocene sandstones, Gulf Coast	0.668	0.011	93
Lower Cretaceous sandstones, Denver basin, Colorado	0.261	0.005	42
Lower Cretaceous sandstones, Denver basin, Nebraska	0.230	0.003	94
Lower Pennsylvanian Atoka sandstones, Arkoma basin, Arkansas	0.284	0.022	30
Greenbrier Limestone, middle Mississippian, West Virginia	0.305	0.023	14
Big Injun sandstone and limestone, Lower and middle Mississippian, West Virginia	0.242	0.016	12
All Mississippian sandstones, excluding Big Injun, West Virginia	0.222	0.019	13
Upper Devonian sandstones, exclusive of reservoirs in "Brown shale," West Virginia	0.358	0.021	16
"Brown shale," Devonian, West Virginia	0.152	0.016	19
Oriskany Sandstone, Lower Devonian, slightly deformed, New York	0.491	0.023	17
Oriskany Sandstone and Onondaga chert, Lower and Middle Devonian, slightly deformed, Pennsylvania	0.491	0.069	11
Oriskany Sandstone and Onondaga chert, Lower and Middle Devonian, strongly deformed, Pennsylvania	0.550	0.014	79
Oriskany Sandstone and Huntersville Chert, Lower and Middle Devonian, West Virginia	0.418	0.020	28
Newburg sandstone, Upper Silurian, West Virginia	0.412	0.019	6
Lower Silurian sandstones, New York	0.299	0.019	15
Lower Silurian sandstones, Ohio	0.293	0.013	22
Lower Silurian sandstones, Pennsylvania	0.298	0.014	16
Lower Silurian sandstones, West Virginia	0.423	0.021	6

* Pressure data from papers by Alkire (1954, 1961), Alkire *et al.* (1959), Cottingham (1944, 1945, 1947–1949), Crowell and Bailey (1943), Dickinson (1953), Fettke (1945, 1950–1954), Fettke and Lytle (1955, 1956), Ingham (1954), International Oil Scouts Association Yearbook (1940, 1966), Jordan and De Brosse (1970), Kreidler (1959, 1963), Latimer (1967, 1968), Lytle (1957–1965, 1967–1970), Lytle *et al.* (1966), Mackay and Loughlin (1952), Reger (1938, 1940–1946, 1948, 1949), Rocky Mountain Association of Geologists (1954, 1955), Van Tyne (1961, 1964–1966, 1968, 1970), and Woodfork (1969).

Continuity of Reservoirs in Appalachian Region

The determination of the extent of permeable reservoirs is essential for solving problems relating to the origin of fluid pressures and the movements of reservoir water. Sandstones that are continuous but only locally sufficiently permeable to permit the free flow of water into boreholes are common in the Appalachian region. Because flows of fluids into boreholes are readily detected where cable tools or rotaries using air as a drilling fluid are used, it is possible to determine as each sandstone is penetrated whether or not it yields free fluids. As a result, it is known that some reservoirs, such as the Lower Silurian sandstones of New York, Ohio, and northwestern Pennsylvania, consist of locally permeable zones in continuous sandstones of a permeability too low to permit yielding free water into boreholes The Lower Silurian sandstones of the Appalachian region have been described by Hardaway (1968), Heald (1965), Heald and Anderegg (1960), Knight (1969), Patchem (1968a), Pepper *et al.* (1953), and Yeakel (1962). These sandstones reach a thickness of more than 500 ft on the east. Toward the west they thin, become interbedded with shale, and terminate in the subsurface. Maps by Kreidler (1959, 1963) and Lytle (1964) show in New York and Pennsylvania east-west-trending zones in which the Oriskany Sandstone is absent. These zones would prevent any northward flow of water through this sandstone from the higher outcrops near the Appalachian Front to the lower outcrops on the north.

The large irregular scattering of the piezometric elevations of reservoirs at the same geologic interval suggests that the reservoirs are lenticular. Table 1 shows that the mean pressure-depth ratios of several reservoirs are definitely subnormal, and there are also gas fields in the Oriskany Sandstone and Onondaga chert of Pennsylvania which have abnormally high fluid pressures. The Appalachian structural basin was formed before the beginning of the Mesozoic Era. In basins as old as this, abnormal fluid pressures indicate permeability barriers between the outcrops of the reservoirs and the locations of the abnormal fluid pressures. If these barriers were not present, the pressures would have been normalized by the flow of water to or from the outcrops. According to Heck *et al.* (1964) and Poth (1962), waters from pre-Late Devonian reservoirs contain around 300,000 ppm solids, of which about half is chlorine. The percentage of dissolved solids generally increases with age. Knight (1969) stated that the Lower Silurian sandstones in northeastern Ohio contain thin lenses of halite, and Schrider *et al.* (1970) reported that the concentration of the connate water in the Lower Silurian sandstones in the East Canton field, northeast Ohio, is near saturation. The chemical character of the waters in the pre-Mississippian reservoirs indicates that there has been no extensive invasion of the reservoirs by fresh surface waters. This supports the idea that the reservoirs are discontinuous.

Expression of Results

The geologic significance of subsurface fluid pressures is generally best brought out by piezometric elevations or by pressure-depth ratios. Piezometric elevations are more useful than pressure-depth ratios for studying conditions in reservoirs which are continuously permeable across great distances, for establishing the presence or absence of hydrodynamic flow, for indicating the pressure-depth relations at shallow depths in hilly country, and for showing the irregular scattering of the values of the pressures in disconnected reservoirs at the same stratigraphic position. The most serious disadvantage of pressure-depth ratios is that they are affected by differences in the elevations of boreholes. However, this disadvantage may be reduced to negligible proportions by using the means of the pressure-depth ratios of many pools. One advantage of pressure-depth ratios is that they may be used if the elevations of the wells are not given with the pressures and depths. Pressure-depth ratios may also be used to compare conditions in different basins. Piezometric elevations are not suitable for this use because of differences in the general elevations of the basins. Moreover, pressure-depth ratios are more suitable for expressing the results if the reservoirs are lenticular and if the structural basins are tilted. Tilting may lower the elevations of the piezometric surfaces in the direction of the tilt without affecting the pressure-depth ratios. The Appalachian structural basin has been tilted northwestward, causing a general northwest slope of the land surface.

Errors or uncertainties may be introduced into the determinations of piezometric elevations by variations in water densities. Poth (1962) has published a table showing the dissolved solids content of waters from various reservoirs in Pennsylvania, and a chart showing the density of waters of varying solid content. According to this information, the density of the water in pre-Pennsylvanian reservoirs would range from 1.02 to 1.22. In some areas the density of the water in the pore space of shales between the reservoir and the surface may affect the fluid pressures in lenticular reservoirs. In the Appalachian region this density is not known. Furthermore, the large variations shown in Table 1 between the pressure-depth ratios of reservoirs in different stratigraphic positions in the same region suggest that these lenticular reservoirs are sealed off from each other. Where this occurs, neither the density of the reservoir water nor the density of the water in the pore space between the reservoirs and the surface may affect the piezometric elevations. Under these conditions piezometric elevations are of doubtful geologic significance.

Field Relations

Relation of Pressure-Depth Ratios to Stratigraphy

Table 1 brings out the fact that there are large variations of the mean pressure-depth ratios of reservoirs in different stratigraphic positions. Furthermore, reservoirs at the same stratigraphic position tend to have a characteristic mean pressure-depth ratio in each region. This characteristic mean value extends over considerable areas. An example is the abnormally low mean pressure-depth ratio of the Lower Silurian sandstones in New York, Ohio, and Pennsylvania, and the much higher ratios of the Lower and Middle Devonian reservoirs in the same region. Table 1 shows that the mean pressure-depth ratios of the Lower Silurian sandstones in New York, Ohio, and Pennsylvania are 0.299, 0.293, and 0.298. The corresponding ratio of the Middle and Lower Devonian reservoirs in this region is 0.491. Table 1 also shows large contrasts between the pressure-depth ratios of different reservoirs in West Virginia. The means range from 0.152 to 0.423. This tendency for the departures of the ratios from normal to be characteristic of each reservoir over large areas is the most important field relation disclosed by the study of fluid pressures in the Appalachian region.

Variations within Basins

The mean pressure-depth ratios do, however, vary over great distances within the same structural basin. For example, in New York, Ohio, and Pennsylvania the ratios of the Lower Silurian sandstones are between 0.293 and 0.299. In West Virginia the ratio of these sandstones is 0.423. The mean pressure-depth ratio of the Lower and Middle Devonian reservoirs is 0.491 in the relatively undeformed region of New York and Pennsylvania, 0.550 in the deformed area of Pennsylvania, and 0.418 still farther south in West Virginia.

Scattering

The piezometric surfaces of lenticular reservoirs in the same area and at the same stratigraphic positions show an irregular variation that probably is greater than would be produced by errors in pressure measurement. Russell (1961) has described a similar scattering of the elevations of the piezometric surfaces of the Lower Cretaceous sandstones of the Denver basin.

Relation to Deformation

The mean pressure-depth ratio of the Oriskany Sandstone and Onondaga chert in the strongly deformed region of Pennsylvania is 0.550, which is 0.059 higher than in the adjacent relatively undeformed region on the northwest. Of the 79 determinations in the strongly deformed region, 20 are above 0.6, 6 above 0.7, and 1 above 0.8. Of the 28 ratios in the relatively undeformed area, only 3 are above 0.6 and none higher than 0.7. These abnormally high values for the fluid pressure are not due to errors in measurement. In some cases abnormally high fluid pressures were measured in several wells in a field before the production of gas reduced the fluid pressure.

Relation of Pressure-Depth Ratios to Depth

The available data show no distinct increase or decrease of pressure-depth ratios with depth in the Appalachian region, but no pressure-depth data are available for depths over 9,000 ft (2,743 m). However, data from many areas outside the Appalachian region show that pressure-depth ratios considerably in excess of 0.5 are very common at depths of over 15,000 ft (4,572 m).

Relation of Fluid Pressures to Deposition and Erosion

Field evidence showing the effects of deposition and erosion on fluid pressures consists in the comparison of the relative magnitudes of the mean pressure-depth ratios of lenticular reservoirs in areas of erosion and deposition. This evidence becomes more convincing as the number of reservoirs and basins studied increases. For this reason, data from three areas outside the Appalachian region are included in Table 1. Dickinson (1953) stated that the reservoirs from which the Gulf Coast pressure data were derived are lenticular. Russell (1961) has presented evidence that the Lower Cretaceous reservoirs of the Denver basin are lenticular, and according to Crowell and Bailey (1943) the Atoka sandstones which furnished the Arkansas pressure data are also lenticular.

The mean pressure-depth ratio of the Gulf Coast sediments, 0.668, is far above normal. The area in which these sediments are present has received thick deposits of sediments very late in geologic time. The Denver basin, the Arkoma basin, and the Appalachian region have been eroded in the late geologic past and are now being eroded. The means of the pressure-depth ratios of the Denver and Arkoma basins are low (0.261, 0.230, and 0.284). Table 1 shows the means of eight reservoirs or groups of reservoirs in the Appalachian region. Only one of these, the Lower and Middle Devonian Oriskany Sandstone and Onondaga chert, has a mean pressure-depth ratio over 0.5. Five reservoirs or groups of reservoirs in West Virginia have mean ratios considerably less than 0.4 and are abnormally low. The mean pressure-depth ratio of the Lower Silurian sandstones in New York, Ohio, and Pennsylvania is slightly less than 0.3, although in West Virginia the ratio of these sandstones is 0.423. In spite of the normal and abnormally high fluid pressures in the Oriskany Sandstone and Onondaga chert, the fluid pressures in the Appalachian region are in general abnormally low. The field evidence therefore supports the idea that abnormally low fluid pressures tend to occur in areas undergoing erosion or which have been eroded late in geologic time. However, evidence from many more reservoirs and structural basins is needed to establish this relation.

Relations of Fluid Pressures to Shales

The reservoirs of the Appalachian region can be divided into two classes, according to the degree of their association with shales. Sandstones within shales form the class with closer association with shales. The other class consists of reservoirs in carbonates, or in sandstones enclosed in carbonates. The effect of association of reservoirs with shales on fluid pressures may be determined by comparing the mean pressure-depth ratios of the two groups. This comparison should be made between reservoirs in the same region. Some of the oil and gas reservoirs of West Virginia have been described by Patchem (1968a, b, c, and d). In that state the reservoirs in carbonates or in sandstones enclosed in carbonates are the Greenbrier Limestone with a mean pressure-depth ratio of 0.305, the Oriskany Sandstone and Huntersville Chert with a ratio of 0.418, and the Newburg sandstone with a ratio of 0.412. In West Virginia the sandstones or siltstones enclosed in shales are the Mississippian sandstones, excluding the Big Injun reservoirs with a mean pressure-depth ratio of 0.222, the Upper Devonian sandstones with a ratio of 0.358, the "Brown shale" with a ratio of 0.152, and the Lower Silurian sandstones with a ratio of 0.423. In the slightly deformed region in New York, northwestern Pennsylvania, and eastern Ohio the Lower Silurian sandstones are enclosed by, or closely associated with, shales. The Oriskany Sandstone and reservoirs in the Onondaga chert are surrounded by carbonates. The mean pressure-depth ratios of the Lower Silurian sandstones in this region are 0.299, 0.293, and 0.298. The mean ratio of the Oriskany Sandstone and Onondaga chert is 0.491. The ratios given in Table 1 for the Denver and Arkoma basins are 0.261, 0.230, and 0.284. The sandstones which furnish these ratios are enclosed by shales. Field data at present available show lower pressure-depth ratios with increasing association with shales, though the ratios of the Greenbrier Limestone and Lower Silurian sandstones of West Virginia appear to be exceptions. Pressure-depth ratios from lenticular reservoirs from many eroded regions are needed to establish or disprove this suggested relation.

Factors Affecting Fluid Pressures

Although the available information does not show conclusively how abnormal fluid pressures on the Appalachian region originated, a discussion of the possible causes should aid in understanding the problems involved. Field evidence as to the role of osmosis, osmotic pressure, osmotic sealing, osmotic straining, and similar processes is largely lacking in the Appalachian region. These phenomena are not discussed because of lack of information, and not because they are known to be unimportant.

The pressure-depth ratios of lenticular reservoirs are controlled by such factors as variations in overburden pressure due to erosion and deposition, variations in the permeability of the reservoirs and associated rocks, compressibility of the reservoirs and contained fluids, ratio of volume of gas to volume of water in reservoirs, release and absorption of water by clay minerals as the pressure changes, mineral transformations which release or absorb water, and variations in temperature with changes in depths or in geothermal gradients. The variations in the permeability of shales associated with reservoirs is of particular importance. It is known from the large variations in mean pressure-depth ratios that the permeability of the pre-Pennsylvanian shales in the Appalachian region is too low to permit sufficient flow between reservoirs to equalize the ratios. There are, therefore, two possible values of the permeability of the nonreservoir rocks. One is extremely low but appreciable; it permits the extremely slow exchange of fluids between the reservoirs and the closely associated shales. The other value is so low that no flow or no important flow of fluids takes place between the reservoirs and the closely associated shales.

Inelastic Loss of Porosity

Sandstones that have been buried to great depths generally have lost much more porosity than unconsolidated sands would lose if subjected in the laboratory to pressures equivalent to the overburden pressures at these great depths. This loss of porosity must be chiefly inelastic, for much of it is not recovered when the overburden pressure is removed. Inelastic loss of porosity occurs as a result of such processes as pressure solution, deposition of cement in pores, closer packing of grains, fracturing, and plastic deformation. The loss might occur at great depths while erosion is taking place at the surface. The large loss and the uncertainty as to when it occurred make it impossible to determine accurately from the compressibility of sandstones in the laboratory the amount of increase in pore space caused by a given loss of overburden pressure. However, the compressibility of sandstones can be used to place an upper limit on the increase in porosity due to erosion.

Deposition with Compaction

As shales containing lenticular reservoirs are deposited, water expelled by compaction percolates through the shales and reservoirs on its way to the surface. While this continues, the fluid pressures in the reservoirs must be at least slightly above normal to force the water upward through the shales. Whether the fluid pressures become slightly or much above normal depends on the rate of deposition and on the thickness and permeability of the shales. After deposition ceases, upward percolation may cause the fluid pressures to decline to normal, but percolation without erosion cannot produce subnormal fluid pressures.

Deposition with Sealing

Deposition and erosion may have important effects on the pressure-depth ratios of lenticular reservoirs. Two factors which affect these ratios are changes in temperature and in pore volume. Calculations have been made to determine the effects of deposition and erosion under certain assumed conditions. The actual conditions in reservoirs in the Appalachian region may differ from those assumed, but the assumed and actual conditons appear to be sufficiently close so that the calculations give useful information. One assumption is that the reservoirs are effectively sealed. The term "effectively sealed" is used here to indicate that the volumes of fluid which enter or leave the reservoirs are too small to control the fluid pressures within the reservoirs. Hence the fluid pressures are controlled by conditions within the reservoirs. Although it is not known whether effectively sealed reservoirs actually exist, the assumption that they do exist is useful in making calculations. It also is assumed that the pore space of the reservoirs is filled with gas, that during deposition there is no inelastic loss of porosity, that there are no clay minerals or solid calcium sulfate in the reservoir, that the geothermal gradient is 1 or 2°F/100 ft (1.82 or 3.64°C/100 m) and remains constant during deposition or erosion, that the depth of the effectively sealed reservoirs is 3,000 ft (914 m) before deposition and 6,000 ft (1,829 m) after deposition, that the pressure depth ratio before deposition was 0.50 and that the compressibility of the reservoir is 3×10^{-6} volumes/psi. This value for the compressibility is within the range of compressibilities given by Fatt (1958). With these assumptions the effect of deposition would be to increase the fluid pressure by about 6–11 percent, with most of the increase due to the rise in temperature. Because the depth is increased by 100 percent, the pressure-depth ratios would be decreased from 0.50 to about 0.27 or 0.28.

As deposition with effective sealing could lower pressure-depth ratios, it might be supposed that this process has produced the abnormally low pressure-depth ratios in the Appalachian region. However, some considerations suggest that deposition with sealing has not been important in affecting subsurface fluid pressures. The abnormally high pressures which are common at great depths imply that a great thickness of deposits above lenticular reservoirs tends to produce high, rather than low, pressure-depth ratios. Moreover, erosion rather than deposition appears to have been dominant in the Appalachian region since the beginning of the Mesozoic. Furthermore, it seems doubtful whether reservoirs can remain effectively sealed during the deposition of considerable thicknesses of sediments. Free water released by the resultant compaction would be expected to pass upward through lenticular reservoirs, preventing effective sealing. This suggests that deposition with effective sealing can decrease pressure-depth ratios only in sediments which have been previously buried to greater depths or exposed to higher temperatures than those present during deposition.

Erosion with Sealing

If it is assumed that erosion takes place above an effectively sealed reservoir, that the depth of the reservoir is 6,000 ft (1,829 m) before erosion and 3,000 ft (914 m) after it, and that before erosion the pressure-depth ratio at

6,000 ft (1,829 m) is 0.50, and other assumptions are the same as in the preceding discussion of deposition with effectively sealed reservoirs, erosion would decrease the fluid pressure by about 6 to 11 percent, with most of the decrease due to cooling of the gas. Because the depth would be decreased by 50 percent, the pressure-depth ratio would be increased from 0.50 to about 0.90 to 0.94. It is very unlikely that all of the conditions assumed for effectively sealed reservoirs would exist in the field, but enough of them might be present to cause the pressure-depth ratios to rise or remain the same with erosion. Whether this has occurred can best be determined from the field evidence. The dominance of erosion in the Appalachian region and the prevailing low pressure-depth ratios suggest that in most of the region this process is unimportant. However, the abnormally high fluid pressures in the Oriskany Sandstone and Onondaga chert in the strongly deformed region of Pennsylvania may have formed in this way. The occurrence of these abnormally high fluid pressures in the strongly deformed zone suggests that they are in some way caused by deformation. Nevertheless, if they were formed during deformation in the Paleozoic Era, they must have been effectively sealed since then to prevent the escape of enough fluid to bring the pressures down to normal. The presence of abnormally high fluid pressures in one reservoir, but not in another, could be explained by differences in sealing. The sealing may have been due to a reduction in the permeability of the surrounding rocks caused by exposure to heat or stresses accompanying deformation, to heat and overburden pressure caused by depth of burial, or to higher temperatures caused by higher geothermal gradients in the geologic past.

The effect of erosion on the fluid pressures of sealed reservoirs containing water is not discussed because of the lack of sufficiently accurate data.

Erosion and Very Low Permeability

The abnormally low fluid pressures in the Appalachian region may be explained as the result of a reduction in fluid pressure in the pore space of shales, combined with an appreciable but very low permeability of the rocks between the shales and the reservoirs. The permeability required is just enough for fluids to move from the reservoirs into the closely associated shales, but not sufficient to equalize the pressure-depth ratios between reservoirs 500–1,000 ft (150–300 m) apart. The lowering of the fluid pressure in the pore space of shale during erosion may be due to the increase in pore volume as the overburden pressure is reduced, to the absorption of water by clay minerals as the overburden pressure decreases, and to the withdrawal of water to enter into mineral transformations. These mineral transformations would be facilitated by the lowering of the temperature caused by erosion or by a change to a lower geothermal gradient even without erosion. Two mineral transformations which might yield or absorb water are montmorillonite-illite and gypsum-anhydrite. Chilingar and Knight (1960) have described the loss of water from clay minerals which occurs as a result of increasing pressure. For a change from 3,000 to 6,000 psi (211 to 422 kg/sq cm) the water loss is about 2 percent for kaolinite, 3 percent for illite, and 5 percent for montmorillonite. The available data on the transformations of clay minerals and the expulsion of water from them apply to increasing temperatures, pressures, or depths. However, if the water absorbed with decreasing temperatures and pressures were even a small part of the water released as they rise, a large effect on fluid pressures could be produced.

Whether erosion has lowered the fluid pressure in shales and this lowering has been transmitted to the adjacent reservoir rocks must be determined from the field evidence. Field conditions which would be expected if these processes are of importance are an association of abnormally low pressure-depth ratios with erosion and a decrease in the pressure-depth ratios with increasing association with shales. The field evidence described previously suggests that both these conditions occur. However, much additional evidence from many regions is needed to determine definitely the effect of erosion on fluid pressures.

The decrease of fluid pressures in shales as a result of erosion also could account for the absence or scarcity of free water in several reservoirs which produce hydrocarbons in the Appalachian region. The free water in these reservoirs could have been drawn into the shales as the fluid pressure in their pore space decreased. The hydrocarbons then could have expanded to fill the pore space in the reservoirs which was formerly occupied by water.

Conclusions

1. In the Appalachian region the pre-Pennsylvanian reservoirs are prevailingly lenticular.

2. Departures of fluid pressures from normal

are, over large areas, characteristic of reservoirs in the same stratigraphic position.

3. Over great distances in the same structural basin, the mean pressure-depth ratios of reservoirs at the same stratigraphic position may differ.

4. The fluid pressures in the Appalachian region are commonly low, but some are normal and a few are above normal.

5. The abnormally high fluid pressures in reservoirs containing gas may be due to sealing and erosion.

6. The most promising explanation for the abnormally low fluid pressures is that they are due to reduction in fluid pressure in shales caused by erosion. This reduction is transmitted to the closely associated reservoir rocks.

7. The reduction of fluid pressure in shales may cause water to move from reservoir rocks into the shales. This could account for the absence of water in many reservoir rocks which produce oil or gas in the Appalachian reigon.

References Cited

Alkire, R. L., 1954, Ohio oil and gas development 1953: Internat. Oil Scouts Assoc. Yearbook, v. 24, p. 510–516.

——— 1961, Ohio oil and gas development 1960: Internat. Oil Scouts Assoc. Yearbook, v. 31, p. 227–231.

——— B. A. Floto, and A. W. Johnson, 1959, Ohio oil and gas development 1958: Internat. Oil Scouts Assoc. Yearbook, v. 29, p. 556–563.

Chilingar, G. V., and L. Knight, 1960, Relationship between pressure and moisture content of kaolinite, illite, and montmorillonite clays: Am. Assoc. Petroleum Geologists Bull., v. 44, p. 101–106.

Cottingham, K., 1944, Ohio oil and gas field development 1943: Internat. Oil Scouts Assoc. Yearbook, v. 14, p. 431–440.

——— 1945, Ohio oil and gas development in 1944: Internat. Oil Scouts Assoc. Yearbook, v. 15, p. 479–486.

——— 1947, Ohio oil and gas development 1946: Internat. Oil Scouts Assoc. Yearbook, v. 17, p. 459–468.

——— 1948, Ohio oil and gas development 1947: Internat. Oil Scouts Assoc. Yearbook, v. 18, p. 418–426.

——— 1949, Ohio oil and gas development 1948: Internat. Oil Scouts Assoc. Yearbook, v. 19, p. 443–449.

Crawford, P. B., R. I. Bradford, A. C. England, and H. Ince, 1954, Static bottom-hole pressures of natural gas wells from surface measurements: Texas Petroleum Research Committee Bull. No. 40, 22 p.

Crowell, A. M., and T. D. Bailey, 1943, Natural gas in northwest Arkansas: AIME Trans., v. 151, p. 278–288.

Dickinson, G., 1953, Geological aspects of abnormal pressures in Gulf Coast Louisiana: Am. Assoc. Petroleum Geologists Bull., v. 37, p. 410–432.

Fatt, I., 1958, Compressibility of sandstones at low to moderate pressures: Am. Assoc. Petroleum Geologists Bull., v. 42, p. 1924–1927.

Fettke, C. R., 1945, Pennsylvania oil and gas activities during 1944: Internat. Oil Scouts Assoc. Yearbook, v. 15, p. 547–559.

——— 1950, Developments in Pennsylvania in 1949: Am. Assoc. Petroleum Geologists Bull., v. 34, p. 1048–1057.

——— 1951, Developments in Pennsylvania in 1950: Am. Assoc. Petroleum Geologists Bull., v. 35, p. 1177–1187.

——— 1952, Developments in Pennsylvania in 1951: Am. Assoc. Petroleum Geologists Bull., v. 36, p. 1047–1057.

——— 1953, Developments in Pennsylvania in 1952: Am. Assoc. Petroleum Geologists Bull., v. 37, p. 1268–1279.

——— 1954, Developments in Pennsylvania in 1953: Am. Assoc. Petroleum Geologists Bull., v. 38, p. 1046–1956.

——— and W. S. Lytle, 1955, Developments in Pennsylvania in 1954: Am. Assoc. Petroleum Geologists Bull., v. 39, p. 811–821.

——— and ——— 1956, Developments in Pennsylvania in 1955: Am. Assoc. Petroleum Geologists Bull., v. 40, p. 1081–1091.

Hardaway, J. E., 1968, Possibilities for subsurface waste disposal in a structural syncline in Pennsylvania, p. 93–127 *in* Subsurface waste disposal in geologic basins—a study of reservoir strata: Am. Assoc. Petroleum Geologists Mem. 10, 253 p.

Heald, M. T., 1965, Lithification of sandstones in West Virginia: West Virginia Geol. and Econ. Survey Bull. 30, 28 p.

——— and R. C. Anderegg, 1960, Differential cementation in Tuscarora Sandstone: Jour. Sed. Petrology, v. 30, p. 568–597.

Heck, E. T., C. E. Hare, and H. A. Hoskins, 1964, Appalachian connate water: West Virginia Geol. and Econ. Survey Bull. 28, 42 p.

Ingham, A. I., 1954, The geology and development of the Leidy-South Leidy gas fields, Clinton County, Pennsylvania: Producers Monthly, v. 18, no. 10, p. 22–28.

International Oil Scouts Association Yearbook, 1940, Pennsylvania: v. 10, p. 345–346.

——— 1966, West Virginia: v. 36, p. 495–496.

Jordan, K., and T. A. De Brosse, 1970, Ohio oil and gas development 1969: Internat. Oil Scouts Assoc. Yearbook, v. 40, p. 236–241.

Knight, W. V., 1969, Historical and economic geology of Lower Silurian Clinton sandstone of northeastern Ohio: Am. Assoc. Petroleum Geologists Bull., v. 53, p. 1421–1452.

Kreidler, W. L., 1959, Selected deep wells and areas of production in eastern and central New York: New York State Mus. and Sci. Serv. Bull. 373, 243 p.

——— 1963, Selected deep wells and areas of production in western New York: New York State Mus. and Sci. Serv. Bull. 390, 404 p.

Latimer, I. S., 1967, West Virginia oil and gas development 1966: Internat. Oil Scouts Assoc. Yearbook, v. 37, p. 512–518.

——— 1968, West Virginia oil and gas development 1967: Internat. Oil Scouts Assoc. Yearbook, v. 38, p. 455–461.

Lytle, W. S., 1957, Developments in Pennsylvania in 1956: Am. Assoc. Petroleum Geologists Bull., v. 41, p. 1010–1020.

——— 1958, Developments in Pennsylvania in 1957: Am. Assoc. Petroleum Geologists Bull., v. 42, p. 1147–1158.

——— 1959, Developments in Pennsylvania in 1958;

Am. Assoc. Petroleum Geologists Bull., v. 43, p. 1144–1160.
——— 1960, Developments in Pennsylvania in 1959: Am. Assoc. Petroleum Geologists Bull., v. 44, p. 688–703.
——— 1961, Developments in Pennsylvania in 1960: Am. Assoc. Petroleum Geologists Bull., v. 45, p. 734–748.
——— 1962, Developments in Pennsylvania in 1961: Am. Assoc. Petroleum Geologists Bull., v. 46, p. 778–795.
——— 1963, Developments in Pennsylvania in 1962: Am. Assoc. Petroleum Geologists Bull., v. 47, p. 940–955.
——— 1964, Developments in Pennsylvania in 1963: Am. Assoc. Petroleum Geologists Bull., v. 48, p. 784–800.
——— 1965, Pennsylvania oil and gas development 1964: Internat. Oil Scouts Assoc. Yearbook, v. 35, p. 279–283.
——— 1967, Pennsylvania oil and gas development 1966: Internat. Oil Scouts Assoc. Yearbook, v. 37, p. 298–304.
——— 1968, Pennsylvania oil and gas development 1967: Internat. Oil Scouts Assoc. Yearbook, v. 38, p. 261–266.
——— 1969, Pennsylvania oil and gas development 1968: Internat. Oil Scouts Assoc. Yearbook, v. 39, p. 258–263.
——— 1970, Pennsylvania oil and gas development 1969: Internat. Oil Scouts Assoc. Yearbook, v. 40, p. 263–266.
——— W. G. McGlade, and W. R. Wagner, 1966, Pennsylvania oil and gas development 1965: Internat. Oil Scouts Assoc. Yearbook, v. 36, p. 269–273.
Mackay, D. K., and P. M. Laughlin, 1952, Oil and gas developments in Arkansas in 1950, *in* Statistics of oil and gas development and production: AIME, v. 5, p. 1–22.
Patchem, D. C., 1968a, A summary of Tuscarora "Clinton sand" and pre-Silurian test wells in West Virginia: West Virginia Geol. and Econ. Survey Circ. 8, 34 p.
——— 1968b, Keefer sandstone gas development and potential in West Virginia: West Virginia Geol. and Econ. Survey Circ. 7, 19 p.
——— 1968c, Newburg gas development in West Virginia: West Virginia Geol. and Econ. Survey Circ. 6, 46 p.
——— 1968d, Oriskany Sandstone-Huntersville Chert production in the eastern half of West Virginia: West Virginia Geol. and Econ. Survey Circ. 9, 47 p.
Pepper, J. F., W. de Witt, Jr., and G. M. Everhart, 1953, The "Clinton" sands in Canton, Dover, Massillon and Navarre quadrangles, Ohio: U.S. Geol. Survey Bull. 1003-A, 15 p.
Poth, C. W., 1962, The occurrence of brine in western Pennsylvania: Pennsylvania Geol. Survey Bull. M 47, 53 p.
Reger, D. B., 1938, Oil and gas development in West Virginia during 1937: AIME Trans., v. 127, p. 593–602.
——— 1940, Oil and gas development in West Virginia during 1939: AIME Trans., v. 136, p. 511–521.
——— 1941, Oil and gas development in West Virginia during 1940: AIME Trans., v. 142, p. 512–521.
——— 1942, Oil and gas development in West Virginia during 1941: AIME Trans., v. 146, p. 521–531.
——— 1943, Oil and gas development in West Virginia during 1942: AIME Trans., v. 151, p. 573–580.
——— 1944, Oil and gas development in West Virginia during 1943: AIME Trans., v. 155, p. 553–560.
——— 1945, Oil and gas development in West Virginia during 1944: AIME Trans., v. 160, p. 593–600.
——— 1946, Oil and gas development in West Virginia during 1945, *in* Statistics of oil and gas development and production: AIME, p. 369–377.
——— 1948, Oil and gas development in West Virginia and the southwestern part of Virginia during 1947: Internat. Oil Scouts. Assoc. Yearbook, v. 18, p. 848–862.
——— 1949, Oil and gas development in West Virginia and the southwestern portion of Virginia during 1948, *in* Statistics of oil and gas development and production: AIME, p. 457–466.
Rocky Mountain Association of Geologists, 1954, Oil and gas fields of Colorado: 302 p.
——— 1955, The oil and gas fields of Nebraska: 264 p.
Russell, W. L., 1961, Reservoir water resistivities and possible hydrodynamic flow in Denver basin: Am. Assoc. Petroleum Geologists Bull., v. 45, p. 1925–1940.
Schrider, L. A., R. J. Watts, and J. A. Wasson, 1970, An evaluation of the East Canton oil field waterflood: Jour. Petroleum Technology, v. 22, p. 1371–1378.
Van Tyne, A. M., 1961, New York oil and gas development 1960: Internat. Oil Scouts Assoc. Yearbook, v. 31, p. 212–220.
——— 1964, New York oil and gas development 1963: Internat. Oil Scouts Assoc. Yearbook, v. 34, p. 230–236.
——— 1965, New York oil and gas development 1964: Internat. Oil Scouts Assoc. Yearbook, v. 35, p. 221–228.
——— 1966, New York oil and gas development 1965: Internat. Oil Scouts Assoc. Yearbook, v. 36, p. 213–221.
——— 1968, New York oil and gas development 1967: Internat. Oil Scouts Assoc. Yearbook, v. 38, p. 219–224.
——— 1970, New York oil and gas development 1969: Internat. Oil Scouts Assoc. Yearbook, v. 40, p. 225–228.
Woodfork, L. D., 1969, West Virginia oil and gas development 1968: Internat. Oil Scouts Assoc. Yearbook, v. 39, p. 411–418.
Yeakel, L. S., Jr., 1962, Tuscarora, Juniata, and Bald Eagle paleocurrents and paleogeography in the central Appalachians: Geol. Soc. America Bull., v. 73, p. 1515–1539.

Reprinted from:
BULLETIN OF THE AMERICAN ASSOCIATION OF PETROLEUM GEOLOGISTS
VOL. 56, NO. 4 (APRIL, 1972), PP. 790–795, 3 FIGS.

Clays with Abnormal Interstitial Fluid Pressures[1]

RICHARD E. CHAPMAN[2]
Brisbane, Queensland, Australia

Abstract Consideration of the stresses in clay rocks, and their compaction processes, leads to the conclusion that abnormally pressured clay sequences generally have been abnormally pressured since burial to a shallow depth—2,000 ft (600 m) or even less—in areas of gravity loading under sediments.

Rubey and Hubbert's concept of equilibrium depth (z_e) is reexpressed

$$z_e = \frac{(1 - \lambda)}{(1 - \lambda_e)} z$$

in which the ratio $(1 - \lambda)/(1 - \lambda_e)$ is a dimensionless measure of the expulsion of fluids from the formation.

Zones of abnormal fluid pressure are caused more by low hydraulic conductivity in a rock unit or sequence than by load or rate of loading. Hence the transition zone from normal to abnormal pressures at the top of a relatively impermeable sequence will have its mirror image at the bottom of the unit, if it overlies a sequence of sufficient hydraulic conductivity. The petroleum potential below an abnormally pressured clay sequence is similar to that above it.

Introduction

Dickinson (1953) showed that abnormal pressures in the Louisiana Gulf Coast are associated with facies, and only incidentally are associated with depth. Hubbert and Rubey (1959) and Rubey and Hubbert (1959) showed that fluid pressure in porous rocks is a significant geologic factor because it affects the mechanical properties of rocks. Bredehoeft and Hanshaw (1968) showed that abnormal pressures can be maintained over geologically significant spans of time if the hydraulic conductivity of the rock is low enough. Other authors have described and discussed the general topic of abnormal pressures, and some have suggested processes other than mechanical compaction under load for their generation (*e.g.*, Hanshaw and Zen, 1965; Heard and Rubey, 1966; Powers, 1967; Hanshaw and Bredehoeft, 1968).

Without prejudice to the other possible causes of abnormal pressures, it can be said that the mechanical aspects of the compaction of clay rocks under the gravity load of overlying sediments are sufficient to explain the essential features observed in such areas as the U.S. Gulf Coast, southern Nigeria, and Borneo. The importance of contributary processes has not yet been assessed satisfactorily.

Compaction and Interstitial Fluid Pressures

Compaction is a diagenetic process that begins with burial and may continue over a long span of time. It increases the bulk density of a rock, and reduces its porosity. It can take place only with the expulsion of interstitial liquid or the compression of interstitial gas.

Different lithologies compact in different ways. Some materials, such as quartz sands, suffer elastic compaction under load. This is largely reversible if the load is removed. Others, notably clays, suffer plastic deformation that is largely irreversible and leads to permanent loss of porosity and permeability. Thus the rate at which compaction takes place is governed not only by the load and the rate of loading, but also by the hydraulic conductivity of the rocks, or the rate at which the fluid can be dispersed through them.

Normal hydrostatic gradient—If the potential for fluid dispersal exceeds the requirements for compaction under load, the interstitial fluid pressures will be approximately normal hydrostatic,

$$p_n = \gamma_w z \tag{1}$$

where γ_w is the specific weight of the fluid in psi per foot (kg/cm²/m) and z the depth in feet (meters) below static water level (usually taken as the surface of the ground). The normal hydrostatic gradient varies from about 0.43 psi/ft of depth to about 0.47 psi/ft (0.10 to 0.11 kg/sq cm/m) according to the water salinity. In many areas it is logical to regard the sediments as being contained in a water reservoir, rather than the water as being contained in the rocks.

Geostatic gradient—If the rate of loading exceeds the corresponding rate of fluid dispersal, abnormal interstitial fluid pressures result. These may approach a limiting pressure that may be expressed

$$p_{max} \rightarrow \bar{\gamma}_{bw} z \tag{2}$$

where $\bar{\gamma}_{bw}$ is the mean specific weight of the water-saturated overburden. The geostatic gradient is

[1] Manuscript received, June 7, 1971; accepted, August 4, 1971.

[2] Department of Geology and Mineralogy, University of Queensland.

commonly taken as 1 psi/ft of depth (0.23 kg/sq cm/m). It is logical to regard the water as being contained within the rock when fluid pressures are abnormal.

Mechanical aspects of pore pressure—Hubbert and Rubey (1959) and others have developed earlier work on soils and shown that the total stress in a liquid-filled rock is divided between the effective grain-to-grain stress and the interstitial fluid pressure,

$$S = \sigma + p \tag{3}$$

where S is the vertical component of total stress, σ the effective stress in solids, and p the liquid pressure.[3] As a water-saturated porous rock is loaded, the load is taken first by the liquid, and then transferred to the grain-to-grain contacts as liquid is expelled from the rock. An abnormally pressured rock is not in stress equilibrium if liquid can leak from it, but it will achieve stress equilibrium when the fluid has been reduced to normal hydrostatic pressure.

The compaction processes set up fluid potential gradients from the more compactible rocks to the more permeable. A fluid potential gradient exists between two points in the subsurface at depths z_1 and z_2 when the pressure difference between them is *not* equal to $\gamma_w(z_1 - z_2)$. Fluid will tend to move down the potential gradient in the direction that tends to restore hydrostatic equilibrium. This is generally upward or downward (Hedberg, 1926), but locally may be lateral. Magara (1968) has discussed an interesting example of downward migration of petroleum from a Miocene mudstone in Japan.

Clays are not ideally impermeable. The continuation of compaction trends with depth indicates that they retain some permeability. A clay in stress equilibrium under a particular load will allow the passage of water if there is a potential difference across the bed. However, during the compaction of a clay that overlies a permeable bed, the downward fluid potential gradient converts the clay into a perfect barrier to the upward migration of fluid.

Effective Compaction

Hubbert and Rubey (1959, p. 142) introduced a useful parameter, λ, which can be written

$$\lambda = \frac{p}{\bar{\gamma}_{bw} z} \tag{4}$$

[3] It would be useful if the term "lithostatic" was not used synonymously with "geostatic," but was confined to the vertical component of stress in solids (unless some other component is specified). The terms "geostatic," "lithostatic," and "hydrostatic" would then refer to S, σ, and p, respectively.

and is the ratio of fluid pressure to geostatic pressure at depth z. For many practical purposes this may be approximated by

$$\lambda' = p \text{ (in psi)/depth (in ft)}.$$

λ is a nondimensional parameter that commonly takes values between 0.43 and 1.0. When $\lambda = 1$, the overburden is supported on water, not on rock, and the effective stress in the solids is zero.

The stress distribution between fluid and solids is in equilibrium under a load when the interstitial fluid pressure is normal, and λ takes its equilibrium value (usually between 0.43 and 0.47, depending mainly on water salinity and overburden lithology). When the fluid pressure is abnormal, the solid stress in the rock is equivalent to the equilibrium stress of an identical rock under a smaller load or, in the context of this paper, at a shallower depth of burial. Rubey and Hubbert (1959, p. 174–175) showed that the equilibrium depth, or effective compaction depth, is given by

$$z_e = \frac{\bar{\gamma}_{bw} z - p}{\bar{\gamma}_{bw} - \gamma_w}$$

where z is the depth in feet, $\bar{\gamma}_{bw}$ the mean specific weight of the overburden in psi/ft, γ_w the specific weight of the pore water in psi/ft, and p the fluid pressure in psi.

However, the equilibrium depth is more simply expressed as a function of λ. As $p = \lambda \bar{\gamma}_{bw} z$ and $\gamma_w = \lambda_e \cdot \bar{\gamma}_{bw}$, the equilibrium depth may be re-expressed

$$z_e = \frac{(1 - \lambda)}{(1 - \lambda_e)} z \tag{5}$$

where z_e and z are in any consistent unit of length, and λ_e is the equilibrium value of λ. Let us denote the expression $(1-\lambda)/(1-\lambda_e)$ by δ.

$$\delta = \frac{z_e}{z} = \frac{(1 - \lambda)}{(1 - \lambda_e)} \tag{5a}$$

δ is nondimensional, and commonly will take values between zero (when $\lambda = 1$) and one (when $\lambda = \lambda_e$). It may be interpreted as a measure of the expulsion of fluids since deposition. When stress equilibrium has been maintained by leakage from the rock, δ is equal to unity, and the equilibrium depth equals the actual depth ($z = z_e$). When no leakage has occurred, δ is equal to zero, and the equilibrium depth is the surface.

Zone of Abnormal Pressures

Drilling from zones of normal pressures into zones of abnormal pressures has shown that there is a transition zone through which the fluid pressure increases at a rate greater than the rate of increase of the geostatic pressure. The rate of in-

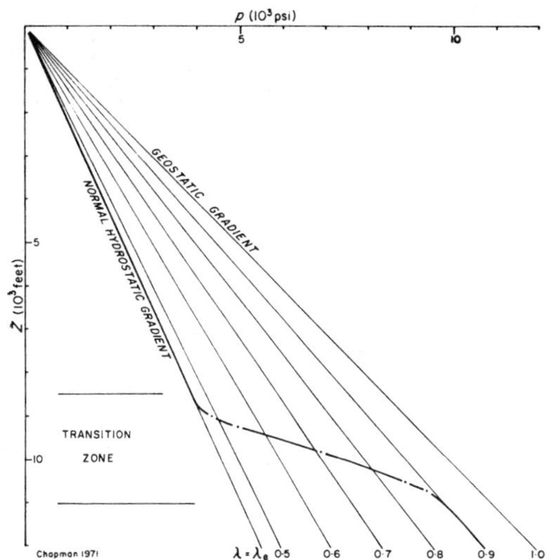

FIG. 1—Plot of interstitial fluid pressure versus depth.

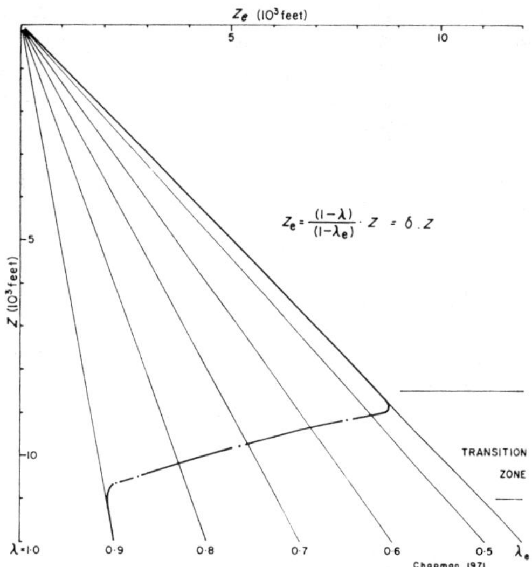

FIG. 2—Plot of equilibrium depth versus depth.

crease in the transition zone is not constant from well to well, nor always in one well. The increase in fluid pressure is accompanied by loss of mechanical strength in the rock (Eq. 3), which contributes to the drilling "break" that usually accompanies the penetration of the zone of abnormal pressures. There also appears to be a maximum value of λ that is rather less than the limiting value of one that corresponds to a fluid pressure equal to the geostatic pressure.

Formation density logging and acoustic logging have shown that the bulk density of clays in the transition zone commonly changes its trend with depth, so that as the pressure increases at a rate greater than the normal hydrostatic, the bulk density of the clays tends to decrease. Therefore, the porosity of the clay rocks tends to increase with depth in the transition zone.

The transition zone is commonly represented graphically by plotting the observed or estimated pressures against depth (Fig. 1). This is sometimes supplemented by a plot of λ against depth, or by including the λ-gradients on the pressure-depth plot (Eq. 4).

Plotting the equilibrium depth (z_e) against depth (z) from the same data (Fig. 2; Eq. 5) shows that the transition zone also is characterized by a reduction in the equilibrium depth with depth. At the bottom of the transition zone (locally, constant or maximum λ), z_e tends to increase again with depth, but on a gradient determined by the value of λ. This is consistent with the bulk density observations from formation density logging and acoustic logging.[4]

The geologic significance of the transition zone, in which so many rock parameters change, is of some interest. The changing rate of increase of pressure with depth cannot be attributed to loading, or rate of loading, because the gradient generally exceeds the geostatic gradient. It therefore must be regarded simply as a zone of leakage from the abnormal to the normal—from the rocks that are not in stress equilibrium to those that are. The top of the transition zone is the bottom of the equilibrium zone, and the transition zone is compacting from the top downward, tending as Rubey and Hubbert (1959) indicated, to seal the pressures in the clay unit. The bottom of the transition zone is marked by constant or maximum λ, minimum z_e, and by a change in the bulk density trends. The thickness of the transition zone is a measure of its hydraulic conductivity. This deter-

[4] Indeed, by use of the arguments of Magara (1969), log anomalies can be used to estimate fluid pressures. If z_e' and z are two depths with the same shale density, the ratio $z_e'/z \approx z_e/z = \delta$. Knowing λ_e, λ can be calculated for depth z, and hence an estimate of the pressure at depth z obtained.

It is perhaps worth noting that Rubey and Hubbert's equations relating porosity, fluid pressure, and depth (Rubey and Hubbert, 1959, p. 176, eqs. 17, 18) can be simplified by the introduction of δ:

$$f = f_o e^{-c\delta z}$$

and

$$\delta = \frac{\log (f_o/f)}{cz \log e} \cdot$$

mines whether the rate of change of pressure in the transition zone shall be large or small.

The concept of equilibrium depth as defined by Rubey and Hubbert (1959, p. 174–175) is a notional depth, z_e, at which the effective stress in solids, σ, in a rock with normal hydrostatic fluid pressure, p, would be equal to the effective stress in solids in a rock at depth z with abnormal fluid pressure. In the special case where the load that gives rise to the abnormal pressure is applied after burial of the rock to depth z, the equilibrium depth is indeed notional. However, in the common case where the compaction of a clay rock under a sedimentary load has been retarded by the limited escape of pore water, this notional depth acquires some degree of reality.

If we accept that the compaction of clay is essentially plastic and irreversible, we cannot suppose that the compaction history of an abnormally pressured clay rock includes a reversal of the trend of increasing bulk density and decreasing porosity with depth. Still less can we suppose that the history of the $z - z_e$ relation in the clay rock now at z_1 has followed the heavy line in Figure 3, passing through z_4, z_3, and z_2. This history must rather be represented qualitatively by the arrows for each point in Figure 3. A rock now abnormally pressured at depth z can only have had normal hydrostatic fluid pressure when it was at depth z_e if there has been no leakage of fluid during burial from depth z_e to depth z.

It therefore must be concluded that the equilibrium depth, z_e, has a real meaning in the compaction history of the rock that has been buried to depth z under a sedimentary load; that is, the depth z_e is an estimate of the *maximum* depth of burial at which that rock ceased to be in stress equilibrium under the load of overlying sediments.

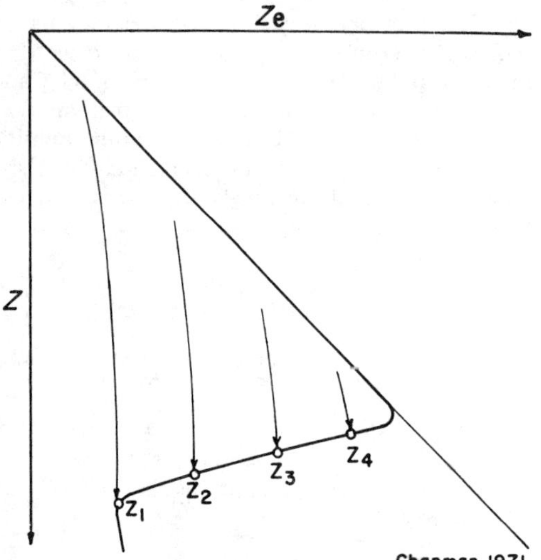

FIG. 3—Diagram suggesting evolution of equilibrium depth curve.

The maximum value of λ would appear to be controlled by the real value of z_e, the depth at which equilibrium was lost. If the true value of z_e is about 2,000 ft (600 m), λ will only approach unity at depths below about 20,000 ft (6,000 m). The maximum value of λ at the base of the transition zone in the Louisiana Gulf Coast appears to be about 0.9 at 10,000–12,000 ft (3,000–3,700 m). The corresponding value of δ is $(1-0.9)/(1-0.465) = 0.19$. This indicates that the abnormally pressured zones at present between 10,000 and 12,000 ft acquired their abnormality at depths of burial of the order of 2,000–2,500 ft (600–800 m) or shallower. Such shallow depths for abnormal pressures are not unknown; Spinks (1970), for example, reported abnormal pressures in the Gulf of Papua at depths between 2,100 and 5,000 ft (640 and 1,500 m) in Pliocene plastic mudstones.

This equilibrium depth may be significant in the diagenesis of clays (Burst, 1969), and relate to the problem of "heaving shales" at shallow depths.

It is also of interest to view the equilibrium depth in another light—that of the lithologies in the sequence of thickness z_e above the bottom of the transition zone. This sequence may be regarded as approximating the sedimentary cover when the rocks at the bottom of the transition zone were buried to depth z_e below the surface. If this sequence is essentially of low hydraulic conductivity, early acquisition of abnormal pressures under a sedimentary load is likely.

A further implication is that the early loss of equilibrium implies retarded fluid expulsion as well as retarded compaction. If the clay unit is a source rock, delayed expulsion of fluids must mean improved chances of entrapment. Indeed, a thick clay may itself contribute to contemporaneous structural growth, because clay diapirism (or incipient diapirism) is understood more readily in the context of abnormal interstitial fluid pressures, which not only lead to relatively low mechanical strength of the material, but also to a density inversion with depth.

Beneath the transition zone—Few wells penetrate far below the transition zone, and although it generally is tacitly accepted that once abnormal pressures have been reached, they will continue in depth, this is not necessarily so. There are some published examples of reversion to lower fluid pressure gradients (Frederick, 1967; Spinks, 1970). If abnormal pressures are inherent in a rock unit, they well may be confined to it. Continuation of abnormal pressures with depth depends mainly on the continuation of low hydraulic conductivity with depth. If the abnormally pressured formation overlies a formation of high hydraulic conductivity and continuity of pore space to the surface, fluid pressures may revert to normality, or near normality.

Where the zone of abnormal pressures is confined to a rock unit of low hydraulic conductivity that overlies one of high hydraulic conductivity and normal fluid pressures, the transition zone at the top of the unit will have its mirror image at the bottom.

In the lower transition zone, the rock parameters will tend toward equilibrium with depth, the fluid potential gradient will be downward, and the migration of fluids will be downward. The zone will be characterized by decreasing fluid potential, increasing mechanical strength, and increasing bulk density. It will be compacting from the bottom upward. The drilling "break" that accompanies penetration of the upper transition zone will have its counterpart in a sharp reduction of penetration rate, warning of a return to more normal pressures.

In brief, the petroleum potential above abnormal pressures also may have its mirror image below, if structures and potential reservoirs exist.

Drilling through a zone of abnormal pressures is accompanied by material and financial risks, involving difficulties of seismic and gravity interpretation, mechanical drilling problems, and well design. The mechanical strength of the rocks in the zone of abnormal pressures may be equivalent to those at depths above 3,000 ft (1,000 m). These rocks are incompetent by any standards. Casing is only secure above the transition zone; liners in abnormally pressured sequences must rely on the cement bond in the annulus below the hanger.

Tectonic or Gravity Load?

Along the western seaboard of the Americas, in Trinidad, and near the Himalayas, abnormal pressures—in places at very shallow depths—have been attributed in part to tectonic loading, rather than to gravity loading.

If the tectonic loading is a continuous process during sediment accumulation (as appears to be the case in California), the abnormal pressures generated may be indistinguishable from those of gravity loading. However, it is suggested that tectonic loading of a sedimentary column that has achieved stress equilibrium will differ in that the equilibrium depth curve will not reflect the bulk density trends of the clay rocks. If the compaction of clays is essentially irreversible, subsequent loading cannot materially alter the density or porosity trends with depth, in these rocks.

Conclusions

1. The rate of sediment accumulation is a significant factor in petroleum generation and accumulation. During the evolution of any sedimentary basin, abnormal interstitial fluid pressures are generated to some extent in the thicker clay units, and these units have a profound effect on the paths of fluid migration by virtue of their downward fluid potential gradient. Major zones of abnormal pressures, far from being economic basement, may overlie petroleum potential equal to that above them.

2. Pooling or averaging pressure-depth data from an area may have some practical value, but it may also obscure the geologic relations. Lithologic columns, equilibrium depth curves, and bulk density curves (if reliable!) should be plotted with the pressure-depth data.

3. The specific weight of the overburden should not be accepted uncritically as 1 psi/ft (0.23 kg/sq cm/m) in areas of thick clay sequences that are not normally pressured, not in stress equilibrium.

Selected References

Athy, L. F., 1930a, Density, porosity, and compaction of sedimentary rocks: Am. Assoc. Petroleum Geologists Bull., v. 14, p. 1–24.

———1930b, Compaction and oil migration: Am. Assoc. Petroleum Geologists Bull., v. 14, p. 25–35.

Bredehoeft, J. D., and B. B. Hanshaw, 1968, On the maintenance of anomalous fluid pressures; 1. Thick sedimentary sequences: Geol. Soc. America Bull., v. 79, p. 1097–1106.

Burst, J. F., 1969, Diagenesis of Gulf Coast clayey sediments and its possible relation to petroleum migration: Am. Assoc. Petroleum Geologists Bull., v. 53, p. 73–93.

Dickinson, G., 1953, Geological aspects of abnormal reservoir pressures in Gulf Coast Louisiana: Am. Assoc. Petroleum Geologists Bull., v. 37, p. 410–432.

Frederick, W. S., 1967, Planning a must in abnormally pressured areas: World Oil, v. 164, no. 4, p. 73–77.

Gretener, P. E., 1969, Fluid pressure in porous media, its importance in geology; a review: Bull. Canadian Petroleum Geology, v. 17, p. 255–295.

Hanshaw, B. B., and J. D. Bredehoeft, 1968, On the maintenance of anomalous fluid pressures—2. Source layer at depth: Geol. Soc. America Bull., v. 79, p. 1107–1122.

——— and E.-A. Zen, 1965, Osmotic equilibrium and overthrust faulting: Geol. Soc. America Bull., v. 76, p. 1379–1386.

Heard, H. C., and W. W. Rubey, 1966, Tectonic implications of gypsum dehydration: Geol. Soc. America Bull., v. 77, p. 741–760.

Hedberg, H. D., 1926, The effect of gravitational compaction on the structure of sedimentary rocks: Am. Assoc. Petroleum Geologists Bull., v. 10, p. 1035–1072.

——— 1936, Gravitational compaction of clays and shales: Am. Jour. Sci., 5th ser., v. 31, p. 241–287.

Hubbert, M. K., and W. W. Rubey, 1959, Role of fluid pressure in mechanics of overthrust faulting; 1. Mechanics of fluid-filled porous solids and its application to overthrust faulting: Geol. Soc. America Bull., v. 70, p. 115–166.

Keep, C. E., and H. L. Ward, 1934, Drilling against high rock pressures with particular reference to operations conducted in the Khaur field, Punjab: Inst. Petroleum Technologists Jour., v. 20, p. 990–1026.

Kok, P. C., and J. H. M. A. Thomeer, 1955, Abnormal pressures in oil- and gas reservoirs: Geologie en Mijnbouw (new ser.), 17e jaargang (August), p. 207–216.

Magara, Kinji, 1968, Compaction and migration of fluids in Miocene mudstone, Nagaoka Plain, Japan: Am. Assoc. Petroleum Geologists Bull., v. 52, p. 2466–2501.

——— 1969, Upward and downward migrations of fluids in the subsurface: Bull. Canadian Petroleum Geology, v. 17, p. 20–46.

Powers, M. C., 1967, Fluid-release mechanisms in compacting marine mudrocks and their importance in oil exploration: Am. Assoc. Petroleum Geologists Bull., v. 51, p. 1240–1254.

Rubey, W. W., and M. K. Hubbert, 1959, Role of fluid pressure in mechanics of overthrust faulting; 2. Overthrust belt in geosynclinal area of western Wyoming in light of fluid-pressure hypothesis: Geol. Soc. America Bull., v. 70, p. 167–206.

Spinks, R. B., 1970, Offshore drilling operations in the Gulf of Papua: Australian Petroleum Exploration Assoc. Jour., v. 10, p. 108–114.

Thomeer, J. H. M. A., and J. A. Bothema, 1961, Increasing occurrence of abnormally high reservoir pressures in boreholes, and drilling problems resulting therefrom: Am. Assoc. Petroleum Geologists Bull., v. 45, p. 1721–1730.

Weller, J. M., 1959, Compaction of sediments: Am. Assoc. Petroleum Geologists Bull., v. 43, p. 273–310.

Reprinted from:
BULLETIN OF THE AMERICAN ASSOCIATION OF PETROLEUM GEOLOGISTS
VOL. 56, NO. 10 (OCTOBER, 1972), PP. 2068–2071, 3 FIGS.

Aquathermal Pressuring—Role of Temperature in Development of Abnormal-Pressure Zones[1]

COLIN BARKER[2]
Tulsa, Oklahoma 74104

Abstract Zones of abnormal subsurface pressures, both above and below hydrostatic, have been described in many areas. The very abrupt changes in pressures and salinities, together with the undercompacted nature of the high-pressure zones, indicate that they are effectively isolated from their surroundings. If this isolation occurred at a shallower depth than the present one, the isolated volume would have been subjected to increasing temperatures as it moved downward. The *P-T*-density diagram for water shows that, for any geothermal gradient greater than about 15°C/km, the pressure in an isolated volume increases with increasing temperature more rapidly than that in the surrounding fluids. This mechanism for producing excess pressures will operate in addition to most of the other processes that have been suggested, but the overall influence in any given area will depend on how well the system remains isolated. If a normally pressured system becomes isolated and is then subjected to a decrease in temperature (for example, if erosion removes considerable quantities of overburden) the pressure in the system will fall below the external hydrostatic pressure. This may have happened in some areas which now have abnormally low pressures.

Zones of anomalous subsurface pressures, both above and below hydrostatic, have been described in many parts of the world. The high-pressure zones along the northern coast of the Gulf of Mexico have been particularly well studied, and there pressures as much as 5,000 psi above the normal hydrostatic values are not uncommon. These excess pressures can be expressed conveniently in terms of the geostatic ratio, which is the ratio of observed fluid pressure to the lithostatic pressure at the same depth. Normal hydrostatic pressures give a ratio close to 0.47, whereas lithostatic pressure corresponds to 1.00. High pressure zones have been encountered with geostatic ratios in excess of 0.95. Generally high-pressure zones are characterized by undercompaction in the shales, and they also commonly contain water which is less saline than the corresponding fluids outside.

In many places there is a very abrupt change from normal pressure to high pressure, indicating that the pressurized volumes are isolated from their surroundings. Dickinson (1953) concluded that "a reservoir containing high pressure must be effectively isolated from any other porous formation which contains normal hydrostatic pressure, otherwise the pressure would be dissipated."

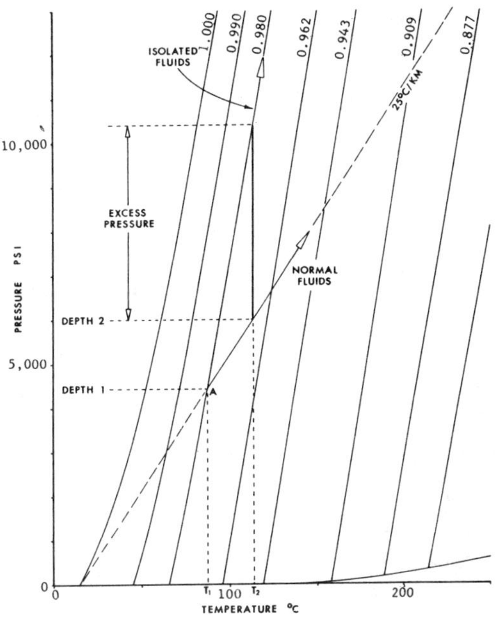

FIG. 1—Pressure-temperature-density diagram for water with superimposed geothermal gradient of 25° C/km, showing *P-T* relation for normal and isolated fluids as temperature rises.

More recently, Jones (1969) has written that, "In theory geopressured reservoirs are closed compartments having fixed or constant volume." Dickey *et al.* (1972) wrote that in Louisiana "the abnormal pressures are found only in sands completely enclosed in shale, with no permeable connection to the outcrop." Most other writers also have reached the conclusion that high pressure zones are effectively isolated from their immediate surroundings.

It is instructive to consider the time when this isolation occurred, and to investigate the consequences of such an isolation. The northern coast of the Gulf of Mexico has been an area of continuous deposition, and any isolation of a potential high pressure zone at a time significantly before

[1] Manuscript received, October 29, 1971; revised and accepted, March 9, 1972.

[2] Department of Chemistry, University of Tulsa.

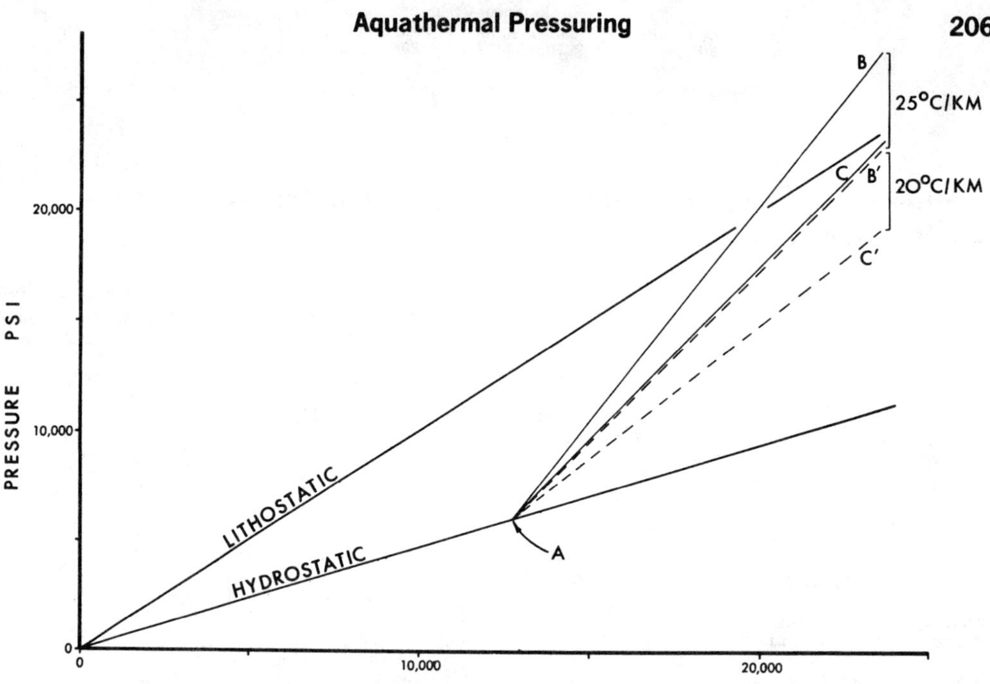

FIG. 2—Increase in excess pressure with increasing depth of burial after initial isolation at point *A*. For both geothermal gradients shown, uppermost of pair of lines *(A-B, A-B′)* is for pure water and lower one (A-C, A-C′) is for 30 percent sodium chloride solution.

the present means that the isolation occurred at a shallower depth than the present one. This corresponds to less compacted shales and to formation waters of lower salinity, for salinity generally increases with depth (Dickey, 1969). It also implies isolation at a temperature lower than that observed now. The effect of increasing temperature on the contents of such an isolated volume is the main topic of this note.

The pressure increase which results from increasing temperature is best considered by reference to the pressure-temperature-density diagram for water. Figure 1 shows a pressure-temperature diagram with selected isodensity lines plotted from the data given by Kennedy and Holser (1966). A geothermal gradient of 25°C/km is superimposed. This value is fairly typical for the Louisiana Gulf Coast, where reported gradients range from 18 to 36°C/km with most values lying between 22 and 24°C/km (Jam L. *et al.*, 1969). Because depth and pressure are related, the pressure and temperature of the formation fluids which follow the normal hydrostatic gradient are related by the geothermal gradient, so that for any given temperature the pressure is fixed. If, however, some of the rocks containing the fluid become isolated by material of low permeability, for example, during faulting, the isolated system becomes one of constant density because a fixed mass of material is trapped in an essentially constant volume. The pressure and temperature of the contents of the sealed volume then will be related, not by the geothermal gradient, but by one of the isodensity lines of the pressure-temperature diagram. For a given rise in temperature the pressure in the isolated volume will increase more than that in the open system. If isolation occurs at Depth 1 and T_1 (point A) and the system is then buried to Depth 2 corresponding to T_2 for this 50°-rise in temperature, the pressure in the normal formation fluids will rise to 6,000 psi. But the same temperature rise increases the pressure in the isolated volume to 10,400 psi. Thus a high-pressure zone with a pressure 4,400 psi above the normal hydrostatic value is created by a temperature increase of 50°C. This corresponds to an increased depth of burial of 3,500 ft if the geothermal gradient is 25°C/km. For higher geothermal gradients smaller increases in depth are required to cause the same increase in pressure. The term "aquathermal pressuring" is proposed for this mechanism of generating high-pressure zones.

The pressure generated aquathermally depends on how much the depth of burial increases after the volume becomes isolated, as well as on the geothermal gradient. Line A-B (Fig. 2) shows the way in which the excess pressure P increases

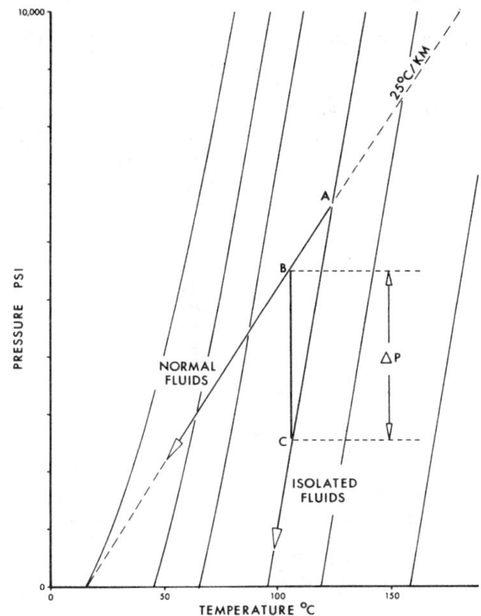

Fig. 3—Pressure-temperature-density diagram for water showing the *P-T* relations for normal and isolated fluids as temperature falls.

relative to the hydrostatic and lithostatic values. In this illustration trapping occurs at 12,500 ft, which is roughly 6,000 psi, and the geothermal gradient is 25°C/km. This shows that aquathermally generated pressures can exceed the lithostatic value at geologically reasonable depths and give a geostatic ratio greater than one. This is a geologically unstable condition which would be relieved by movement of some kind, possibly faulting or diapirism. Up to this point we have been using data from the phase diagram for pure water, but most waters in the subsurface are saline. The pressure-temperature-density diagrams for sodium chloride solutions closely resemble the diagram for water, but the isodensity lines are displaced to higher temperatures. With increasing depth of burial the excess pressure does not rise as rapidly for a saline solution as for pure water, and line A-C (Fig. 2) shows how the pressure increases for a 30-percent solution of sodium chloride. Real subsurface fluids lie between the extremes represented by lines A-B and A-C.

So far we have been discussing the ideal case in which the contents of the high-pressure zone remain isolated and of constant density. However, in nature it is possible that the volume, or the mass of trapped liquid, or indeed both, may change. In the first case the contents of the high-pressure zone may be isolated in the way that the contents of a balloon are isolated, so that a volume increase can occur without loss of material. If the volume increases, it will act to reduce the density and hence reduce the excess pressure. An increase in volume of about 5 percent would be enough to reduce an excess pressure of 4,000 psi back to the normal hydrostatic value. The isolated zones probably are imperfectly sealed so that some of the contents may be leaking out. This also would cause a reduction in the pressure, and an excess pressure of 4,000 psi would be reduced to the normal hydrostatic value by a 5-percent loss of material.

The shales in high-pressure zones are undercompacted compared with other shales at the same depth, so that the isolated volumes contain a greater percentage of water and are poorer conductors of heat. Therefore, it has been suggested that they can have somewhat higher temperatures than the surrounding rocks (Lewis and Rose, 1970). If this is the case, the higher temperatures will favor aquathermal pressuring within the isolated volume. The increase is approximately 200 psi per °C, so that a temperature 5° above normal is needed to generate an excess pressure of 1,000 psi.

Mineral transformations which occur with the release of water also can increase the pressure, or alternatively can provide a means of offsetting losses due to leakage. The gypsum-anhydrite conversion may be important at shallow depths, but the dehydration of montmorillonite probably becomes the most important diagenetic process as depth increases. The amount of water released is substantial and may be as much as 10 percent of the volume of the isolated sediment (Burst, 1966). The dehydration process is endothermic and needs temperatures in excess of 100°C. It would be favored by a slightly higher temperature in the high-pressure zone. In the presence of potassium ions, montmorillonite converts to illite and this process releases a volume of water equal to about half of the volume of the altered montmorillonite (Powers, 1967). As the water released by mineral transformations is fresh, it reduces the salinity of the trapped liquids.

In summary, aquathermal pressuring is a possible mechanism for generating abnormal subsurface pressures, and the effect will be enhanced by slightly higher temperatures in the high pressure zone and by mineral transformations which occur with the release of water. An increase in volume or leakage of material from the high-pressure zone will reduce the excess pressure. It is not suggested that aquathermal pressuring is responsible for all abnormally pressured zones, but it will operate whenever an isolated volume is moved down a geothermal gradient. Whether aquathermal pressuring modified by the factors discussed will produce a net increase in pressure must be evaluated separately in each area.

In addition to abnormally high pressures, areas have been described where the pressures are abnormally low, as for example in the San Juan basin. The different response of the contents of isolated and open systems to changing temperatures again may provide the explanation. If a normally pressured zone becomes effectively isolated from its surroundings (Fig. 3, point A), and if this zone then is cooled (*e.g.*, by uplift, or removal of overburden during erosion), the pressure in the isolated volume as given by A-C will fall below the normal hydrostatic value, because the contents of the isolated volume correspond to conditions of constant density and follow an isodensity line, whereas the fluids in the open system follow the geothermal gradient.

The discussion presented here makes no reference to scale. The properties discussed are independent of the volume of the system, and the treatment applies equally well to a cubic mile of fluid trapped in a sand lens as it does to a fraction of a milliliter of fluid trapped between clay particles.

References Cited

Burst, J. F., 1966, Diagenesis of Gulf Coast clayey sediments and its possible relation to petroleum migration (abs.): Am. Assoc. Petroleum Geologists Bull., v. 50, no. 3, p. 607.

Dickey, P. A., 1969, Increase in concentration of subsurface brines with depth: Chem. Geology, v. 4, p. 361-370.

——— A. G. Collins, and I. Fajardo, 1972, Chemical composition of deep formation waters in southwestern Louisiana: Am. Assoc. Petroleum Geologists Bull., v. 56, no. 8 (in press).

Dickinson, G., 1953, Geological aspects of abnormal reservoir pressures in Gulf Coast Louisiana: Am. Assoc. Petroleum Geologists Bull., v. 37, no. 2, p. 410-432.

Jam L., P., P. A. Dickey, and E. Tryggvason, 1969, Subsurface temperature in south Louisiana: Am. Assoc. Petroleum Geologists Bull., v. 53, no. 10, p. 2141-2149.

Jones, P. H., 1969, Hydrodynamics of geopressure in the northern Gulf of Mexico basin: Jour. Petroleum Technology, v. 21, p. 803-810.

Kennedy, G. C., and W. T. Holser, 1966, Pressure-volume-temperature and phase relations of water and carbon dioxide, Sec. 16 *in* Handbook of physical constants (rev. ed.): Geol. Soc. America Mem. 97, p. 371-383.

Lewis, C. R., and S. C. Rose, 1970, A theory relating high temperatures and overpressures: Jour. Petroleum Technology, v. 22, p. 11-16.

Powers, M. C., 1967, Fluid-release mechanisms in compacting marine mudrocks and their importance in oil exploration: Am. Assoc. Petroleum Geologists Bull., v. 51, no. 7, p. 1240-1254.

Reprinted from:
BULLETIN OF THE AMERICAN ASSOCIATION OF PETROLEUM GEOLOGISTS
VOL. 56, NO. 11 (NOVEMBER, 1972), PP. 2185-2191, 5 FIGS.

Primary Migration of Petroleum from Clay Source Rocks[1]

RICHARD E. CHAPMAN[2]
Brisbane, Queensland, Australia

Abstract An important difficulty in explaining primary migration from compacting source rocks concerns timing. The commonly accepted compaction curves for clay rocks imply maximum fluid expulsion early in a clay's compaction history, whereas most accumulations clearly have been formed later, after burial to depths of several thousands of feet.

It was shown recently that abnormally pressured clays have been abnormally pressured since burial to a shallow depth. It was concluded that the thicker clays in most sedimentary basins have been abnormally pressured to some extent.

If this conclusion is correct, the expulsion of petroleum becomes reconcilable with the time of entrapment because (a) thicker source clays, e.g., more than 200 ft, or 60 m, initial thickness) are probably more important than thinner source clays, and (b) mean fluid expulsion from the thicker clays is retarded.

In addition, most of the fluid in a thin clay is expelled under geothermal temperatures and essentially hydrostatic pressures (implied in early migration) over a small depth range, whereas fluid in a thick clay is exposed to geothermal temperatures and near-geostatic pressures over a much greater depth range.

The protopetroleum thus may be subjected to higher temperatures and much higher pressures for a longer time, thereby facilitating its solution in the pore water. The pressure decrease that necessarily accompanies expulsion may lead to the release of petroleum from solution in the outer, compacting layer of clay, and the accumulation of a slug that could be expelled by capillary force aided by the fluid potential gradient.

Introduction

Primary migration from clay source rocks long has interested petroleum geologists, and the subject occupies an important part of the literature. It has been extensively reviewed recently by Dott and Reynolds (1969).

This paper considers the primary migration of petroleum from compacting clay rocks in a dominantly regressive sequence, such as that shown schematically in Figure 1. This is a common and significant geologic setting of petroleum (see Chapman, 1972a), and is applicable to areas such as the U.S. Gulf Coast, Nigeria, and many fields in southeast Asia.

The premises of this paper are as follows.

1. Clay is the principal source rock of petroleum in regressive sequences, but not all clays, nor all parts of clays, are source rocks. Thus a clay unit may consist of source clays and nonsource clays.

2. Petroleum is generated by the diagenesis of organic matter during burial, and some of the products of this diagenesis are soluble in water.

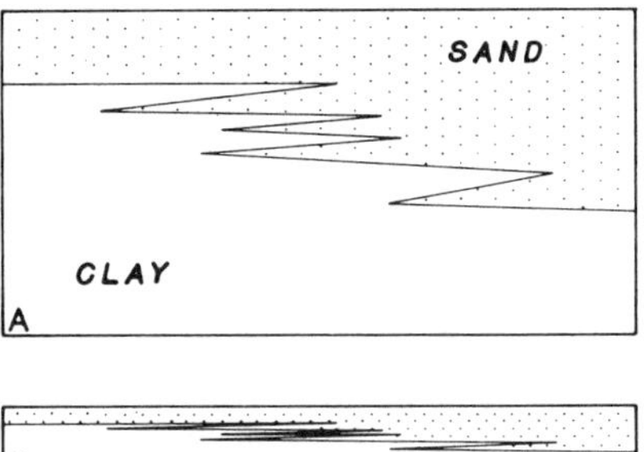

FIG. 1—Schematic cross section through regressive sand-clay sequence. **A,** vertical scale exaggerated. **B,** more natural vertical scale.

3. The composition of the petroleum in the reservoir reflects the temperatures and pressures under which the diagenesis took place, the composition of the original organic matter, and probably the catalytic effects of clay minerals. Other effects are the mixing of petroleums from different source rocks, and changing temperatures and pressures in the reservoirs during subsidence (or elevation).

When petroleum generation and migration are understood, the full process clearly will involve considerations of geology, biology, chemistry, and physics. The justification for simplifying the treatment here is largely that the mechanical aspects of compaction appear sufficient to explain the essential features of primary migration and to indicate the physical environment of the interstitial fluids buried with, and subsequently expelled from, a clay. Processes invoking clay diagenesis, as proposed by Powers (1967) and Burst (1969), accept general clay stability above depths of 6,000 to 8,500 ft (2,000–2,700 m), and their conclusions in this respect were supported by Perry and Hower (1970, p. 171). Athy's (1930a) empirical data are considered valid to about 7,000 ft (2,000 m).

[1] Manuscript received, October 18, 1971; revised and accepted, February 14, 1972.

[2] Department of Geology and Mineralogy, University of Queensland. The writer thanks his colleagues J. M. Fitz-Gerald and D. C. Green for critical reading of the manuscript.

In the many disciplines involved, the writer notes with regret his ignorance, and so restricts himself as far as possible to geologic processes.

Primary Migration

Compaction

The compaction of clay rocks usually is thought of in terms of one of the clay compaction curves (Hedberg, 1926, 1936; Athy, 1930a; Dickinson, 1953). These compaction curves indicate the present distribution of clay densities with depth. They have been interpreted as representing the compaction history of a clay during burial. Perhaps because they are quantifiable, these curves, on occasion, have been seriously abused. It is readily forgotten that they summarize widely scattered data, and it has even been forgotten that they relate to clays only, for they have been used to represent densities of overburden that includes important proportions of other lithologies.

From such curves it is apparent that the rate of compaction of clays decreases with depth, thus it is easy to conclude that the rate of fluid expulsion from a clay during burial decreases with depth. In other words, more water apparently is expelled during the first thousand feet (or meters) of burial than during the second, more during the second than the third, and so on until the pore space virtually is eliminated.

Largely for this reason primary migration due to compaction has been seen as an inefficient process; most of the transporting medium for petroleum apparently is lost at shallow depth, and the petroleum is irrevocably lost unless a trap exists at this early stage.

The conclusion that petroleum was generated and began primary migration early has been supported by research into recent sediments (Smith, 1952, 1954; Emery and Rittenberg, 1952; Kidwell and Hunt, 1958). This conclusion is not disputed here as a statement of fact, but the matter of emphasis requires clarification.

The work of Dickinson (1953), Hubbert and Rubey (1959), and Rubey and Hubbert (1959) has shown that clay rocks will be abnormally pressured under load if the hydraulic conductivity is so low that the rate of expulsion of liquids is insufficient for the maintenance of stress equilibrium between liquids and solids in the rock. Bredehoeft and Hanshaw (1968) showed that such pressures can be created and maintained in thick sedimentary sequences during significant spans of geologic time. Chapman (1972b) interpreted the geologic and physical data of abnormally pressured clays to mean that such clays have been abnormally pressured since burial to shallow depths, and suggested that the thicker clays in most sedimentary basins have been abnormally pressured to some extent during their burial.

A clay unit compacts from the outer surfaces towards the center, by virtue of the fluid potential gradient[3] induced by the processes of compaction. This compacting outer layer is not only the site of decreasing fluid potential toward the outer surface, but also of decreasing porosity and permeability. It therefore tends to seal the interstitial fluids in the clay. In general the mean fluid potential gradient between the center of a compacting clay and one of its outer surfaces will be less in a thick clay than in a thin clay under the same load. The rate of compaction of a clay unit is partly a function of the load, partly a function of its hydraulic conductivity (and that of the rocks to the surface), and partly a function of its thickness (see Appendix). A thick clay will compact more slowly per unit thickness than a thin clay under identical loads if both have equal initial amounts of interstitial liquid per unit volume; and retarded average compaction implies retarded average liquid expulsion.

Fluid Expulsion

Let us accept Athy's clay compaction curve as representing the compaction history of a thin clay that maintains approximate stress equilibrium between liquids and solids during burial. Rubey and Hubbert (1959, p. 175, 176) developed functional relations between depth, porosity, and fluid pressure based on Athy's empirical curve.

It can be shown that Rubey and Hubbert's equation 17 (p. 176) can be written

$$f = f_o e^{-c\delta z}, \qquad (1)$$

where f is the porosity at depth z, f_o the initial porosity with the numerical value of 0.48 (from Athy's data), c a factor with the numerical value of 4.33×10^{-4} per ft (1.42×10^{-3} per m), and e the base of Napierian logarithms. δ is a dimensionless measure of fluid expulsion defined by Chapman (1972b) as $(1-\lambda)/(1-\lambda_e)$, where λ is the ratio of fluid pressure to geostatic pressure (Hubbert and Rubey, 1959, p. 142) and λ_e is the equilibrium value of λ. The product δz is the equilibrium depth, z_e, of Rubey and Hubbert (1959, p. 175).

Under the special conditions of stress equilibrium being maintained between liquids and

[3] A fluid potential gradient exists between two points in the subsurface at depths z_1 and z_2 when the pressure difference between them is *not* equal to $\gamma_\omega(z_1 - z_2)$, where γ_ω is the specific weight of the pore fluid. Fluid will tend to move down the potential gradient in the direction that tends to restore hydrostatic equilibrium.

solids in the clay (*i.e.*, maintenance of normal hydrostatic liquid pressures) $\delta = 1$, and the equation reduces to

$$f = f_o e^{-cz}, \tag{2}$$

which is Rubey and Hubbert's equation 15 (p. 176).

The volume of liquid expelled from a rock during compaction is equal to the difference between the initial bulk volume of the rock and its compacted bulk volume (assuming that the volume of solids remains constant). It may be regarded alternatively as the difference between the initial pore volume and the "compacted" pore volume.

If we consider unit initial *bulk* volume of clay, the initial porosity, f_o, is a measure of the initial liquid volume. The corresponding measure of solid volume is therefore $(1-f_o)$. On compaction, the void ratio, $f/(1-f)$, decreases. The ratio of liquid to solid decreases, but the actual volume of solids is considered to remain constant. It can be shown that when f_o is the measure of initial fluid volume, $f(1-f_o)/(1-f)$ is the corresponding measure of fluid volume on compaction to porosity f. The expulsion of liquid during compaction of a clay from an initial porosity of f_0 to a porosity of f is therefore proportional to

$$\hat{q} = f_o - \frac{f}{1-f}(1-f_o) = \frac{f_o - f}{1-f}, \tag{3}$$

where $\hat{q}$ is a measure of the liquid expelled from unit initial volume of clay.

Substituting Equation 1 into Equation 3 and re-arranging, we obtain

$$\hat{q} = \frac{f_o(1 - e^{-c\delta z})}{1 - f_o e^{-c\delta z}} \cdot \tag{4}$$

Hence the ratio of liquid expelled to initial interstitial liquid is

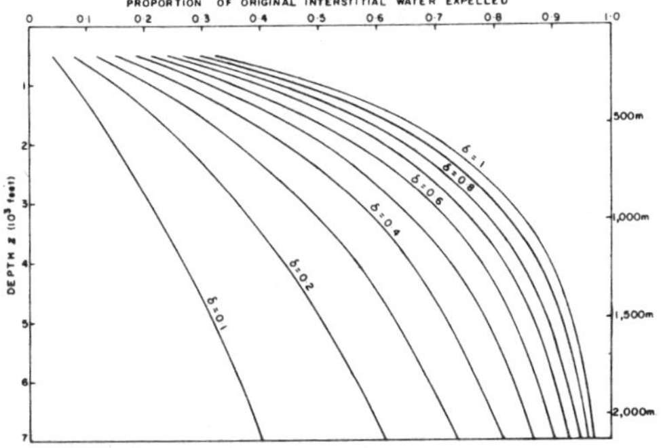

FIG. 2—Relations between proportion of initial interstitial liquid expelled, depth, and δ.

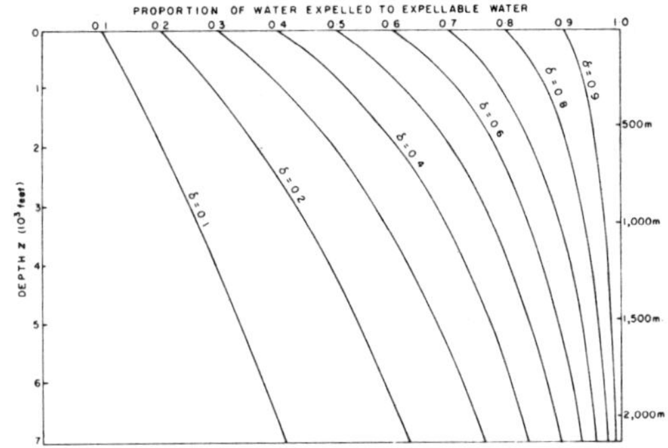

FIG. 3—Ratio of liquid expelled to liquid that would have been expelled under equilibrium conditions, and its relations with depth and δ.

$$\hat{q}/f_o = \frac{1 - e^{-c\delta z}}{1 - f_o e^{-c\delta z}} \cdot \tag{5}$$

This ratio is plotted in Figure 2 for various values of δ over the depth range of Athy's clay compaction curve. It shows that at any given depth z, the higher the interstitial liquid pressure, the higher the proportion of original liquid retained in the clay's pore space.

Also of interest is the ratio of liquid expelled during burial to depth z to *liquid that would have been expelled had the clay been in stress equilibrium* at depth z, *i.e.*

$$\frac{\hat{q}}{\hat{q}_e} = \frac{f_o(1 - e^{-c\delta z})/(1 - f_o e^{-c\delta z})}{f_o(1 - e^{-cz})/(1 - f_o e^{-cz})} = \frac{(1 - e^{-c\delta z})(1 - f_o e^{-cz})}{(1 - e^{-cz})(1 - f_o e^{-c\delta z})} \cdot \tag{6}$$

This ratio is plotted in Figure 3. It takes the value of unity when $\delta = 1$, and the liquid pressures are normal hydrostatic. It approaches zero as δ approaches zero, and the liquid pressures approach the geostatic. It approaches the value of δ as z approaches zero; in other words, δ is only a *linear* measure of fluid expulsion at very shallow depths.

If Figure 3 is a reasonable approximation to reality, it shows that an abnormally pressured clay at 7,000 ft (2,300 m) with a mean δ-value of 0.2 ($\lambda = 0.9$) has expelled about 60 percent of its liquids, but nearly 40 percent remains.

To gain some idea of the quantitative effect of retarded liquid expulsion from unit volume of clay, consider the compaction of a clay by burial in such a way that δ remains constant at 0.2. This does not seem unreasonable as there are clays with this value to depths well below the range considered here. Figure 4 shows the proportion of initial interstitial liquid that has been expelled during burial to depth z, with maintenance of

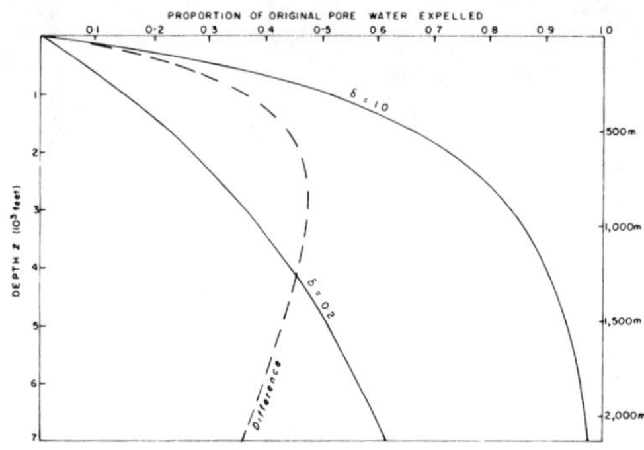

FIG. 4—Comparison between proportions of initial interstitial liquid expelled from normally pressured and abnormally pressured clays during burial.

$\delta=0.2$ and with maintenance of $\delta=1.0$ (equilibrium conditions). The difference between these two reaches a maximum at about 3,000 ft (1,000 m) where the abnormally pressured clay has an excess of initial liquid over that of the clay in equilibrium amounting to about 45 percent of the initial unit interstitial liquid volume. In oil field units, an initial acre-foot of clay with 3,700 bbl of initial liquid retains 2,300 bbl on burial to 3,000 ft with maintenance of $\delta=0.2$. This is 1,700 bbl more than that retained when $\delta=1.0$.

It must be stressed, however, that these relations are intended only to illustrate the orders of magnitude. Nevertheless, the results are consistent with Weeks' observation that most petroleum is found between depths of 2,500 and 9,000 ft (750–2,700 m), bearing in mind that burial may continue after accumulation (Weeks, 1958, p. 22).

Temperatures and Pressures

A thin clay unit that compacts under a gravitational load in such a manner that there is little loss of stress equilibrium will be subjected during burial to geothermal temperatures, and its interstitial liquids will be at pressures corresponding to the normal hydrostatic gradient—or marginally in excess of these (Fig. 5, line $\delta=1$).

A thick clay also will be subjected to geothermal temperatures, but the pressures in the pore liquids in the middle of the clay may approach the geostatic pressures, *i.e.*, about twice the normal hydrostatic pressures at a given depth (Fig. 5, line $\delta=0.2$).

Figures 4 and 5 show that a clay compacting under equilibrium conditions ($\delta=1$) retains 50 percent of its initial pore liquids at about 1,000 ft (300 m) of burial. The temperature in these liquids will be about 85° F (30° C), and the pressure will be about 460 psi (32 kg/sq cm). However, if the mean value of δ is 0.2, 50 percent of the initial pore liquids will be retained until buried to nearly 5,000 ft (1,500 m) where the temperature will be about 160° F (70° C) and the liquid pressure will be about 4,300 psi (300 kg/sq cm). In the latter case, only about 15 percent of the initial pore liquids will be expelled under the temperature and pressure conditions of the normally pressured clay.

Thus, retarded mean compaction exposes more pore liquids to higher temperatures and much higher pressures for a longer time.

EXPULSION OF PETROLEUM

Quantitative evaluation of the rate of liquid expulsion from a clay unit must await evaluation of the permeability changes in a compacting clay unit, and the way these changes change with depth and with bed thickness (see Smith, 1971, for a mathematical model of these). Nevertheless, the example described by Magara (1968) of downward migration from a compacting mudstone leaves no doubt that the process is quantitatively adequate for the expulsion of significant volumes of petroleum.

If protopetroleum is finely disseminated through a source rock, and if some of the components of this protopetroleum or the products of its diagenesis are soluble in water, the combined effects of higher temperatures and higher pressures, and

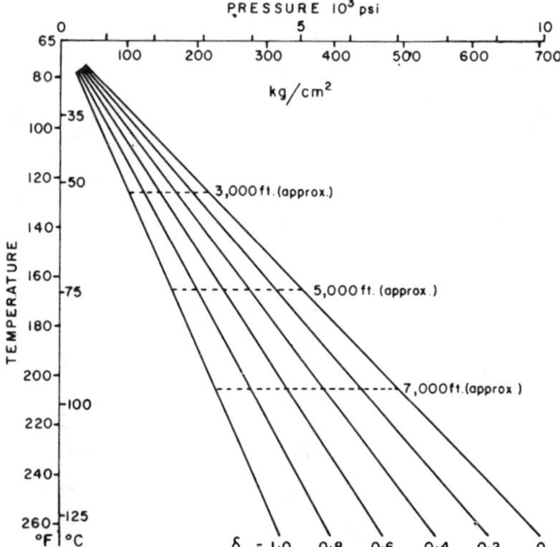

FIG. 5—Temperature-pressure field during burial and compaction. In this figure, geothermal gradient has been taken as 20°F/1,000 ft (36°C/1,000 m), the normal hydrostatic gradient as 460 psi/1,000 ft (106 kg/sq cm per 1,000 m) and λ_e as 0.46.

delayed expulsion are capable of causing a larger proportion of the protopetroleum or its derivatives to be dissolved in the pore water. These components would be expelled selectively during burial, and the pattern of expulsion and the nature of the petroleums expelled would depend on the factors already discussed.

During expulsion, however, the solution moves essentially isothermally down a fluid potential gradient from abnormal towards normal hydrostatic pressures in the outer, compacting, zone of the clay. This change in pressure will have one of two effects: (1) if the water is sufficiently undersaturated with petroleum, the petroleum will be expelled in solution, or, (2) if the saturation of any or all of the components is reached during this movement, petroleum will be released from solution *before* expulsion from the clay.

In the first case, the petroleum is free to move (in solution) in the permeable bed into which it is expelled, in the hydrodynamic field present, which will be lateral in alternating sequences of compactible/permeable beds because of the downward fluid potential gradient of the base of the compacting clays (Chapman, 1972b, p. 791).

In the second case, one conclusion is clear—the effective permeability to water will be reduced in the compacting outer layer of the clay, thus tending further to seal the fluids in the clay. If the release of petroleum from aqueous solution is physically controlled, its release will take place in a limited zone of the clay determined by these physical conditions, and a slug of sufficient dimensions may accumulate in the outer layer of the clay to be expelled by capillary force aided by the fluid potential gradient (Hubbert, 1953, p. 1979).

To a geologist such a process would appear capable of producing a wide range of petroleums, the nature of which would depend on the nature of the organic matter in the source rock, the thickness of the clay unit that is (or contains) the source rock, the catalytic properties of the clay, the time of expulsion, and the temperature and pressure conditions up to, and during, expulsion. Further selectivity may be found in the preferential expulsion of more saline pore water from clays (von Engelhardt and Gaida, 1963).

Therefore there is considered to be no serious conflict between the evidence of early migration and accumulation in the Orinoco delta (Kidwell and Hunt, 1958) for example, and late in the Los Angeles basin (Philippi, 1965). The important distinction concerns the time of bulk migration, and the conditions prevailing in the interstitial fluids at that time. These factors cannot be known for recent sediments. There is also no *need* to postulate a "flushing" of petroleum (Powers, 1967; Burst, 1969), although this may well be contributory. Indeed, the interstitial fluid pressure in a clay probably affects clay diagenesis; but abnormal fluid pressures may, as Burst suggested, inhibit the release of layered water.

Petroleum seemingly cannot be expelled in quantity from a clay source rock if the diagenesis of most of the organic matter takes place after the expulsion of the bulk of the interstitial liquid. If a source clay is buried in such a manner that stress equilibrium is maintained between liquids and solids, most of the liquid is expelled above depths of 2,000–3,000 ft (700–1,000 m), and the petroleums expelled with it are those generated under these conditions of temperature and pressure. Subsequent deeper burial may lead to further diagenesis of the organic matter or protopetroleum, the products of which will remain largely in the clay. Some of the so-called "bituminous shales" may have been formed in this manner.

Conclusions

1. Significant petroleum accumulations in regressive sequences probably are to be related to thicker source clays that now are nearly in stress equilibrium, with near-normal interstitial fluid pressures. Such source rocks would be expected to show little petroleum that is soluble in water. Unusually prolific source rocks may expel significant quantities of petroleum before stress equilibrium is reached.

2. There is a need to establish a clay compaction curve based on thin clay units and the tops (and bottoms, where appropriate) of thicker units, in order to confirm or disprove the common assumption that Hedberg's, Athy's, and Dickinson's curves differ only in the dimension of time.

Log evaluation of pressures similarly should be referred to a base line drawn on thin units and tops and bottoms of thicker units, not on the average of all points.

3. There is a need to establish for clay units under a gravity load the effect of thickness on fluid expulsion, the permeability patterns across the beds, and the volumetric relations between known reservoirs and their presumed source clays.

Appendix

Specific Storage of Clays

Bredehoeft and Hanshaw (1968) listed *specific storage* of clay layers as a parameter in the creation and maintenance of abnormal interstitial fluid pressures of secondary importance after hydraulic conductivity. In their discussion (p. 1100) they stated

$$S_s = \rho_\omega g \left(\frac{1}{E_s} + \frac{n}{E_\omega}\right),$$

where S_s is specific storage;
$\rho_\omega g$ is weight of unit volume of water;
$E_s = \sigma_{zz}/\epsilon_{zz}$ is modulus of compression of the rock skeleton confined *in situ* (σ_{zz} is vertical stress; ϵ_{zz} is vertical strain);
n is porosity ($=f$ in the present paper); and
E_ω is bulk modulus of elasticity of water.

The specific storage (S_s) may be regarded as the sum of two components:

$$\rho_\omega g/E_s = \rho_\omega g \epsilon_{zz}/\sigma_{zz},$$

relating to the water in storage by virtue of the compressibility of the rock skeleton; and

$$\rho_\omega g n/E_\omega,$$

relating to the water in storage by virtue of the compressibility of water.

For *clays*, σ_{zz} and n are functions of depth, fluid pressure and geostatic pressure. Through Terzaghi's relation, $\sigma_{zz} = S - p$ (where S is the vertical component of total stress, the geostatic pressure, and p is the fluid pressure; see Hubbert and Rubey, 1959, p. 115, and Rubey and Hubbert, 1959, p. 173), and $n = n_0 e^{-cbz}$ (this paper, Equation 1, writing n for f). The role of ϵ_{zz} in clays is not clear. Thus for constant ϵ_{zz} (for the sake of argument), the higher the fluid pressure at depth z, the lower σ_{zz} and the more water retained in storage by virtue of the compressibility of the rock skeleton. The higher the fluid pressure at depth z, the greater the porosity, and the more water retained by virtue of its compressibility.

Specific storage in clays seems to be a concept that may be regarded as a general statement of the problem—a variable dependent largely on load, hydraulic conductivity (also a function of porosity, presumably), and bed thickness. The specific storage of sands (elastic, permeable) is more readily understood.

These observations do not alter materially the validity of Bredehoeft and Hanshaw's conclusions unless the value for specific storage assumed in their computations is widely divergent from reality. This does not seem to be the case.

Selected References

Athy, L. F., 1930a, Density, porosity, and compaction of sedimentary rocks: Am. Assoc. Petroleum Geologists Bull., v. 14, no. 1, p. 1–24.

——— 1930b, Compaction and oil migration: Am. Assoc. Petroleum Geologists Bull., v. 14, no. 1, p. 25–35.

Bredehoeft, J. D., and B. B. Hanshaw, 1968, On the maintenance of anomalous fluid pressures—I. Thick sedimentary sequences: Geol. Soc. America Bull., v. 79, no. 9, p. 1097–1106.

Buckley, S. E., C. R. Hocott, and M. S. Taggart, Jr., 1958, Distribution of dissolved hydrocarbons in subsurface waters, *in* Habitat of oil: Am. Assoc. Petroleum Geologists, p. 850–882.

Burst, J. F., 1969, Diagenesis of Gulf Coast clayey sediments and its possible relation to petroleum migration: Am. Assoc. Petroleum Geologists Bull., v. 53, no. 1, p. 73–93.

Chapman, R. E., 1972a, Petroleum and geology: a synthesis: Australian Petroleum Exploration Assoc. Jour., v. 12, no. 1, p. 36–38.

——— 1972b, Clays with abnormal interstitial fluid pressures: Am. Assoc. Petroleum Geologists Bull., v. 56, no. 4, p. 790–795.

Conybeare, C. E. B., 1970, Solubility and mobility of petroleum under hydrodynamic conditions, Surat basin, Queensland: Geol. Soc. Australia Jour., v. 16, p. 667–681.

Dickinson, G., 1953, Geological aspects of abnormal reservoir pressures in Gulf Coast Louisiana: Am. Assoc. Petroleum Geologists Bull., v. 37, no. 2, p. 410–432.

Dott, R. H., and M. J. Reynolds, eds., 1969, Sourcebook for petroleum geology: Am. Assoc. Petroleum Geologists Mem. 5, 471 p.

Emery, K. O., and S. C. Rittenberg, 1952, Early diagenesis of California basin sediments in relation to origin of oil: Am. Assoc. Petroleum Geologists Bull., v. 36, no. 5, p. 735–806.

Gibson, R. E., 1958, The progress of consolidation in a clay layer increasing in thickness with time: Géotechnique, v. 8, p. 171–182.

Gussow, W. C., 1955, Time of migration of oil and gas: Am. Assoc. Petroleum Geologists Bull., v. 39, no. 5, p. 547–574.

Hedberg, H. D., 1926, The effect of gravitational compaction on the structure of sedimentary rocks: Am. Assoc. Petroleum Geologists Bull., v. 10, no. 11, p. 1035–1072.

——— 1936, Gravitational compaction of clays and shales: Am. Jour. Sci., 5th ser., v. 31, p. 241–287.

——— 1964, Geologic aspects of origin of petroleum: Am. Assoc. Petroleum Geologists Bull., v. 48, no. 11, p. 1755–1803.

Hubbert, M. K., 1953, Entrapment of petroleum under hydrodynamic conditions: Am. Assoc. Petroleum Geologists Bull., v. 37, no. 8, p. 1954–2026.

——— and W. W. Rubey, 1959, Role of fluid pressure in mechanics of overthrust faulting; pt. 1, mechanics of fluid-filled porous solids and its application to overthrust faulting: Geol. Soc. America Bull., v. 70, no. 2, p. 115–166.

Kidwell, A. L., and J. M. Hunt, 1958, Migration of oil in Recent sediments of Pedernales, Venezuela, *in* Habitat of oil: Am. Assoc. Petroleum Geologists, p. 790–817.

Magara, K., 1968, Compaction and migration of fluids in Miocene mudstone, Nagaoka Plain, Japan: Am. Assoc. Petroleum Geologists Bull., v. 52, no. 12, p. 2466–2501.

Perry, E., and J. Hower, 1970, Burial diagenesis in Gulf Coast pelitic sediments: Clays and Clay Minerals, v. 18, p. 165–177.

Philippi, G. T., 1965, On the depth, time and mechanism of petroleum generation: Geochim. et Cosmochim. Acta, v. 29, no. 9, p. 1021–1049.

Powers, M. C., 1967, Fluid-release mechanisms in compacting marine mudrocks and their importance in oil exploration: Am. Assoc. Petroleum Geologists Bull., v. 51, no. 7, p. 1240–1254.

Roof, J. G., and W. M. Rutherford, 1958, Rate of migration of petroleum by proposed mechanisms: Am. Assoc. Petroleum Geologists Bull., v. 42, no. 5, p. 963–980.

Rubey, W. W., 1927, The effect of gravitational compaction on the structure of sedimentary rocks; a discussion: Am. Assoc. Petroleum Geologists Bull., v. 11, no. 6, p. 621–632.

——— and M. K. Hubbert, 1959, Role of fluid pressure in mechanics of overthrust faulting, Pt. 2, Overthrust belt in geosynclinal area of western Wyoming in light of fluid-pressure hypothesis: Geol. Soc. America Bull., v. 70, no. 2, p. 167–205.

Smith, J. E., 1971, The dynamics of shale compaction and evolution of pore-fluid pressures: Internat. Assoc. Mathematical Geology Jour., v. 3, p. 239–263.

Smith, P. V., 1952, The occurrence of hydrocarbons in Recent sediments from the Gulf of Mexico: Science, v. 116, no. 3017, p. 437–439.

——— 1954, Studies on origin of petroleum—occurrence of hydrocarbons in Recent sediments: Am. Assoc. Petroleum Geologists Bull., v. 38, no. 3, p. 377–404.

von Engelhardt, W., and K. H. Gaida, 1963, Concentration changes of pore solutions during the compaction of clay sediments: Jour. Sed. Petrology, v. 33, no. 4, p. 919–930.

Weeks, L. G., 1958, Habitat of oil and some factors that control it, *in* Habitat of oil: Am. Assoc. Petroleum Geologists, p. 1–61.

Reprinted from:
BULLETIN OF THE AMERICAN ASSOCIATION OF PETROLEUM GEOLOGISTS
VOL. 57, NO. 2 (FEBRUARY, 1973), PP. 321–337, 11 FIGS., 2 TABLES

Interstitial Water Composition and Geochemistry of Deep Gulf Coast Shales and Sandstones[1]

GENE W. SCHMIDT[2]
Tulsa, Oklahoma 74102

Abstract Interstitial water from shales and sandstones shows a contrast in concentration and composition. Sidewall cores of shales were taken every 500 ft between 3,000 and 14,000 ft in a well in Calcasieu Parish, Louisiana, which encountered abnormally high fluid pressures just below 10,000 ft. Significant differences between the total dissolved solids concentrations in waters from normally pressured sandstones (600–180,000 mg/l) and highly pressured sandstone (16,000–26,000 mg/l) were noted. Shale pore water has a lower salinity than the water in the adjacent normally pressured sandstones, but the concentrations are more similar in the high pressure zone. Shale water generally has a concentration order of $SO_4^{=} > HCO_3^{-} > Cl^{-}$, whereas water in normally pressured sandstone has a reversed concentration order.

Conversion from predominantly expandable to nonexpandable clays accelerates near the top of the high pressure zone, which appears correlative with a major temperature gradient change, an increase in shale porosity (decrease in shale density), a lithology change to a massive shale, an increase in shale conductivity, an increase in fluid pressure, and a decrease in the salinity of the interstitial waters.

The data presented suggest that the clays subjected to diagenetic change release two layers of deionized water and that this released water may be responsible for the lower salinity of the water found in the high pressure section.

Introduction

In the past few years, many papers have been published relating to the occurrence, causes, and geology of normally and abnormally pressured shales in the Gulf Coast basin (Wallace, 1969; Hottman and Johnson, 1965; Dickey *et al.*, 1968; Burst, 1959, 1966; Powers, 1967; Jones, 1968; Harkins and Baugher, 1969). Very few studies have given any attention to the geochemistry and electrochemical properties of the sandstone and shale sequences, and particularly to the chemistry of the interstitial waters associated with normally and abnormally pressured sediments. Attempts have been made to determine the salinity of these subsurface waters by electric log calculations (Hottman and Johnson, 1965) and by analyzing the water-soluble constituents leached and squeezed from shale samples (Chilingar and Rieke, 1969; Hedberg, 1968; Weaver and Beck, 1969). However, no literature is available reliably describing the chemical composition of the interstitial waters of shales from deep wells. The information available about interstitial waters in the subsurface is mainly about water produced with hydrocarbons from porous sandstones. It has been reported that the deeper waters commonly were characterized by lower than normal salinity (Myers and VanSiclen, 1964; Timm and Maricelli, 1953)

To obtain reliable new data concerning the interstitial water in shales, sidewall cores were taken in shale sections in a well in Louisiana. These shale cores were analyzed for their content of soluble salts, the cation-exchange capacity of the clays, exchangeable ions, and clay mineralogy. Electric logs of this well and others in the area, subsurface temperature and pressure measurements, shale density measurements, and analyses of produced water provided additional information.

Shales with abnormal physical properties commonly are encountered in deep wells of the Gulf Coast. Their porosity is abnormally high considering the depth of burial. Apparently, the normal processes of compaction which require the expulsion of the interstitial water were arrested. The waters in the sandstones enclosed by the abnormal shales usually have abnormally high pressures. Waters in the normal sandstones increase in concentration of total dissolved solids with depth, but those in the abnormally pressured sandstones have abnormally low concentrations. If compaction caused the normal increase in concentration, the water, like the shale porosity, appears to be immature. Not only the concentration, but the ionic ratios of the deeper water resemble those of waters found at much shallower depths in the normal shales. A remarkable fact is that the interstitial waters in the normal shales are much less concentrated in total dissolved solids than those in the adjacent sand-

[1] Manuscript received, February 18, 1972; accepted, May 24, 1972.

[2] Staff Research Scientist, Amoco Production Company.

Amoco Production Company supported this research and gave permission for its publication. The writer is also indebted to P. A. Dickey, John Hower, and J. T. Robison for their helpful contributions during this work. This study represents part of a thesis submitted in partial fulfillment for the M.S. degree at the University of Tulsa, 1971.

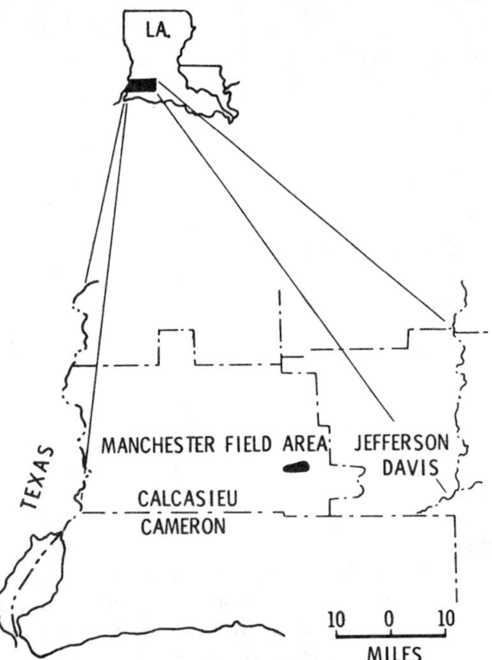

Fig. 1—Index map showing location of Manchester field area.

stones. However, the waters in the abnormally pressured sandstones have concentrations similar to those in the adjacent shales. More remarkable are the high concentrations of $SO_4^=$ and HCO_3^-, relative to Cl^-, that are found in the interstitial shale water. Mature subsurface sandstone waters generally contain low concentrations of $SO_4^=$ and HCO_3^-, relative to Cl^-.

Sidewall Core Samples

Sidewall core samples were obtained from as "pure" shale sections as possible in the interval from 3,496 to 13,970 ft in the Pan American, Farmers Land and Canal Co., Inc. A-5 well in Sec. 17, T10S, R7W, Calcasieu Parish, Louisiana. The well, a dry hole, is in the Manchester field (see Fig. 1). The 3/4-in. sidewall cores were collected every 500 ft and kept as fresh as possible by immediately wrapping each in Saran Wrap and placing it in tightly sealed Schlumberger sample bottles. At the laboratory, the outer parts of each core were trimmed away with a knife so that only a clean part of the core remained for analysis. Four parts of each sample were selected: one for the X-ray-mineralogy examination and the other three to determine the soluble salts, the exchangeable cations, and the cationic exchange capacity. All the samples were air dried at room temperature for 72 hours prior to additional handling.

Regional Stratigraphy

The rocks discussed in this paper range in age from Pleistocene to middle Oligocene; the sidewall cores were taken from the Oligocene Anahuac and Frio Formations. The Tertiary section is the product of a long-duration marine regression. The rate of sediment supply exceeded that of subsidence, resulting in progradation seaward. The Tertiary beds are mostly land-derived clastics deposited by processes similar to those which control the present deposition along the Gulf Coast. Facies of dominantly nonmarine origin grade southward into a continuous sequence of marine facies which represent progressively deeper water environments. The initial time-rock units tended to migrate southward and overlie older beds of deeper marine deposition.

The result of continuous regression throughout the Tertiary was the building of a sedimentary column of interbedded sandstone and shale of continental and shallow-marine origin which overlies thick, massive, deep-water marine shale. The upper part of the section, which averages 10,000 ft in thickness in the Manchester field study area, is referred to as the "sand-shale sequence"; the lower part is called the "major shale sequence." Distinction between these two major sedimentary sequences is important to an understanding of the geochemical properties of the sandstones and shales.

High-Pressure Zone

Abnormally high formation pressures are common in the post-Cretaceous sediments of Calcasieu Parish, Louisiana (Dickey *et al.*, 1968), and particularly in the Manchester field area. The abnormal pressures in the Manchester area appear to be correlative with a thick, massive shale section with relatively low sandstone content, which is present slightly below 10,000 ft. Shale conductivities (Fig. 2) determined from the A-5 well electric log were used to make plots of a normal conductivity divided by the observed conductivity, against the fluid-pressure gradient, which in turn was used to obtain the normal and abnormal pressure plot (Fig. 3). Measured formation pressures from the A-5 and five other wells in the same field agree reasonably closely with the estimated plot.

The origin of abnormal pressures appears to be caused by compaction of rapidly deposited sediments. As sediments are deposited in a water environment, the grains are packed closer together with added weight of the overburden. As compaction continues, the water within the pore spaces is squeezed. If the water is free to move out of the shale and is in communication with the surface, the pressure of the water in the pores is hydrostatic (normal). If the water is not able to move freely from the pores as the weight of the overburden is increased, then it cannot be squeezed out. Compaction of the sediment grains will not occur, and the water in the pores will begin to assume the extra weight of the overburden. When this occurs, the formation fluid pressure is not hydrostatic but abnormally high (increases from the normal 0.465 psi/ft pressure gradient). The abnormally high pressures apparently are associated with shale deposits of large extent and thickness, not completely impervious, but with such low permeability that they largely retard fluid movement.

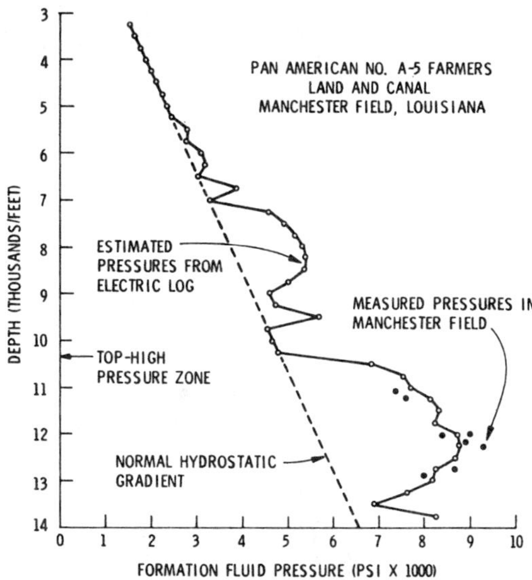

Fig. 3—Estimated formation fluid pressures from electric-log derived shale properties compared to measured pressures from Manchester field.

Density and Porosity of Shales

Most Gulf Coast shales have a systematic increase in density with depth, because shale density is a function of compaction. However, shales below a certain depth in some areas have densities that are lower than normal. Because of the limited size of sidewall-core samples available from the A-5 well, density measurements were made on shale cuttings from a well in Sec. 21, T10S, R7W, in the Manchester field, approximately 1 mi from the A-5 well. These densities were measured by a weight-differential method utilizing a pycnometer, dry and varsol-saturated weights. The varsol-saturated weights are measured and the dry densities calculated by dividing the dry weights in grams by the total volume in cubic centimeters.

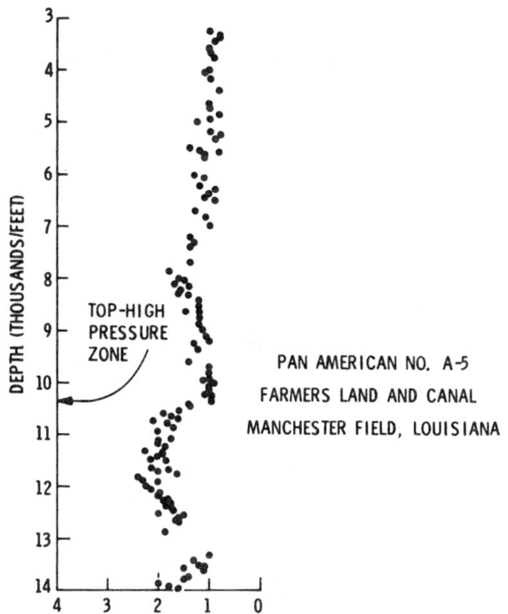

Fig. 2—Shale conductivity plot showing major conductivity increase at top of high-pressure zone.

The porosities were computed from the densities by use of a plot relating rock porosity and dry-bulk density, and a constant grain density of 2.65 g/cc. These densities and porosities plotted against depth were adjusted stratigraphically to the A-5 well (Fig. 4). The density and porosity plots were extrapolated to cover the interval from which the sidewall cores were obtained. Shale porosities determined from the A-5 well formation-density log essentially agree with those calculated from the measured densities. The small porosity differences would not alter greatly the resulting salinities of the calculated interstitial water of the shales.

For normally compacted rocks, the shale porosity ranges from about 22 percent at 2,000 ft to about 12 percent at 10,000 ft, where the major shale sequence begins. In the abnormal section the porosity shows a sudden increase to about 17 percent at 11,000 ft and then begins a decrease to about 12 percent at 14,000 ft in the A-5 well.

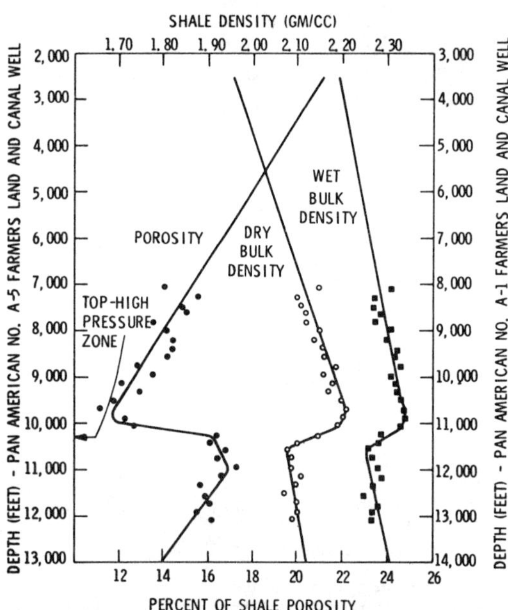

Fig. 4—Determination of shale porosity for Farmers A-5 well from shale density data from nearby Farmers A-1 well.

Lithology

Sand-shale percentage plots for the A-5 well and five other wells in the Manchester field show a thick, massive shale section which correlates with the high-pressure and low-density shale zones (Fig. 5). Near the same depth there appears to be a significant decrease in the salinity gradient, an increase in the temperature gradient, and a definite decrease in percent of expandable clays.

Increases in shale conductivities correlate with the thicker shale sections in the 10,000-13,000-ft and 8,000-9,000-ft intervals. The upper shale section is not so massive or thick as the lower section and has only a slight increase in conductivity. The close correlation between the thick shale section directly below the main sand-shale sequence, the high-pressure zone, the permeability of the rocks, the chemical changes in the interstitial waters in the shales and sandstones, and the chemical and electrochemical change of the shales is not just coincidence. It is a fundamental aspect of the interrelation of the chemical and electrochemical properties of the clay-water system.

Temperatures

More than 45 temperature measurements were obtained in and near the Manchester field. They are plotted in Figure 6. Approximately 40 of the

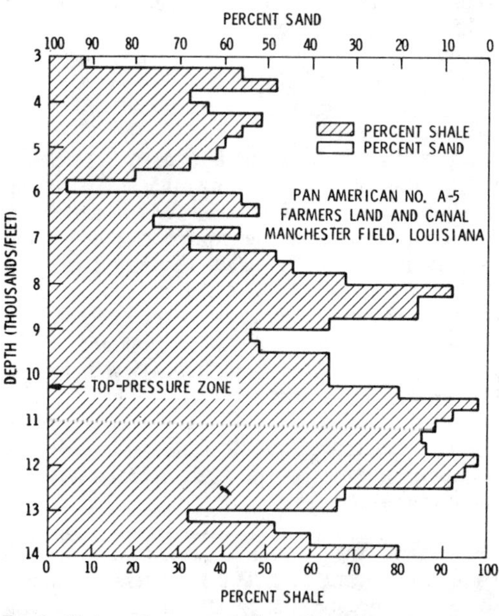

Fig. 5—Shale-sand percentage plot estimated over 250-ft intervals from SP shale base line, short normal, and induction curves of electric log.

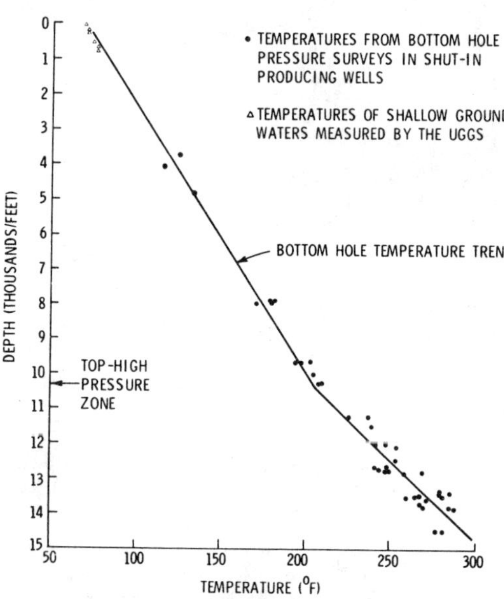

Fig. 6—Plot of measured temperatures versus depth which shows increased temperature gradient at top of high-pressure zone.

temperatures were taken during shut-in bottom-hole-pressure measurements and should be fairly representative of the earth temperatures. Temperatures of waters between 200 and 800 ft were obtained by the Louisiana Geological Survey (Harder, 1960; Hodges *et al.*, 1963).

A computer-plotted least-squares fit of the shut-in temperatures between 7,500 and 10,500 ft gives a geothermal gradient of about 1.3°F/100 ft of depth. This gradient, when extended to the surface, fits the shallow-temperature data. The lithology down to 10,000 ft is mostly alternating sandstones and shales.

The least-squares fit of the shut-in temperature between 10,500 and 14,000 ft gives a gradient of about 2.1°F per 100 ft. Over this interval the lithology is predominantly massive, under-compacted shale. The depth where the increase in gradient occurs cannot be picked, but it appears to be between 10,000 and 11,000 ft.

Lewis and Rose (1969) suggested that the change from normal to abnormally high shale porosity should cause a decrease in thermal conductivity. This would cause an increase in geothermal gradient because the overpressured zone acts as an insulator. The increase in gradient from 1.3 to 2.1°F/100 ft is undoubtedly real. It must be caused by a change in thermal conductivity and must be related to the change in lithology around 10,000 ft.

Mineralogy

During the past several years discussion regarding the diagenesis of clay minerals has concluded that mixed layering of montmorillonite and illite is caused by increasing temperature as the sediments are buried to greater depths. Burst (1966) indicated that clay compaction by interlayer-water discharge is a temperature-dependent phase change in which montmorillonite begins to dehydrate at a critical temperature in the 200-220°F range. Temperature is held to be a more effective cause of dehydration than pressure or age.

The clay mineralogy of the less than 2.0-micron fractions of the sidewall cores was determined by X-ray diffraction. Montmorillonite or mixed-layer montmorillonite-illite was detected in major amounts (more than 25 percent) in all the clay fractions analyzed; however, no montmorillonite was found below 11,000 ft, nor was any mixed-layer montmorillonite-illite found above 11,000 ft. Kaolinite is present in all the clay-fraction samples in minor amounts (25-5 percent) except in the 13,504-ft sample, where none was detected. Illite was present in trace amounts (less than 5 percent) down to 7,015 ft

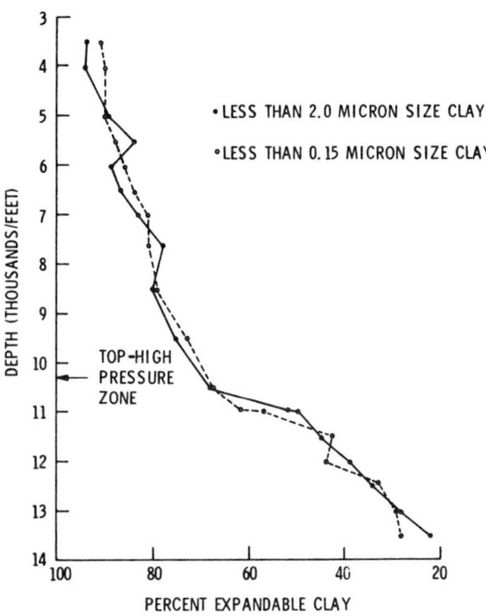

Fig. 7—Plot of percent expandable clay versus depth showing accelerated increase in diagenesis of montmorillonite to mixed layer montmorillonite-illite at top of high-pressure zone.

and in minor amounts below 7,584 ft. Quartz and calcite were present in all clay samples in trace amounts.

The X-ray-diffraction examination of the shale samples indicates that the clay mineral montmorillonite (expandable) alters to illite (nonexpandable) with depth. A plot (Fig. 7) of the percent of expandable clay (less than 2.0-micron and less than 0.15-micron fractions) versus depth clearly demonstrates that the amount of expandable clay is decreased greatly below the top of the highly pressured or abnormal shale. The change from the more random montmorillonite to the more orderly illite system occurs by a mixed-layer montmorillonite-illite sequence. The illite part increases at the expense of the montmorillonite part of the clay structure.

The data demonstrate that the increase in nonexpandable clay with depth is progressive. The accelerated increase beginning at about 10,300 ft correlates with the top of the high-pressure zone and with an increase in the temperature gradient. This accelerated transition from predominantly expandable to nonexpandable clays has a significant role in the fluid-expulsion mechanism of shales. Expandable clays contain large amounts of intracrystalline water, which is squeezed into intergranular spaces during diagenesis to a non-

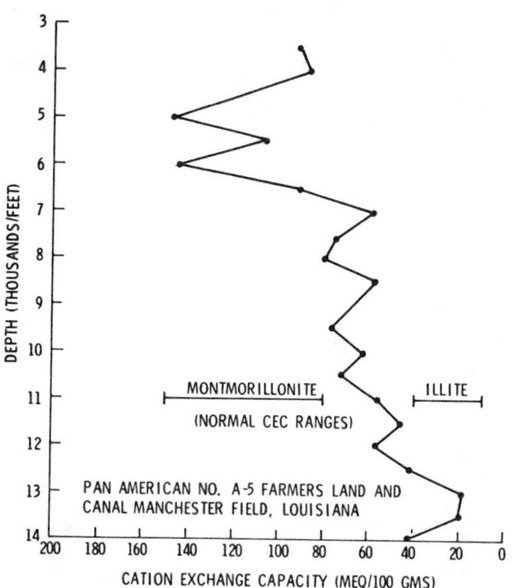

FIG. 8—Plot of variation of cation-exchange capacity (CEC) of shale with depth.

expandable state. Data from this study indicate that the mechanism allows the water to be released at a temperature around 200–220°F and to continue progressively at higher temperatures.

Exchange Capacity and Exchangeable Cations of Shales

The cation-exchange capacity (CEC) of the sidewall cores decreases with depth (Fig. 8), correlatively with the mineralogy change. Because the montmorillonite has a higher CEC than the illite clays and decreases with depth, the CEC also should decrease with depth. A plot of CEC versus the percent expandable clay shows that the CEC generally reflects the exchange capacities of the montmorillonite fraction of the shale samples. X-ray examination detected no montmorillonite in any of the shale samples below 11,000 ft. These samples all have a CEC much lower than 80 meq/100 g. This is generally the lower limit of exchange capacities for montmorillonite clays.

As the CEC increases, the shale conductivity (corrected to 100°F) generally increases. This suggests that the exchangeable cations on the clay particles may be partly responsible for the conduction of electricity. The conductivity is relatively high in the abnormal shale section where the CEC and the interstitial water concentrations are relatively low compared to the normal shale section. This may be related to the sudden increase in temperature, which could increase the mobility of the dissolved and exchangeable ions causing the higher conductivity. The increased mobility of the ions may be related to the sudden mineralogy change at the high-pressure zone.

A decrease of exchangeable Na^+ with depth follows the mineralogy change. This decrease is due to the decrease of CEC with depth which follows the mineralogy change. The exchangeable Na^+ is the major exchangeable cation found in the exchange sites of the clay. Sodium accounts for an average of almost 80 percent of the total exchangeable ions on the clays in the shales.

The exchangeable K^+ shows a trend similar to that of the exchangeable Na^+ which also is explained by the mineralogy. The exchangeable K^+ slightly increases from 3,500 to 9,500 ft and then progressively decreases to 14,000 ft. The decrease below 10,000 ft is correlative with the increase of illite. The exchangeable K^+ probably is being used up in the fixation of K^+ in interlayer position of illite.

The exchangeable Li^+ of the less than 2.0-micron clay fraction of the shale samples appears to correlate with the environment of deposition. None was detected in any of the shales above 7,000 ft. These shallower shales are mostly of continental and shallow-marine origin, so an abundance of Li^+ would not be expected. However, below 7,000 ft, where the shales are considered to be of deep-marine origin, exchangeable Li^+ was detected in all the samples.

Kinds and amounts of the various exchangeable cations are useful because they can influence markedly the physical and chemical properties of shales. A shale that contains sufficient exchangeable Na^+ and water with a low amount of soluble salts may exhibit low permeability to aqueous solutions. Likewise, a sodium clay in contact with a water containing a low concentration of soluble salts is water-sensitive. This suggests an additional mechanism for the abnormal shales having a high water volume. These explanations are only suggestions because sufficient supporting evidence is not available at this time. The exchangeable cations also may have a bearing on the shale properties being measured during electric logging and may be used as correlation, environmental, and diagenetic indicators.

Interstitial Water in Sandstones

Subsurface waters in South Louisiana have a wide range in concentration of total dissolved solids (Timm and Maricelli, 1953). According to

Dickey (1969), sandstone waters from normally compacted sediments usually increase in total dissolved solids with increasing depth at a rate of 20,000-50,000 mg/l per 1,000 ft. However, the total dissolved solid concentrations of the sandstone waters from the Manchester area (Fig. 9) increase at a rate of about 45,000 mg/l per 1,000 ft down to 3,500 ft depth, and about 5,000 mg/l per 1,000 ft from 3,500 ft to the high-pressure zone. The freshness of the waters from the surface to about 3,500 ft is probably the result of recharge by meteoric water. Waters from the abnormally pressured sandstones have very low total dissolved solid concentrations for the depth at which they are encountered. These deep fresher waters also show evidence of "immaturity," as suggested by Fajardo (1969).

The salinities of the formation waters in the sandstones of the A-5 well were calculated from the spontaneous potential (SP) curve of the electric log (Fig. 10). The calculations were confirmed by analyses of 56 water samples (Table 1) from producing zones, taken from the normally pressured zones (700-13,000 ft) in the nearby Hackberry field and from the highly pressured zones (11,000-13,000 ft) from the Manchester field area (*cf.* Figs. 9, 10). The salinity agreement between the actual water analyses and the SP-

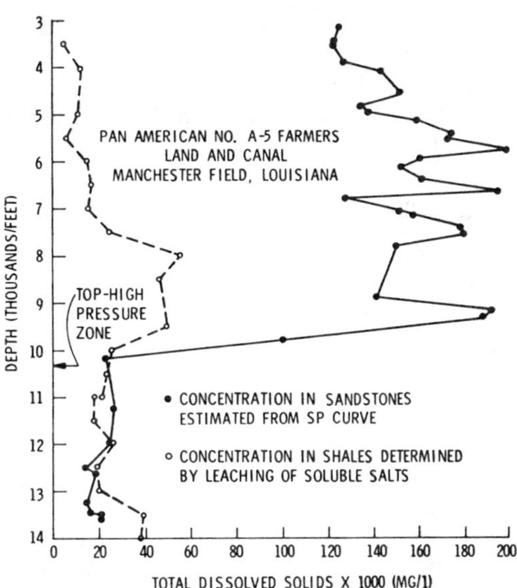

Fig. 10—Changes in concentration of interstitial water with depth in sandstones and shales. Total dissolved solids concentration in normally-pressured shale water is much less than the adjacent sand, whereas, similar concentrations are found in abnormally pressured shale and sandstone waters.

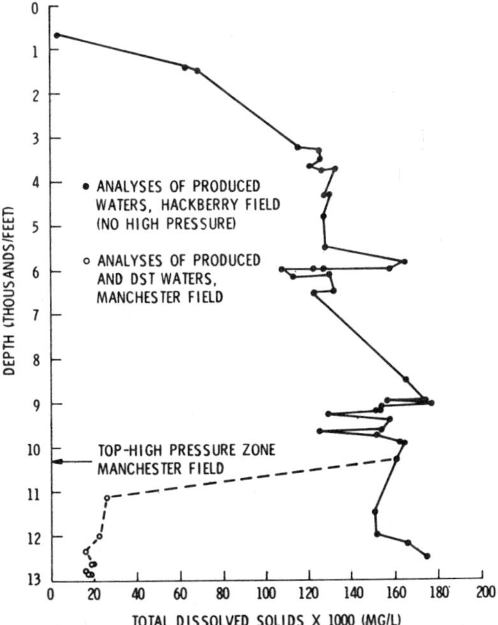

Fig. 9—Changes in total dissolved solids concentrations of produced waters from sandstones with depth, from wells where high pressure is present and not present.

calculated formation water in the sandstones is remarkably close. Water samples were taken from the Hackberry field because (1) water samples are not available from the Manchester field above 11,000 ft, (2) the Hackberry field is near the Manchester field, and (3) the lack of high pressure zones in the Hackberry field suggests that water contains abnormally low TDS (about 20,000 mg/l) only where abnormally high pressure is present.

The major cations and anions in the sandstone waters are Na^+ and Cl^-. Except for the samples from the abnormally pressured zones, sodium and chloride contents increase with depth. The rate of increase is approximately 30,000 mg/l Na^+ per 1,000 ft and 45,000 mg/l Cl^- per 1,000 ft down to 3,000 ft, and only about one-tenth of those rates between 3,000 and 10,000 ft. The Na^+ and Cl^- concentrations (about 7,000 mg/l Na^+ and 10,000 mg/l Cl^-) are approximately constant from 10,000 to 14,000 ft. The proportion of potassium in the total dissolved solids also increases with depth. The ratio of sodium to potassium is low (about 80-130) in the fresher abnormally pressured water. However, the ratio is much higher (about 150-400) in the more concentrated normally pressured water.

The ratio of sodium to magnesium remains constant (about 60) with depth until the high-pressure zone is encountered. Then Mg^{++} decreases rapidly relative to Na^{+} with the freshening of the abnormally pressured waters. A similar relation holds between the sodium and calcium. This relative increase in calcium with increasing concentration has been observed frequently (de Sitter, 1947; Dickey, 1966). The deep waters seem to be "immature."

When plotted against depth, the bicarbonate content of the normally pressured waters increases gradually from less than 100 to 300 mg/l. However, the bicarbonate concentrations of the abnormally pressured waters increase rapidly to almost 2,000 mg/l. The waters from the abnormally pressured sandstones have relatively high bicarbonate concentrations compared to shallower, normally pressured waters.

This study shows that barium is absent or very low in the waters from highly pressured sandstones. This is expected because the sulfate concentrations in the high-pressure zone are relatively high (more than 125 mg/l compared to the sulfate concentrations in the water from the normally pressured sandstones (less than 40 mg/l). The barium concentrations are inversely proportional to the sulfate concentration.

The trace ions, such as lithium and rubidium, show a progressive increase with depth to about 7.0 mg/l Li^{+} and 0.7 mg/l Rb^{+} in the normally pressured zones. They are absent or present in only small amounts in waters from abnormally highly pressured sandstones. No cesium was detected in any of the waters from the normal or the high pressure zones. Strontium (up to 6.0 mg/l) and bromine (up to 200 mg/l) increase with increasing total dissolved solid concentrations which, except for the abnormally pressured zone, also increase with depth. The iodide decreases with depth in the normally pressured waters, whereas the abnormally pressured sandstones have abnormally high iodide concentrations (20-30 mg/l) for their depth. The higher iodide concentrations tend to be associated with the less mineralized water or the most recent shale water.

Interstitial Water in Shales

An effort was made to determine the amount and composition of the dissolved solids in the water in the pores of the shales by using the sidewall cores. Although the cores were sealed in bottles at the drilling site, when they arrived in the laboratory they were partly dried. It was necessary to extract the soluble salts by leaching and then to determine the shale-pore volume to find the concentration of the original water in the shale. The composition of the water extracted by leaching depends greatly on the extraction procedure. Consequently, this procedure is described in detail.

A part of each sidewall core is cleaned carefully and oven-dried (105°C). A 4-10-g lightly crushed samples is placed in a stainless-steel centrifuge tube with 30 ml of 50 percent ethanol and distilled water. The alcohol solution is used to reduce hydrolysis, to help prevent dispersion of the clay particles, and to promote flocculation during centrifuging. This is dispersed ultrasonicly for 1 hour and then centrifuged at 16,500 RPM for 20 minutes or until the supernatant fluid is clear. The fluid is decanted and the extraction procedure repeated three more times with the extracts being combined. The solution is evaporated carefully over a steam bath until the residue is just moist. Carefully, the residue is diluted with 30-40 ml of distilled water and centrifuged at 16,500 RPM for 20 minutes. The supernatant fluid is transferred to a 50-ml volumetric flask. The cations (Li^{+}, Na^{+}, K^{+}, Rb^{+}, Cs^{+}, Mg^{++}, Ca^{++}, Sr^{++}) are determined by atomic absorption spectrophotometry. The anions (Cl^{-}, HCO_3^{-}, and $SO_4^{=}$) are determined by conventional wet methods (American Petroleum Institute, 1965).

The analytical data of the soluble salts extracted from the sidewall cores are given in Table 2. The milliequivalents per liter (meq/l) of the cations reasonably agree with the meq/l of the anions for each sample except sample 2, for which the sulfate determination was not possible. The close agreement of the meq/l of the cations and antions suggests that the cations were actually the soluble cations extracted from the solution and not cations on exchange sites of any excess clays in the solution. This indicates that very little, if any, clay material remained in the extraction solution and the extraction procedure is valid for estimating the soluble salts in shales.

The soluble cation and anion data from the shales were used to reconstitute the chemical composition of the interstitial waters originally in the shales. To do this, the bulk volume of the shale samples must be calculated using the mass (dry weight) and the dry density ($V=M/D$). The volume representing the percent porosity is divided into the extracted sample volume for a factor which is multiplied by the mg/l of each constituent.

The interstitial shale waters have the same salinity trends as the analyzed and log-calculated sandstone waters, except that the normally pressured shale waters have much lower total dissolved solid concentrations than the adjacent sandstone waters (Fig. 10). The concentration of the dissolved constituents of the shale waters in

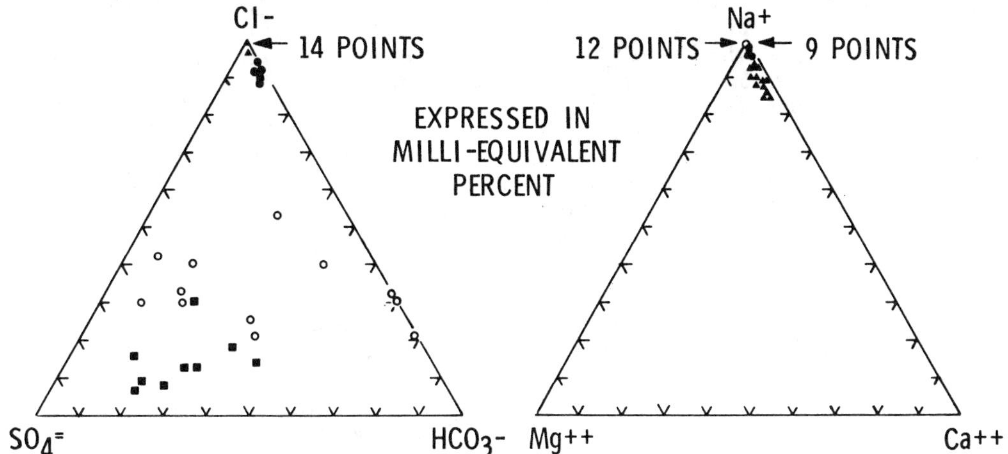

FIG. 11—Triangular coordinate plots showing distinct composition differences of interstitial waters in shales and selected sandstones.

the high-pressure zone is nearly identical with that in the high-pressure sandstones.

The Na^+ and Cl^- of the interstitial waters in the shales show trends similar to those of the waters in the adjacent sandstones, but at lower concentrations. The soluble K^+ shows a systematic increase (from 25 to 425 mg/l) with depth, which may be reflecting an increase of K^+ caused by the destruction of the feldspars. The Ca^{++} in the shale waters increases (up to 70 mg/l) downward to the abnormal shales, where the Ca^{++} concentrations are relatively constant and very low (5.0 to 15 mg/l). The sandstone waters contain as much as 500 times more Ca^{++} than the shale waters. The Mg^{++} shows a reversed relation, in that the Mg^{++} concentrations in the water of the normal shales are relatively low (about 5.0 mg/l) and constant, and have a general increase (20-30 mg/l) in the abnormal shale water. These relations are similar to those in the waters in the sandstones. No soluble Li^+ was detected in the water from shallow-water marine shale, but small amounts were detected in the waters from sediments originally deposited in deeper marine waters. No Rb^+, Cs^+, or Sr^{++} concentrations were detectable in any of the shale waters.

The SO_4^- concentrations in the interstitial waters of the shales have a general increase with depth and are unexpectedly high in the abnormally pressured shales. The SO_4^- ions increase from a trace at 3,500 ft to almost 27,000 mg/l at 14,000 ft. The soluble HCO_3^- concentrations, which are also unexpectedly high, show a less spectacular increase with depth from about 4,000 to 12,000 mg/l. In the interstitial waters of the shales the SO_4^- concentrations are generally greater than the HCO_3^- concentrations, which are generally greater than the Cl^- concentrations.

There is a very marked contrast between the composition of the interstitial waters of the shales and that of the sandstones in both the normally pressured and the abnormally pressured zones. Figure 11 demonstrates the distinct compositional difference between the sandstone and shale waters. Sodium is the only major cation in both the sandstone waters and the shale waters. The shale water generally has low salinity (low in Cl^-) and is high in HCO_3^- and SO_4^-. The normally pressured sandstones contain water of high salinity but low in HCO_3^- and SO_4^-. The abnormally high pressured sandstone waters have abnormally low salinities, with the HCO_3^- concentrations being intermediate between those in the normally pressured sandstone and shale waters.

The anion concentrations in sandstone waters ($Cl^- > HCO_3^- > SO_4^-$) are generally the reverse of those found in the shales ($SO_4^- > HCO_3^- > Cl^-$).

Weaver and Beck (1969) found that HCO_3^- is the dominant anion in the interstitial waters of shales; $SO_4^=$ is next in abundance, followed by Cl^-. White (1965) suggested that sulfate ions may be less mobile than bicarbonate, which is more mobile than chloride ions. These relative "mobilities" may suggest why more sulfate than bicarbonate, which is greater than chloride, is found in the interstitial waters of the shales. The sulfates being the least mobile are retained in the shales to the greatest degree, with bicarbonate next and chlorides last. This mobility also could explain the relative abundance of the ions in the sandstone waters ($Cl^- > HCO_3^- > SO_4^=$) as a result of the escape of the ions to the sandstone waters.

Water-Release Mechanism

During the diagenesis of expandable clay to nonexpandable clay, intracrystalline water is expelled. If the diagenetic water expelled from the shale is deionized water that mixes completely with the sandstone water, then a release mechanism may be used to explain the lower total dissolved solid concentration of the sandstone waters in the abnormally pressured zone. From data obtained in this study, the amount of released water can be calculated. Because the abnormally pressured section contains 5.0 percent sandstone with 25 percent porosity, the sandstone water occupies 1.25 percent of the total pore volume ($0.05 \times 0.25 = 0.0125$). The decrease in sandstone water concentration (175,000 mg/l total dissolved solids) from the normally pressured section (10,000 ft) to the high pressured (25,000 mg/l total dissolved solids) section (13,000 ft) is seven times. Therefore, 7.50 percent ($7 \times 1.25 = 8.75 - 1.25 = 7.50$) by volume of the water needed must come from the shale. As there is 95 percent shale, 7.9 percent by volume of the water needed must be from the shale ($7.50/0.95 = 7.9$). About 12.9 percent total water in the system must be accounted for, because of the 5.0 percent porosity change in the shale from the normal to the high pressure sections ($7.9 + 5.0 = 12.9$). Gulf Coast shales average about 80 percent clay-size (less than 5 microns) material (Hower, personal commun., 1972), of which about 75 percent is montmorillonite-illite. This study shows that 43 percent of the montmorillonite layers release water because montmorillonite decreases from 75 percent at 10,000 ft to 29 percent at 13,000 ft. Therefore, the montmorillonite volume of total shale is 25.8 percent ($0.80 \times 0.75 \times 0.43 = 25.8$). Inasmuch as the total water released in the system is 12.9 percent and the aluminum silicate layer in the clay lattice is 9.5A thick, the thickness of the water layers must be 4.8A ($12.9/25.8 \times 9.5 = 4.8$). This (4.8A) is about the thickness of two layers of water, because one water layer is between 2.25 and 2.50A. It is concluded that this released water may be responsible for the lower salinity of water found in the sandstones and shales in the high pressure zone.

Conclusions

Conclusions reached as a result of this investigation follow.

1. At depths greater than 10,300 ft in the Manchester field, the pressures of the fluids (oil, gas, and water) are abnormally high. Below 10,300 ft, the shales have abnormally high conductivity, which is typical of intervals where abnormally high pressures are found in the sandstones. The high conductivity of the shales may be related to, or caused by, the abnormally high porosity, considering the depth to which they have been buried.
2. The electric logs show a thick, massive-shale section which correlates almost perfectly with the abnormally high-pressure zone and the zone of low-density shales. The massive-shale section also correlates with the decrease in the salinity gradient and the increase in the temperature gradient.
3. The clay-mineral composition of the shales changes from mostly expandable clays (montmorillonite) at depths less than 11,000 ft to mostly nonexpandable (mixed-layer montmorillonite-illite) at depths greater than 11,000 ft. The clay-mineralogy change appears to be caused by increasing temperature as the shales are more deeply buried.
4. The temperature gradient down to 10,000 ft is about 1.3°F/100 ft; below 10,000 ft it changes to about 2.1°F/100 ft. The major temperature gradient change appears to occur at the position of the sudden increase in shale porosity, the mineralogy change, the lithology change to massive shale, and the increase in pressure.
5. The accelerated transition from predominantly expandable to nonexpandable clays occurs near the high-pressure zone and may have a significant role in the fluid-expulsion mechanism. Intracrystalline water of expandable clays may be squeezed into intergranular spaces during diagenesis to the nonexpandable state. Data from this study indicate that the mechanism apparently allows the water to be released at about 200°F and to continue progressively at higher temperatures.
6. The cation exchange capacity of the shales decreases from more than 140 meq/100 g at 5,000 ft to less than 40 meq/100 g at 14,000 ft, which is expected because montmorillonite has a much greater cation-exchange capacity than the mixed-

layer clays. The major exchangeable cation is almost entirely sodium.

7. The formation waters in the sandstones in the Manchester field increase gradually in concentration from about 120,000 mg/l total dissolved solids at 3,000 ft to more than 170,000 at 10,000 ft. The waters in the sandstones from the surface to 3,000 ft probably are affected by meteoric input water, which accounts for the fresher waters found in this interval. Below 10,000 ft the concentration decreases abruptly to around 20,000 mg/l. In the nearby Hackberry field, where there are no abnormal pressures, the concentration of the water in the sandstones continues to increase to almost 180,000 mg/l at 14,000 ft.

8. The total dissolved solids in the interstitial water of the normal shales increase gradually with depth, ranging between 10,000 and 70,000 mg/l. Shallower than 10,000 ft, the water in the adjacent sandstones has a concentration of more than 150,000 mg/l, which is much more concentrated than the shale water. At depths greater than 10,000 ft, the water in the adjacent sandstones has a concentration of about 20,000 mg/l, which is nearly the same as in the shales.

9. The water in the shales contains large amounts of sulfate (8,000-28,000 mg/l) and bicarbonate (averages 8,000 mg/l). This is very remarkable because the water in the adjacent sandstones, like most deep subsurface water, is notably low in sulfate and bicarbonate.

10. The large difference in composition between water in the sandstones ($Cl^- > HCO_3^- > SO_4^=$) and that in the adjacent shales ($SO_4^= > HCO_3^- > Cl^-$) is surprising and possibly would require changes in the current theories regarding the mechanisms for concentrating soluble salts in subsurface waters. Many speculations on the origin of deep subsurface brines generally have not considered the possibility that interstitial waters in shales might have a chemical composition different from those in the sandstones. The data seem to support a selective membrane-filtration process suggested by Billings *et al.*(1969).

11. Data from this work indicate that the amount of water released during the conversion of montmorillonite to illite is equal to two layers of water in the clay system. The released water is deionized and could be responsible for the lower salinity of water found in the high-pressure zone.

These data appear to have profound implications, although more are needed to substantiate the conclusions. The data suggest a chemical pattern within which the processes of sedimentary diagenesis take place, although it is too early to propose a new theory at this time.

References Cited

American Petroleum Institute, 1965, Recommended practice for analysis of oilfield waters: Am. Petroleum Inst., RP 45, 1st ed., p. 1-34.

Billings, G. K., B. Hitchon, and D. R. Shaw, 1969, Geochemistry and origin of formation waters in the western Canada sedimentary basin; 2, Alkali metals, *in* Geochemistry of subsurface brines: Chem. Geology, v. 4, no. 1-2, p. 211-223.

Burst, J. F., Jr., 1959, Postdiagenetic clay-mineral environmental relationships in the Gulf Coast Eocene, *in* Clays and clay minerals: Internat. Ser. Mons. Earth Sci., v. 2, p. 327-341.

——— 1966, Diagenesis of Gulf Coast clayey sediments and its possible relation to petroleum migration (abs.): Am. Assoc. Petroleum Geologists Bull., v. 50, no. 3, p. 607.

Chilingar, G. V., and H. H. Rieke, III, 1969, Some chemical alterations of subsurface waters during diagenesis, *in* Geochemistry of subsurface brines: Chem. Geology, v. 4, no. 1-2, p. 235-252.

de Sitter, L. U., 1947, Diagenesis of oil-field brines: Am. Assoc. Petroleum Geologists Bull., v. 31, no. 11, p. 2030-2040.

Dickey, P. A., 1966, Patterns of chemical composition in deep subsurface waters: Am. Assoc. Petroleum Geologists Bull., v. 50, no. 11, p. 2472-2478.

——— 1969, Increasing concentration of subsurface brines with depth, *in* Geochemistry of subsurface brines: Chem. Geology, v. 4, no. 1-2, p. 361-370.

——— C. R. Shriram, and W. R. Paine, 1968, Abnormal pressures in deep wells of southwestern Louisiana: Science, v. 160, no. 3828, p. 609-615.

Fajardo, I., 1969, A study of the connate water and clay mineralogy in southern Louisiana: Master's thesis, Tulsa Univ.

Harder, A. H., 1960, The geology and ground-water resources of Calcasieu Parish, Louisiana: U.S. Geol. Survey Water-Supply Paper 1488, p. 25.

Harkins, K. S., and J. W. Baugher, 1969, Geological significance of abnormal formation pressures: Jour. Petroleum Technology, v. 21, no. 8, p. 961-966.

Hedberg, W. H., 1968, Pore-water chlorinities of subsurface shales: Ph.D. thesis, Wisconsin Univ. Dissert. Abs., sec. B, v. 28, no. 8, abs. no. 67-12,428, p. 3338B-3339B.

Hodges, A. L., Jr., S. M. Rogers, and A. H. Harder, 1963, Gas and brackish water in fresh-water aquifers, Lake Charles area, Louisiana: Louisiana Geol. Survey, Water Resources Pamph. 13, 35 p.

Hottman, C. E., and R. K. Johnson, 1965, Estimation of formation pressures from log-derived shale properties: Jour. Petroleum Technology, v. 234, no. 6, p. 717-722.

Jones, P. H., 1968, Hydrodynamics of geopressure in the northern Gulf of Mexico bayou: Houston, Texas, Soc. Petroleum Engineers 2207, AIME, 43d Ann. Meeting.

Lewis, C. R., and S. C. Rose, 1969, A theory relating high temperatures and over-pressures: Denver, Colorado, Soc. Petroleum Engineers 2564, AIME, 44th Ann. Meeting.

Myers, R. L., and D. C. VanSiclen, 1964, Dynamic phenomena of sediment compaction in Matagorda County, Texas: Gulf Coast Assoc. Geol. Socs. Trans., v. 14, p. 241-252.

Powers, M. C., 1967, Fluid release mechanisms in compacting marine mudrocks and their importance in oil exploration: Am. Assoc. Petroleum Geologists Bull., v. 51, no. 7, p. 1240-1254.

Timm, B. C., and J. J. Maricelli, 1953, Formation waters in southwest Louisiana: Am. Assoc. Petroleum Geologists Bull., v. 37, no. 2, p. 394-409.

Wallace, W. E., 1969, Water production from abnormally pressured gas reservoirs in South Louisiana: Jour. Petroleum Technology, v. 21, August, p. 969-983.

Weaver, C. E., and K. C. Beck, 1969, Changes in the clay-water system with depth, temperature, and time: Georgia Inst. Technology Water-Resources Center, p. 1-95.

White, D. E., 1965, Saline waters of sedimentary rocks, *in* Fluids in subsurface environments: Am. Assoc. Petroleum Geologists Mem. 4, p. 342-366.

TABLE I. COMPOSITION OF SANDSTONE WATERS FROM CALCASIEU, CAMERON, JEFFERSON DAVIS, AND ACADIA PARISHES, LOUISIANA

Field	Well Name*	Location	Formation	Depth (ft.)	Date Collected	Major Ions (mg/l) Na	Ca	Mg	K	Cl	HCO_3^-	SO_4^-	CO_3	TDS	Trace Ions (mg/l) Li	Rb	Cs	Sr	Br	I	Ba
CALCASIEU PARISH																					
Wildcat	#1 W. J. Boudreau	5-10S-7W		11164-11174	11-12-68	9,280	216	47	65	14,500	1,710		0	25,800	1.9	0	0	8.3			2.9
				12028-12056	11-10-68	8,380	109	22	58	12,400	1,810		0	22,700	1.6	0	0	5.8			1.3
Manchester	#2 Farmers Ld and Canal A	17-10S-7W	Hackberry (12,300' Sand)	12346-12353	2-5-69	5,660	57	13	55	8,200	1,520	175	0	15,700	2.0	0		3.1	26	22	0
W. Manchester	#3 Lake Charles Sugar Co.	13-10S-8W	Hackberry (12,200' Sand)	12396-12404	2-5-69	17,300	728	112	208	29,600	854	183	0	49,000	5.8	0		42	43	28	0
Manchester	#4 Farmers Ld and Canal A	17-10S-7W	Hackberry (12,725' Sand)	12750-12756	2-5-69	5,640	68	15	81	8,450	1,430	232	0	16,000	2.9	0		4.4	29	22	0
Manchester	#1 W. H. McBurney	16-10S-7W	Hackberry (12,650' Sand)	12670-12677	11-30-66	6,700	125	20	84	9,850	1,930	215	0	18,900	3.5	0.30	0	8.8	43	24	1.1
				12670-12677	2-5-69	6,580	138	18	86	9,950	1,330	175	0	18,300	3.1	0		5.3	33	27	0
Manchester	#2 McBurney	16-10S-7W	Hackberry (12,800' Sand)	12862-12870	6-20-67	6,400	158	23	89	9,700	1,710	170	0	18,300	3.6	0.30	0	11	27	20	1.7
				12862-12870	2-5-69	6,330	117	21	89	9,800	1,270	128	0	17,700	3.2	0		8.1	32	25	0
CAMERON PARISH																					
W. Hackberry	School Board water well	16-12S-10W	Shallow fresh water sand	660	6-18-42	131	41	17		142	205	11	48	595							
	Salt water supply well			1400	12-17-64	21,700	1,660	820		38,700	183	0	0	63,100							
	#1-SW School Board			1470-1500	2-7-45	23,800	1,840	664		41,800	207	20	0	68,400							
E. Hackberry	#28 State Land	24-12S-10W	Miocene	2558	10-15-43	53,500	2,900	610		88,200	219	1,560	0	147,000							
W. Hackberry	#2 Mary Duhon	22-12S-10W	Amphistigina	3189-3235	9-23-48	42,400	2,560	834		72,300	92	38		118,000							
	#8 Mary Duhon	22-12S-10W	Miocene	3300	6-18-42	46,000	2,460	1,060		78,400	61	80	0	128,000							
Hackberry	#5 J. R. Watkins	23-12S-10W		3500		44,900	1,540	658	114	73,900	19	0	0	121,000	1.5	0.11	0	135	79	9.1	70
W. Hackberry	#7 M. Duhon	22-12S-10W	Miocene	3696	6-18-42	43,300	2,620	966		74,100	61	62	0	121,000							
				3696	10-7-43	47,900	3,140	724		81,400	213	135	0	133,000							
			Miocene	3800	8-20-43	47,000	2,210	358		77,300	195	17	0	127,000							
	#129 Gulf Land A R/AC	30-12S-10W		4336-4384	9-6-67	46,000	3,320	830		79,100	293	10	0	130,000							
	#137 Gulf Land A R/AC	30-12S-10W		4384-4394	9-13-67	45,800	3,080	830		78,400	305	9	0	128,000							
	Gulf Land AR/A "B"	14-12S-10W		5500		42,500	2,070	693	119	74,600	15	0	0	120,000	1.3	0.23	0	136	90	7.5	81
Hackberry	#22 Gulf Land "B"	14-12S-10W	Miocene	5867-5887	5-16-37	60,600	2,340	846		100,000	134	85	0	164,000							
				5867-5887	5-21-37	61,100	2,370	809		101,000	110	44	0	165,000							
	#"B"-5 Gulf Land R/AC	14-12S-10W		5900	6- -36	46,400	9,010	870		89,900	177	51	0	146,000							
	#18 School Land "A"	16-12S-10W	Miocene	5917-5926	5-12-37	58,700	2,530	809		97,200	134	116	0	159,000							
				5918-5926	5-21-37	58,500	2,440	834		96,800	134	41	0	159,000							

*All wells operated by Amoco Production Company

Table 1. Continued

Field	Well Name*	Location	Formation	Depth (ft.)	Date Collected	Major Ions (mg/l) Na	Ca	Mg	K	Cl	HCO_3^-	SO_4^-	CO_3	TDS	Trace Ions (mg/l) Li	Rb	Cs	Sr	Br	I	Ba
CAMERON PARISH (Cont'd)																					
E. Hackberry	#10 Gulf Land A R/A B	14-12S-10W		5959-5967	9-6-67	44,100	2,460	781		74,500	159	8	0	122,000							
Hackberry	#9 Gulf Land "B"	14-12S-10W	Miocene	5991-6013	6-10-37	39,000	2,300	773		66,500	98	Trace	0	109,000							
	Gulf Land AR/AB #111-D		Amphistigina	6000	4-18-69	46,100	2,420	830	142	78,000	129	0	0	128,000							
W. Hackberry	#28 Gulf Land "A"	30-12S-10W	Discorbis	6119-6129	7-29-40	47,700	2,700	501		79,100	232	759	0	131,000							
Hackberry	#21 Gulf Land "B"	14-12S-10W	Miocene	6140-6151	6-12-37	41,000	2,430	713		69,500	128	Trace	0	114,000							
	#96-D Gulf Land AR/A "B"			6500		44,900	2,410	723	124	77,700	154	13	0	126,000	1.3	0.23	0	141	90	9.1	73
W. Hackberry	#107-D State Lease #42	17-12S-10W	Camerina "C"	6524-6534	1-5-65	44,000	2,720	756		74,800	117	0	0	122,000							
	#125 State Lease #42			7245-7252		78,500	7,790	1,180	280	137,100	85	8	0	225,000	2.9	0.55	0	608	274	0	63
	#136 State Lease #42		Camerina "C"	8500		57,000	6,620	1,140	335	105,400	233	0	0	171,000	4.0	0.50	0	586	221	2.7	84
	#62 State Land	17-12S-10W	Camerina "C"	8935-8955	9-22-48	58,200	7,670	1,210		107,000	207	63	0	174,000							
	#66 State Land	17-12S-10W	Camerina "C"	8940-8962	9-22-48	52,500	6,870	1,090		96,100	244	47	0	157,000							
	#75-D WH NG Cam "C" VU; S.L. 42	17-12S-10W	Camerina "C"	9050.	4-2-65	58,600	8,000	1,570		109,000	137	0	0	177,000							
	#58 State Land	16-12S-10W	Camerina "B"	9082-9110	9-22-48	52,200	6,310	978		94,300	250	49	0	154,000							
	#22 School Board "C"	16-12S-10W	Camerina "A"	9210-9220	9-23-48	51,900	5,500	1,020		92,600	329	70	0	151,000							
	#65-B State Land #42	17-12S-10W	Camerina	9234-9248	6-27-44	51,200	6,470	422		91,500	244	172	0	150,000							
	#24 School Board "C"	16-12S-10W	Camerina "B"	9285-9294	9-22-48	45,000	4,030	797		78,700	305	103	0	129,000							
	#21 School Board "C"	16-12S-10W	Camerina "B"	9291-9318	9-22-48	49,700	7,610	1,020		92,900	262	67	0	152,000							
	#66-D State	17-12S-10W	Marginulina "D"	9374	3-7-45	52,200	7,420	1,210		96,800	305	337	0	158,000							
	#58-B State Land	16-12S-10W	Camerina "B"	9619	1-2-51	52,300	6,250	1,090		94,700	281	26	0	155,000							
	#23-D School Board "C"	16-12S-10W	Camerina "D"	9648-9674	9-22-48	42,100	5,050	1,000		76,600	183	81	0	125,000							
	#65-B State	17-12S-10W	Marginulina "B"	9752	3-7-45	51,400	6,460	1,090		93,700	244	29	0	153,000							
	#24 School Board "C"	16-12S-10W	Marginulina "D"	9955	6-18-42	58,300	3,370	1,030		98,200	293	805	0	162,000							
				9955	5-12-42	59,800	3,400	544		98,900	512	848	0	164,000							
E. Hackberry	#57 Gulfland AR/A "C"			11500		49,100	5,850	903	309	92,100	286	37	0	149,000	4.5	0.60	0	569	183	5.3	23
				11674-11684		50,000	5,590	881	305	90,500	312	41	0	148,000	6.8	0.70	0	492	200	Trace	19
	#6-D C. Elender "A"	12-12S-10W		12078-12088	10-19-66	52,600	5,120	756		92,200	234	66	0	151,000							
	#6 C. Elender "A"	12-12S-10W		12236-12250	10-19-66	58,100	5,200	830		101,000	210	110	0	165,000							
	#3 C. Ellender "A"	12-12S-10W		12500		62,800	5,520	848	421	111,500	243	550	0	182,000	4.8	0.66	0	376	158	5.3	0
				12510-12515	10-19-66	52.700	4,640	756		91,500	195	130	0	150,000							

TABLE II. SOLUBLE SALTS EXTRACTED FROM SIDEWALL CORES OF SHALES

Sample	Constituents	Wt. (grams)	mg/l	meq/l	Cations/ Anions (meq/l)	Bulk Sample meq/l x $\frac{50}{1000}$ x $\frac{100}{\text{spl.wt.}}$ (meq/l)	Bulk NaCl Eq.	Reconstructed Composition of Interstitial Water in Shale mg/l based on Dry Density
1 (3,496')	Li^+	8.7520	0	0		0	0	0
	Na^+		54	2.35		1.34	54	2,970
	K^+		.53	.01		0.01	.53	29
	Rb^+		0	0		0	0	0
	Cs^+		0	0		0	0	0
	Mg^{++}		.10	.01		.01	.20	5.5
	Ca^{++}		.22	.01		.01	.21	12
	Sr^{++}		0	0		0	0	0
	Cl^-		26	.13		.42	26	1,430
	HCO_3^-		95	1.56		.89	25.65	5,230
	$SO_4^=$		Trace	Trace		Trace	Trace	Trace
	Totals				2.38/2.29		107	9,680
2 (4,000')	Li^+	6.3717	0	.01		.01	.04	5.1
	Na^+		83	3.61		2.83	83	10,600
	K^+							
	Rb^+		0	0		0	0	0
	Cs^+		0	0		0	0	0
	Mg^{++}		.27	.02		.02	.54	35
	Ca^{++}		.08	Trace		Trace	.08	10
	Sr^{++}		0	0		0	0	0
	Cl^-		16	.45		.35	16	2,050
	HCO_3^-		64	1.05		.82	17.28	8,190
	$SO_4^=$		*	*		*	118	*
	Totals				3.68/1.50			21,100
3 (5,004')	Li^+	8.8218	0	0		0	0	0
	Na^+		88	3.83		2.17	88	5,370
	K^+		.45	.01		.01	.45	27
	Rb^+		0	0		0	0	0
	Cs^+		0	0		0	0	0
	Mg^{++}		.06	.01		.01	.12	3.7
	Ca^{++}		.25	.01		.01	.24	15
	Sr^{++}		0	0		0	0	0
	Cl^-		65	1.83		1.04	65	3,970
	HCO_3^-		67	1.10		.62	18.09	4,090
	$SO_4^=$		28	.58		.33	14	1,710
	Totals				3.86/3.51		186	15,200
4 (5,494')	Li^+	8.3979	0	0		0	0	0
	Na^+		61	2.65		1.49	61	3,840
	K^+		.54	.01		.01	.54	34
	Rb^+		0	0		0	0	0
	Cs^+		0	0		0	0	0
	Mg^{++}		.07	.01		.01	.14	4.4
	Ca^{++}		.19	.01		.01	.18	12
	Sr^{++}		0	0		0	0	0
	Cl^-		16	.45		.25	16	1,010
	HCO_3^-		103	1.69		.95	27.81	6,490
	$SO_4^=$		Trace	Trace		Trace	Trace	Trace
	Totals				2.68/2.14		106	11,400
5 (6,020')	Li^+	9.0974	0	0		0	0	0
	Na^+		124	5.39		2.96	124	7,810
	K^+		.66	.02		.01	.66	42
	Rb^+		0	0		0	0	0
	Cs^+		0	0		0	0	0
	Mg^{++}		.04	Trace		Trace	.04	25
	Ca^{++}		.28	.01		.01	.28	18
	Sr^{++}		0	0		0	0	0
	Cl^-		37	1.04		.57	37	2,330
	HCO_3^-		122	2.00		1.10	32.94	7,690
	$SO_4^=$		91	1.89		1.04	45.50	5,730
	Totals				5.42/4.93		240	23,600
6 (6,499')	Li^+	9.0294	0	0		0	0	0
	Na^+		120	5.22		2.89	120	8,160
	K^+		.79	.02		.01	.79	54
	Rb^+		0	0		0	0	0
	Cs^+		0	0		0	0	0
	Mg^{++}		.05	Trace		Trace	.05	3.4
	Ca^{++}		.48	.02		.01	.48	33
	Sr^{++}		0	0		0	0	0
	Cl^-		68	1.92		1.06	68	4,620
	HCO_3^-		140	2.30		1.27	37.80	9,520
	$SO_4^=$		28	.58		.32	14	1,900
	Totals				5.26/4.80		241	24,300

*Present, but not able to determine quantitatively.

TABLE 2. Continued

Sample	Constituents	Wt. (grams)	mg/l	meq/l	Cations/ Anions (meq/l)	Bulk Sample meq/l x $\frac{50}{1000}$ x $\frac{100}{\text{sampl.wt.}}$ (meq/l)	Bulk NaCl Eq.	Reconstructed Composition of Interstitial Water in Shale mg/l based on Dry Density
7 (7,015')	Li^+	7.1405	0	0		0	0	0
	Na^+		89	3.87		2.71	89	7,830
	K^+		1.1	.03		.02	1.10	97
	Rb^+		0	0		0	0	0
	Cs^+		0	0		0	0	0
	Mg^{++}		.11	.01		.01	.22	9.7
	Ca^{++}		.35	.02		.01	.33	31
	Sr^{++}		0	0		0	0	0
	Cl^-		34	.96		.67	34	2,990
	HCO_3^-		88	1.44		1.01	23.76	7,740
	$SO_4^=$		69	1.44		1.01	34.50	6,070
	Totals				3.93/3.84		183	24,800
8 (7,584')	Li^+	9.7486	0	0		0	0	0
	Na^+		147	6.39		3.28	147	10,300
	K^+		1.2	.03		.02	1.20	84
	Rb^+		0	0		0	0	0
	Cs^+		0	0		0	0	0
	Mg^{++}		.05	Trace		Trace	.10	3.5
	Ca^{++}		.82	.04		.02	.78	57
	Sr^{++}		0	0		0	0	0
	Cl^-		94	2.65		1.36	94	6,580
	HCO_3^-		88	1.44		.74	23.76	6,160
	$SO_4^=$		187	3.89		2.00	93.5	13,090
	Totals				6.46/7.98		360	36,300
9 (8,000')	Li^+	9.5596	.03	0		0	.03	2.2
	Na^+		322	14.01		7.33	322	23,800
	K^+		1.8	.05		.03	1.80	133
	Rb^+		0	0		0	0	0
	Cs^+		0	0		0	0	0
	Mg^{++}		.07	.01		.01	.14	5.2
	Ca^{++}		.64	.03		.02	.61	47
	Sr^{++}		0	0		0	0	0
	Cl^-		234	6.60		3.45	234	17,300
	HCO_3^-		76	1.25		.65	20.52	5,620
	$SO_4^=$		362	7.53		3.94	181	26,800
	Totals				14.10/15.38		760	73,700
10 (8,504')	Li^+	8.2581	.03	Trace		Trace	.03	2.6
	Na^+		235	10.22		6.19	235	20,700
	K^+		1.6	.04		.02	1.60	141
	Rb^+		0	0		0	0	0
	Cs^+		0	0		0	0	0
	Mg^{++}		.06	.01		.01	.12	5.3
	Ca^{++}		.75	.04		.02	.71	66
	Sr^{++}		0	0		0	0	0
	Cl^-		112	3.16		1.91	112	9,860
	HCO_3^-		67	1.10		.67	18.09	5,900
	$SO_4^=$		312	6.49		3.93	156	27,500
	Totals				10.31/10.75		524	64,100
11 (9,470')	Li^+	6.9404	.04	.01		.01	.04	4.8
	Na^+		220	9.57		6.89	220	26,180
	K^+		1.5	.04		.03	1.50	179
	Rb^+		0	0		0	0	0
	Cs^+		0	0		0	0	0
	Mg^{++}		.07	.01		.01	.14	8.3
	Ca^{++}		.60	.03		.02	.57	71
	Sr^{++}		0.	0		0	0	0
	Cl^-		101	2.85		2.05	101	12,000
	HCO_3^-		76	1.25		.90	20.52	9,040
	$SO_4^=$		150	3.12		2.25	75	17,900
	Totals				9.66/7.22		419	65,400
12 (10,003')	Li^+	4.2120	.04	0		0	0	0
	Na^+		117	5.09		6.04	117	13,600
	K^+		1.4	.04		.05	1.40	162
	Rb^+		0	0		0	0	0
	Cs^+		0	0		0	0	0
	Mg^{++}		.24	.02		.02	0	28
	Ca^{++}		.50	.03		.04	.48	58
	Sr^{++}		0	0		0	.48	0
	Cl^-		55	1.55		1.84	0	6,380
	HCO_3^-		61	1.00		1.19	16.47	7,080
	$SO_4^=$		125	2.60		3.09	62.50	14,500
	Totals				5.18/5.15		198	41,800

(Table 2 continued on next page)

Table 2. Continued

Sample	Constituents	Wt. (grams)	mg/l	meq/l	Cations/ Anions (meq/l)	Bulk Sample meq/l x $\frac{50}{1000} \times \frac{100}{\text{smpl.wt.}}$ (meq/l)	Bulk NaCl Eq.	Reconstructed Composition of Interstitial Water in Shale mg/l based on Dry Density
13	Li^+	4.3716	.03	Trace		Trace	.03	4.1
(10,500')	Na^+		80	3.48		3.98	80	10,800
	K^+		1.1	.03		.03	1.10	149
	Rb^+		0	0		0	0	0
	Cs^+		0	0		0	0	0
	Mg^{++}		.16	.01		.01	.32	22
	Ca^{++}		.12	.01		.01	.11	16
	Sr^{++}		0	0		0	0	0
	Cl^-		18	.51		.58	18	2,430
	HCO_3^-		67	1.10		1.26	18.09	9,050
	$SO_4^=$		112	2.33		2.66	56	15,120
	Totals				3.51/3.94		174	37,600
14	Li^+	6.1424	.02	Trace		Trace	.02	1.9
(10,970')	Na^+		103	4.48		3.65	103	9,680
	K^+		1.1	.03		.02	1.1	103
	Rb^+		0	0		0	0	0
	Cs^+		0	0		0	0	0
	Mg^{++}		.14	.01		.01	.28	13
	Ca^{++}		.12	.01		.01	.11	11
	Sr^{++}		0	0		0	0	0
	Cl^-		52	1.47		1.20	52	4,890
	HCO_3^-		67	1.10		.90	18.09	6,300
	$SO_4^=$		112	2.32		1.89	56	10,500
	Totals				4.53/4.89		231	31,500
15	Li^+	4.3716	0	0		0	0	0
(11,000')	Na^+		59	2.57		2.94	59	7,790
	K^+		.77	.02		.02	.77	102
	Rb^+		0	0		0	0	0
	Cs^+		0	0		0	0	0
	Mg^{++}		.24	.02		.02	.48	32
	Ca^{++}		.18	.01		.01	.17	24
	Sr^{++}		0	0		0	0	0
	Cl^-		16	.45		.51	16	2,110
	HCO_3^-		67	1.10		1.26	18.09	8,840
	$SO_4^=$		94	1.96		2.24	47	12,400
	Totals				2.63/3.51		142	3,300
16	Li^+	6.9304	.01	Trace		Trace	.01	0.9
(11,500')	Na^+		106	4.65		3.33	106	9,330
	K^+		1.9	.05		.04	1.90	167
	Rb^+		0	0		0	0	0
	Cs^+		0	0		0	0	0
	Mg^{++}		.20	.02		.01	.40	18
	Ca^{++}		.11	.01		.01	.10	10
	Sr^{++}		0	0		0	0	0
	Cl^-		28	.79		.57	28	2,460
	HCO_3^-		98	1.61		1.16	26.46	8,620
	$SO_4^=$		94	1.96		1.41	47	8,270
	Totals				4.69/4.36		2.10	28,900
17	Li^+	6.8546	.02	Trace		0	.02	1.8
(12,000')	Na^+		128	5.57		4.06	128	11,650
	K^+		1.2	.03		.02	1.20	109
	Rb^+		0	0		0	0	0
	Cs^+		0	0		0	0	0
	Mg^{++}		.23	.02		.01	.46	21
	Ca^{++}		.08	Trace		Trace	.08	7.3
	Sr^{++}		0	0		0	0	0
	Cl^-		20	.56		.41	20	1,820
	HCO_3^-		79	1.30		.95	21.33	7,190
	$SO_4^=$		219	4.56		3.33	109.50	19,900
	Totals				5.62/6.42		281	40,700
18	Li^+	6.1271	0	0		0	0	0
(12,500')	Na^+		93	4.05		3.30	93	10,137
	K^+		3.2	.08		.07	3.20	349
	Rb^+		0	0		0	0	0
	Cs^+		0	0		0	0	0
	Mg^{++}		.30	.03		.02	.60	33
	Ca^+		.08	Trace		Trace	.08	8.7
	Sr^{++}		0	0		0	0	0
	Cl^-		19	.54		.44	19	2,070
	HCO_3^-		107	1.75		1.43	28.89	11,700
	$SO_4^=$		75	1.56		1.27	37.50	8,180
	Totals				4.16/3.85		182	32,400

(Table 2 continued on next page)

TABLE 2. Continued

Sample	Constituents	Wt. (grams)	mg/l	meq/l	Cations/ Anions (meq/l)	Bulk Sample meq/l x $\frac{50}{1000} \times \frac{100}{\text{smpl.wt.}}$ (meq/l)	Bulk NaCl Eq.	Reconstructed Composition of Interstitial Water in Shale mg/l based on Dry Density
19 (12,998')	Li^+	4.1952	0	0		0	0	0
	Na^+		57	2.48		2.96	57	9,520
	K^+		1.4	.04		.05	14	234
	Rb^+		0	0		0	0	0
	Cs^+		0	0		0	0	0
	Mg^{++}		.08	.01		.01	.16	13
	Ca^{++}		.36	.01		.01	.34	60
	Sr^{++}		0	0		0	0	0
	Cl^-		7.8	.22		.26	7.80	1,300
	HCO_3^-		46	.75		.89	12.42	7,680
	$SO_4^=$		92	1.91		2.28	46	15,400
	Totals				2.54/2.88		125	34,200
20 (13,504')	Li^+	2.6349	0	0		0	0	0
	Na^+		68	2.96		5.62	68	18,900
	K^+		1.5	.04		.08	1.50	417
	Rb^+		0	0		0	0	0
	Cs^+		0	0		0	0	0
	Mg^{++}		.11	.01		.02	.22	31
	Ca^{++}		.74	.04		.08	.70	206
	Sr^{++}		0	0		0	0	0
	Cl^-		7.8	.22		.42	7.8	2,170
	HCO_3^-		40	.66		1.25	10.80	11,100
	$SO_4^=$		112	2.33		4.42	56	31,100
	Totals				3.05/3.21		145	64,000
21 (13,970')	Li^+	6.5211	.02	Trace		Trace	.02	24
	Na^+		146	6.35		4.87	146	17,400
	K^+		1.4	.04		.03	1.40	167
	Rb^+		0	0		0	0	0
	Cs^+		0	0		0	0	0
	Mg^{++}		.14	.01		.01	.28	17
	Ca^{++}		.19	.01		.01	.18	23
	Sr^{++}		0	0		0	0	0
	Cl^-		38	1.07		.82	38	4,520
	HCO_3^-		64	1.05		.81	17.28	7,620
	$SO_4^=$		225	4.68		3.59	112.50	26,800
	Totals				6.41/6.80		316	64,500

Reprinted from:
BULLETIN OF THE AMERICAN ASSOCIATION OF PETROLEUM GEOLOGISTS
VOL. 57, NO. 5 (MAY, 1973), PP. 878–886, 10 FIGS.

Pressured Shale and Related Sediment Deformation: Mechanism for Development of Regional Contemporaneous Faults[1]

CLEMONT H. BRUCE[2]
Houston, Texas 77046

Abstract Regional contemporaneous faults of the Texas coastal area are formed on the seaward flanks of deeply buried linear shale masses characterized by low bulk density and high fluid pressure. From seismic data, these masses, commonly tens of miles in length, have been observed to range in size up to 25 mi in width and 10,000 ft vertically. These features, aligned subparallel with the coast, represent residual masses of undercompacted sediment between sandstone-shale depoaxes in which greater compaction has occurred. Most regional contemporaneous fault systems in the Texas coastal area consist of comparatively simple down-to-basin faults that formed during times of shoreline regression, when periods of fault development were relatively short. In cross-sectional view, faults in these systems flatten and converge at depth to planes related to fluid pressure and form the seaward flanks of underlying shale masses. Data indicate that faults formed during regressive phases of deposition were developed primarily as the result of differential compaction of adjacent sedimentary masses. These faults die out at depth near the depoaxes of the sandstone-shale sections.

Where subsidence exceeded the rate of deposition, gravitational faults developed where basinward seafloor inclination was established in the area of deposition. Some of these faults became bedding-plane type when the inclination of basinward-dipping beds equaled the critical slope angle for gravitational slide. Fault patterns developed in this manner are comparatively complex and consist of one or more gravitational faults with numerous antithetic faults and related rotational blocks.

Postdepositional faults are common on the landward flanks of deeply buried linear shale masses. Many of these faults dip seaward and intersect the underlying low-density shale at relatively steep angles.

Conclusions derived from these observations support the concept of regional contemporaneous fault development through sedimentary processes where thick masses of shale are present and where deep-seated tectonic effects are minimal.

Introduction

Regional contemporaneous faults are structural features that develop on the shoreward sides of major depocenters during deposition. They are characterized by thickening of sediment on the basinward, downthrown side, where rollover (reverse drag) is common. Carver (1968) enumerated possible causes of regional contemporaneous faults, the most significant of which are basement tectonics, deep salt or shale movement, slump across flexures, slump at the shelf edge, differential compaction, response to crustal loading, or combinations of these factors.

In southern Texas, where thick masses of Tertiary shale are present and where salt diapirism is not dominant, extensive contemporaneous fault systems are developed that extend for several miles subparalleled with the coast. Data from deep wells that have encountered low-density (high-pressure) shale, together with information gained through improved seismic methods, now provide information which indicates shale to be a dominant factor in the formation of these fault systems. A diagrammatic cross section (Fig. 1) illustrates how Tertiary sedimentary rocks are related to the underlying pre-Tertiary section in southern Texas. In this area, the Tertiary section is composed primarily of sandstone in the west, which grades into shale in a seaward direction. Contemporaneous faults, indicated by arrows, are shown to be developed where relatively abrupt changes from sandstone to shale occur. The heavy black line extended across this section to separate Tertiary sandstone above from the underlying shale illustrates a "ridge and valley" effect that becomes more pronounced in a seaward direction. These "ridges" represent residual masses of undercompacted shale between sandstone-shale depoaxes along which greater compaction has occurred. These linear shale bodies, commonly referred to as "shale masses," are the subject of this discussion.

Shale Masses

The term "shale mass" as used in this paper is the same as that defined by Musgrave and Hicks (1968), in that it is a body of shale that is not less than 500 ft in the smallest dimension. The shale

[1]Manuscript received, June 9, 1972; accepted, October 21, 1972.

[2]Mobil Oil Corporation.

The writer thanks many co-workers within Mobil Oil Corporation for helpful advice and constructive criticism. Special thanks are extended to D. R. Palmore, J. H. Parsley, and R. W. Aldrich for seismic interpretation. I thank the Gulf Coast Association of Geological Societies for permission to reprint this revised paper from GCAGS Transactions, Volume 22.

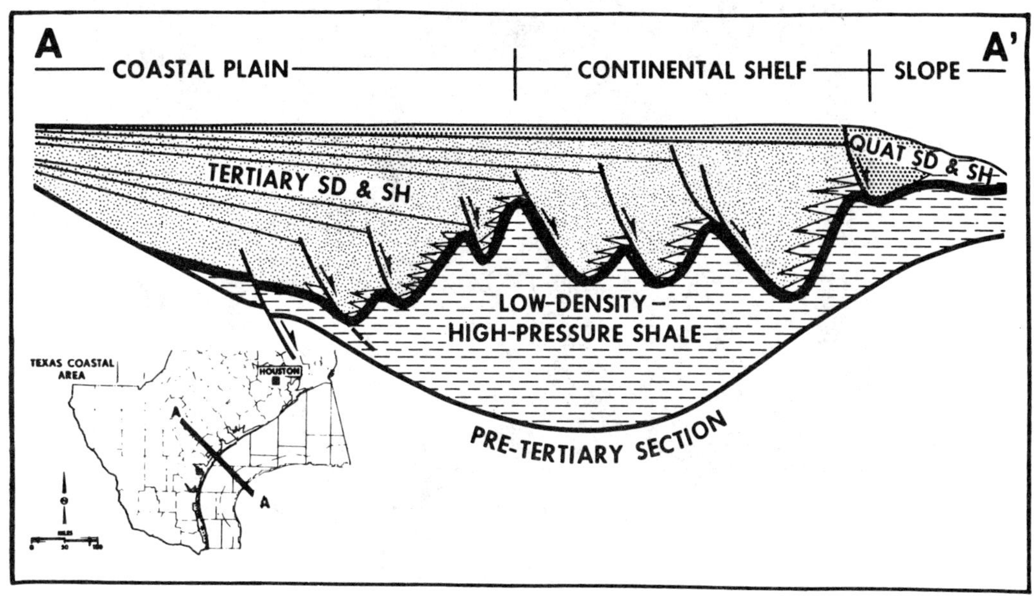

FIG. 1—Diagrammatic cross section across Texas part of northern Gulf of Mexico basin.

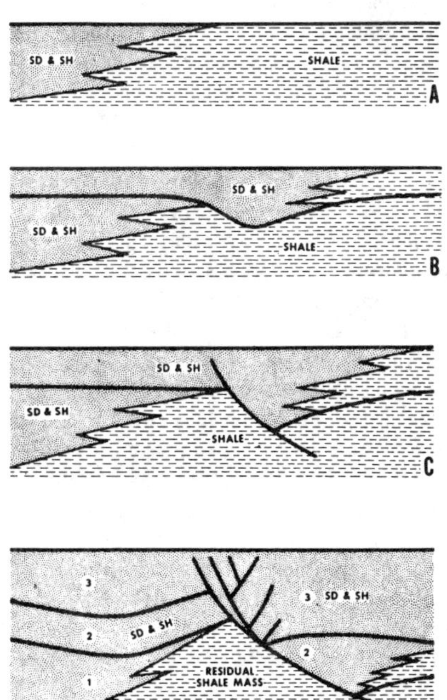

FIG. 2—Diagrammatic illustration showing four stages in development of residual shale mass.

can be in the original state of deposition or may be in any stage of deformation. The simplest forms of shale masses are flat lying and are developed by deposition (Fig. 2A). However, because of the instability of newly deposited clay materials; shale masses of this simple form are uncommon. Initial development of residual shale masses of regional extent occurs where regressive deposition of sandy sediment on a plastic clay surface results in compaction and downwarping of the unconsolidated clay under the heavier sandy section (Fig. 2B). Through continued deposition (Fig. 2C), growth faults develop and remain active as long as the depositional axis is maintained along the same line. Figure 2D shows mature development of a residual shale mass and illustrates the relation of the mass to facies on the landward side and to structure on the seaward side. These relations also are illustrated by a line of seismic data (Fig. 3) that crosses a major contemporaneous fault system where the residual shale mass is more than 10,000 ft vertically and 10 mi across.

The principal process involved in the development of residual shale masses similar to those found in southern Texas is considered to be differential compaction resulting from differential loading and shale diagenesis, both of which con-

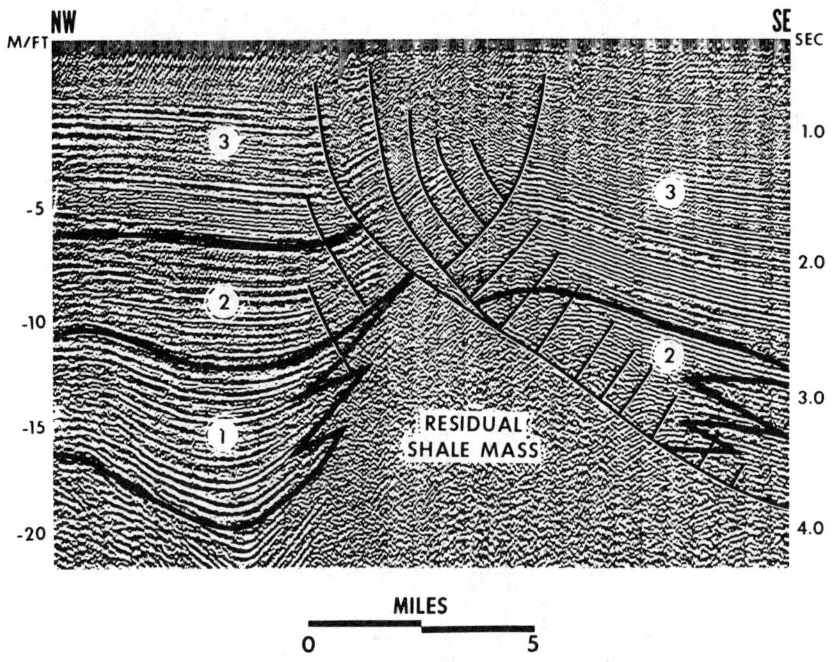

FIG. 3—Seismic illustration of residual shale mass. Numbers on either side of fault system indicate depositional sequence of sand and shale units.

trol subsurface fluid movement. In the initial stages of residual shale-mass development (Fig. 2A, B), water is free to escape from both the shale and the adjacent sandstone-shale sections. At first the water loss is greater from the shales, because of the relatively higher water content. However, as subsidence continues, water loss from the shale decreases progressively, as a result of decreasing permeability, until a critical depth is reached where orderly expulsion of water from the shale section is restricted and abnormally high pore pressure is developed within the shale mass. If subsidence continues to depths where fluid pressure within the shale approaches total overburden pressure, compaction ceases.

Some fault patterns observed on seismic records indicate that masses of this type tend to expand. It is probable, although the subject is not considered in this study, that under these conditions the shale may become diapiric locally along the linear shale mass. During the process of abnormal pressure development within the shale mass, the flanking sections of interbedded sandstones and shales continue to compact normally with water loss through the permeable sandstone layers. Water expulsion through the sandstone layers continues as long as the interlayered permeable strata are in communication with the surface.

Additional compaction occurs within the flanking sediments when the critical temperature-pressure level is reached at which montmorillonite is altered to illite. During this process, large volumes of interlayer water are transferred from the montmorillonite into the pore-water system of the host sedimentary section, as suggested by Powers (1967) and Burst (1969). This released interlayer water also can escape through permeable sandstone layers present within the sedimentary section. The amount of water in motion during this stage of clay-mineral diagenesis is considered by Burst (1969) to be 10-15 percent of the compacted bulk volume of the shale involved. This second-stage water loss does not affect the pressured shale section because permeable strata are not present within the mass to permit the escape of fluids. The extended period of water loss from the interbedded sandstone-shale section reduces the thickness of the flanking sediments, leaving them draped on the landward side of the pressured shale mass, and further accentuates the faulting that has developed on the seaward side (Fig. 2D, 3).

FAULT MECHANISMS

The processes responsible for development of regional contemporaneous faults similar to those

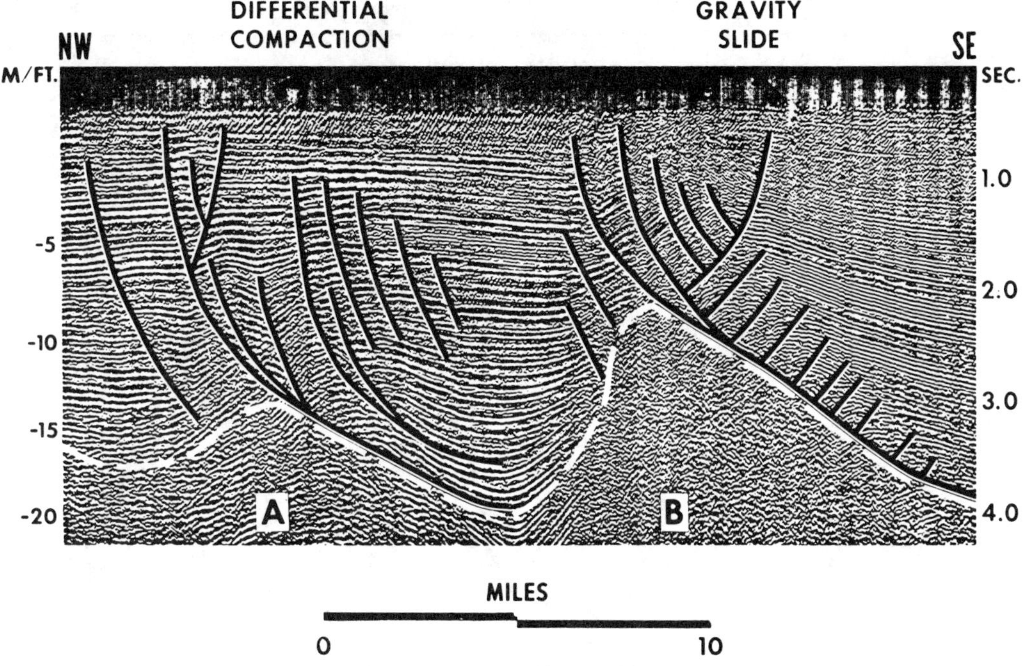

Fig. 4—Seismic illustration showing differences between fault systems formed by differential compaction and gravity slide. Dashed white line shows configuration of shale masses.

present in southern Texas are considered to be differential compaction and gravity slide. Examples are shown on a line of seismic data (Fig. 4) which includes the fault system shown in Figure 3 and a second fault system located approximately 10 mi northwest. The heavy dashed white line extending across this seismic section illustrates how these two fault systems are interpreted to be related to underlying shale masses, labeled "A" and "B." Although fault patterns within these systems are different, faults in both systems flatten and converge at depth to form the southeast (seaward) flanks of the underlying masses. Progressive flattening at depth of major basinward-dipping faults is common in Tertiary sedimentary rocks of the Texas coastal area, regardless of mechanism, and is considered to be related directly to progressive increases in subsurface fluid pressure with depth.

Where gravitational slide faults are developed, the mechanism is considered to follow in general the fluid-pressure hypothesis presented by Hubbert and Rubey (1959). In this hypothesis they propose that development of high fluid pressure greatly reduces internal friction, thus facilitating the formation of low-angle gravitational faults. Faulting associated with shale mass "B" is a typical example of faults formed through gravitational sliding. Fault systems formed in this manner are characterized by one or more major faults that are downthrown basinward and many antithetic (adjustment) faults. Together, these faults produce extremely complex systems.

Faults formed by differential compaction, in contrast, consist predominantly of normal faults that are downthrown basinward. Carver (1968) considered differential compaction faults to represent simple shear failure in sediments that became more compacted on the downthrown side. Faults formed in this manner flatten at depth and die out at near the depositional axis of the adjacent syncline. The fault system associated with shale mass "A" illustrates, in cross-sectional view, faults which die out at depth near the synclinal axis between shale masses "A" and "B." The relatively short distance between these masses confined the sandstone-shale section between them in such a manner that major faulting could occur only by dehydration and compaction. Across several shallow faults, present between the depths of 5,000 and 10,000 ft near the center of the syncline, no displacement is observed below the depth of 10,000 ft. Faults of this type are best explained by differential water loss from adjacent blocks during the process of compaction.

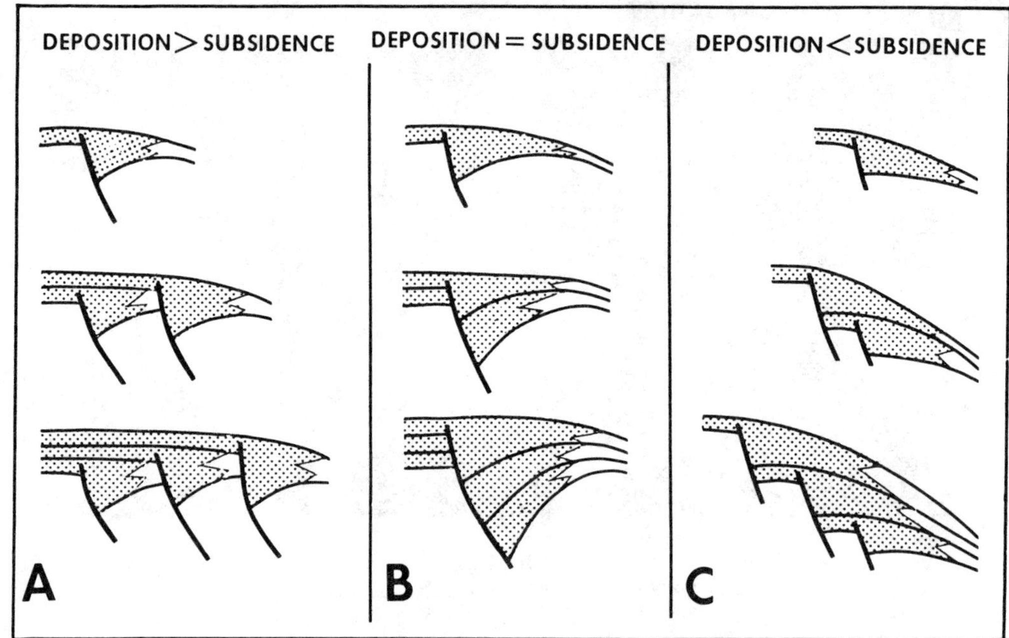

Fig. 5—Diagrammatic illustration showing development of three types of contemporaneous fault systems.

Fault Types

Tertiary deformation in southern Texas was controlled primarily by sediment overburden distribution, rather than deep-seated tectonic activity. Three basic types of regional contemporaneous faults are recognized (Fig. 5), with differentiation based on rates of deposition of sandy sediment upon unconsolidated clay surfaces. Two of these types are considered to be associated with sea floors which were relatively flat at time of deposition, and the third appears to be formed in areas of slope environments where sea floor subsidence exceeded the rate of deposition.

The first example (Fig. 5A) represents faults formed during a regressive sequence of deposition (progradation locally), when the amount of sediment available for deposition was greater than the space available for accumulation. Under these conditions each successive depoaxis was formed seaward from that of the adjacent underlying unit. Antiregional dip, developed adjacent to the downthrown sides of these faults, varies in relation to the amount of sediment deposited. In areas where still-stand depositional conditions prevailed, the rate of faulting was sufficient to accommodate all incoming sediment (Fig. 5B). In these areas, strong antiregional dip developed that increased with depth and time. Contemporaneous faults, formed during still-stand and regressive phases of deposition, are common in southern Texas and are considered to have developed primarily through differential compaction associated with relatively flat sea floors.

Faults formed during transgressive phases of deposition are present in southern Texas; however, they are less common than the other two forms. Where subsidence exceeded the rate of deposition (Fig. 5C), the sea floor is considered to have been inclined basinward at an angle related to the rate of subsidence. The primary cause of sea-floor subsidence and tilting was not dependent on differential compaction and differential loading, as described for faults formed during regressive and still-stand phases of deposition, but was controlled by forces below or outside of the area of deposition. These forces may have been related to either salt movement or basement tectonics. Other manifestations of contemporaneous faulting can be explained when sea-floor inclination and basinward formational dips are considered with rates of deposition. The most significant of these are gravity-slide faults, many of which become bedding-plane types at depth.

Faults of bedding-plane type begin with normal displacement (Fig. 6A) and become bedding-plane faults where, at depth, the dip of the fault plane becomes the same as the dip of the basinward tilted beds (Fig. 6C). After bedding-plane characteristics have been developed, all major subsequent displacement is along one or more planes parallel with the bedding. Quarles (1953) has shown that faulting of this type produces a hypothetical gap in the normal part of the fault zone above the point where the fault becomes bedding-plane type. This hypothetical gap, which is relatively wide in the upper section, does not develop in nature but is filled by collapsed material from both sides of the fault zone. An interpreted seismic section (Fig. 7) illustrates in cross-sectional view a complex fault system that became bedding-plane type. At this location the collapse zone is approximately 7 mi wide near the depth of 5,000 ft; however, at greater depths, the width of the fault zone decreases progressively, reaching the point of nonrecognition near the 15,000 ft level. Reconstructed sections (Fig. 8) prepared from these data indicate the presence of three types of faulting which account for the complicated nature of the system. A phase of faulting associated with still-stand deposition (Fig. 8A) was followed by faulting that formed during a transgressive sequence of deposition (Fig. 8B). During later deposition (Fig. 8C), sea-floor inclination and abnormal subsurface fluid pressure were sufficient to become factors in fault development. All subsequent major down-to-basin faults are shown to flatten and to converge at depth into faults of gravity-slide type. Figure 8D duplicates present structural conditions shown in Figure 7. These illustrations demonstrate, in part, why faults formed during transgressive phases of deposition are relatively difficult to recognize when compared with contemporaneous faults of the other types. Seismic and well data indicate that faults formed during periods of transgressive deposition generally flatten and converge at relatively shallow depths, and that later displacement results in development of other faults which complicate preexisting structure and make recognition of mechanisms difficult.

Faults formed on the landward sides of linear shale masses are usually postdepositional and have relatively small amounts of displacement. Generally, these faults dip basinward and show little indication of flattening at depth. They appear to enter the low-density shale at relatively steep angles. Differences in fault types, relative to landward or seaward locations on linear shale masses, are illustrated on a line of seismic data (Fig. 9) that covers a distance of approximately 30 mi and crosses a broad shale mass with a well-developed contemporaneous fault system. The configuration of the low-density shale mass, as interpreted from seismic, gravimetric, and well data, is represented on the section by a heavy dashed white line. These data indicate a mass with dimensions greater than 10,000 ft vertically and 25 mi across. Contemporaneous faulting on the southeast (seaward) side of the mass is considered to have formed during a phase of still-stand deposition with considerable thickening of sediments on the downthrown side. The angle of faulting is relatively steep above the crest of the underlying shale mass, being 50-60°, but at depth the dip of the fault plane flattens to approximately 10° where it coincides with the seaward slope of the low-density (high-pressure) shale surface. This faulting cannot be traced beyond the adjacent synclinal axis, which is shown to be at a depth of approximately 20,000 ft. On the northwest (landward) part of the shale mass the faults are mostly postdepositional and have relatively small amounts of displacement. They show little indication of flattening at depth, but appear to penetrate the low-density shale at nearly right angles. Fault patterns observed here are common to other fault systems in southern Texas where faulting occurred in regressive and still-stand phases of deposition.

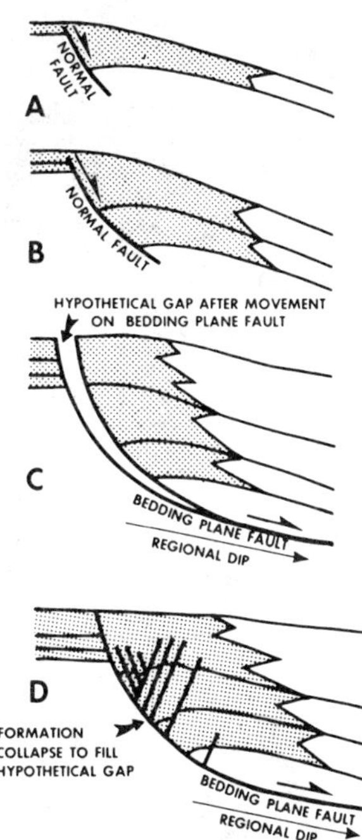

Fig. 6—Diagrammatic illustration showing development of bedding-plane fault.

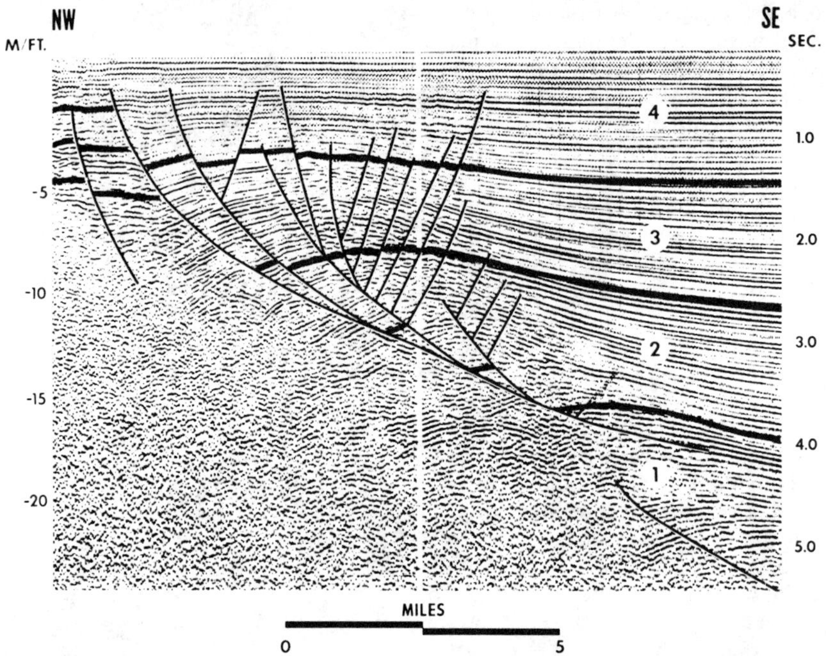

FIG. 7—Seismic illustration of combination differential compaction and bedding-plane fault system. (AQUAPULSE-Courtesy Western Geophysical Company.)

Fluid Pressure-Fault Angle Relation

Flattening at depth is normal for contemporaneous faults in the Texas coastal area. Subsurface data indicate fault flattening to be coincident with progressive increases in fluid pressure with depth. An example of the fluid pressure-fault angle relation is illustrated by a line of seismic data (Fig. 10) that crosses a major contemporaneous fault system where well information is sufficient to establish the relation down to 20,000 ft. The normal vertical pressure gradient for the Gulf Coast Tertiary sedimentary section is considered to be 1.0 psi/ft of depth. The two pressure components are 0.465 psi/ft for fluid and 0.535 psi/ft for sediment. Well data indicate fluid pressure along the line of this section to be normal down to approximately 5,000 ft. The bottom-hole fluid-pressure gradients, as determined from drilling mud weights at total depth, are recorded for each of the four tests shown on the section. These data indicate that, near the depth of 15,000 ft, the fluid pressure/overburden ratio increases to more than 0.90 psi/ft. It can also be observed that the angle of faulting reflects this pressure increase, in that a progressive change in dip occurs from 60° in the upper normally pressured section to 15° where fluid pressure reaches the higher value. Similar fluid pressure-fault angle

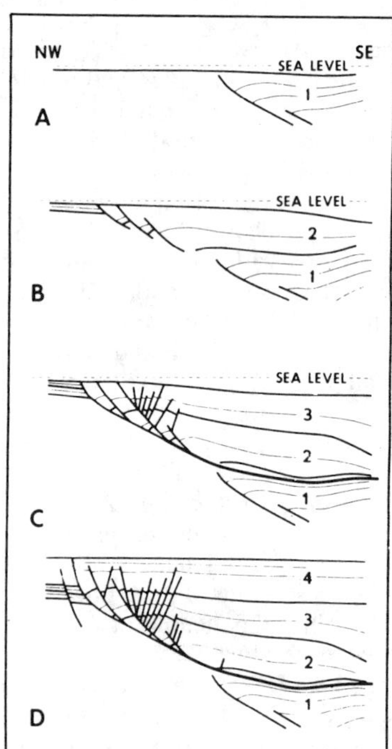

FIG. 8—Diagrammatic illustration showing development of combination differential compaction and bedding-plane fault system.

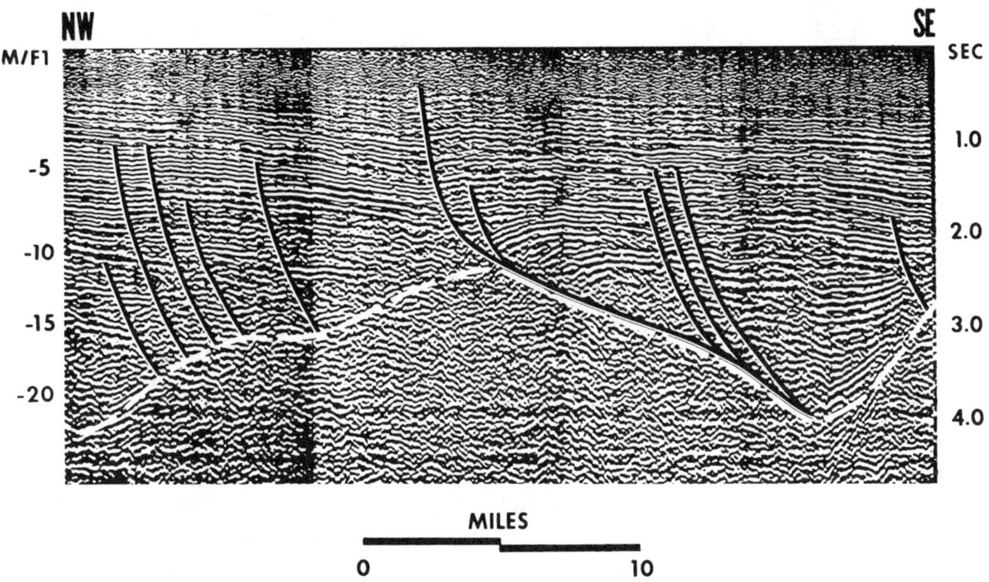

Fig. 9—Seismic illustration showing differences in faults on landward and seaward sides of shale mass. Dashed white line shows configuration of shale mass.

relations have been observed along other contemporaneous fault systems in Texas. The depth at which flattening occurs, and the degree of flattening attained, vary between areas depending on the physical and chemical properties of the shale involved. The principal factors controlling fluid pressure are the number and thickness of sandstone layers in communication with the surface and the clay-mineral composition of the shales.

Summary and Conclusions

In southern Texas, where salt diapirism is minimal, seismic data and well information indicate low-density (high-pressure) shale to be a dominant factor in the formation of regional contemporaneous fault systems. Observations concerning fault development in this area, where thick shale is dominant, are as follows:

1. Regional contemporaneous fault systems are formed on the seaward flanks of underlying shale masses, where all down-to-basin faults flatten and converge at depth. These fault systems may be formed either by differential compaction or gravitational sliding.

2. Faults formed in sediments deposited on the landward flanks of underlying shale masses are primarily postdepositional. These faults, which involve little or no sedimentary thickening, do not converge at depth but intersect the abnormally pressured shale at relatively steep angles.

3. Regional contemporaneous faults associated with regressive and still-stand phases of deposition were the dominant types formed during Tertiary time. Subsurface data indicate that fault systems formed under these conditions developed during relatively short periods of time. These fault systems, with comparatively simple patterns, appear to have formed primarily by differential compaction with some associated gravity adjustments.

4. Complex fault systems are found where gravitational sliding occurred. Faults formed in this manner are in areas of rapid subsidence, where sea-floor inclination was relatively steep (slope environments). Differential compaction also occurred within these systems.

Conclusions derived from these observations support the concept of the establishment of regional contemporaneous faulting through sedimentary processes. The physical and chemical properties of Gulf Coast Tertiary shale were such that abnormal fluid-pressure development occurred within the shale at relatively shallow depths. During periods of regressive deposition, linear masses of abnormally pressured shale developed between depoaxes of sandy sediments in which greater compaction occurred. The role of shale in

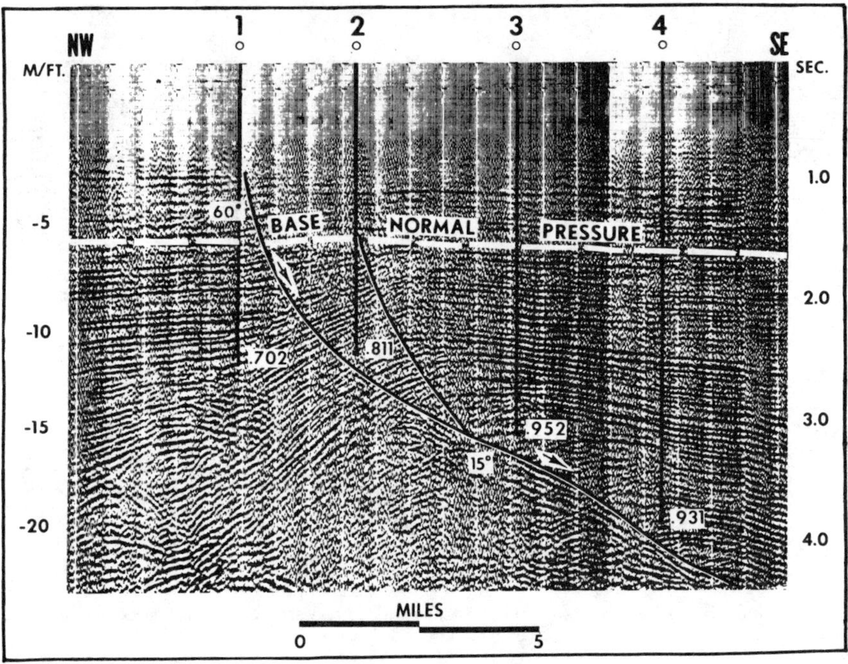

FIG. 10—Seismic illustration with well locations showing fluid pressure/overburden ratio at total depth adjacent to fault plane.

the development of regional contemporaneous faults formed in this manner is generally passive, in that structural uplift is minimal. These mechanisms for fault origin and growth satisfy the observed characteristics of regional contemporaneous fault systems found in the Texas coastal area, where thick masses of shale overlie the pre-Tertiary section and where salt diapirism is not dominant.

References Cited

Bruce, C. H., 1972, Pressured shale and related sediment deformation: a mechanism for development of regional contemporaneous faults: Gulf Coast Assoc. Geol. Socs. Trans., v. 22, p. 23-31.

Burst, J. F., 1969, Diagenesis of Gulf Coast clayey sediments and its possible relation to petroleum migration: Am. Assoc. Petroleum Geologists Bull., v. 53, no. 1, p. 73-93.

Carver, R. E., 1968, Differential compaction as a cause of regional contemporaneous faults: Am. Assoc. Petroleum Geologists Bull., v. 52, no. 3, p. 414-419.

Hubbert, M. K., and W. W. Rubey, 1959, Role of fluid pressure in mechanics of overthrust faulting. 1. Mechanics of fluid-filled porous solids and its application to overthrust faulting: Geol. Soc. America Bull., v. 70, no. 2, p. 115-166.

Musgrave, A. W., and W. G. Hicks, 1968, Outlining shale masses by geophysical methods, *in* Diapirism and diapirs: Am. Assoc. Petroleum Geologists Mem. 8, p. 122-136.

Powers, M. C., 1967, Fluid-release mechanisms in compacting marine mudrocks and their importance in oil exploration: Am. Assoc. Petroleum Geologists Bull., v. 51, no. 7, p. 1240-1254.

Quarles, M. W., Jr., 1953, Salt-ridge hypothesis on origin of Texas Gulf Coast type of faulting: Am. Assoc. Petroleum Geologists Bull., v. 37, no. 3, p. 489-508.

Reprinted from:
BULLETIN OF THE AMERICAN ASSOCIATION OF PETROLEUM GEOLOGISTS
VOL. 57, NO. 7 (JULY, 1973), PP. 1219–1249, 17 FIGS., 1 TABLE

High Fluid Potentials in California Coast Ranges and Their Tectonic Significance[1]

FREDERICK A. F. BERRY[2]
Berkeley, California 94704

Abstract Anomalous high fluid potentials exist within the miogeosynclinal Great Valley and eugeosynclinal Franciscan sequences of Jurassic-Cretaceous age within the Coast Ranges and at depth on the west side of the Central Valley, California. These rocks are dominantly mudstones with low fluid transmissibilities.

Certain problems exist as to the probable regional distribution of these high fluid potentials. Low fluid potential areas such as The Geysers geothermal district are present in the Franciscan of northern California within a region generally characterized by high fluid potentials. The low potential areas are attributed to fracture zones with channel-type flow whose transmissive characteristics exceed those of intergranular flow. It is concluded that the Franciscan of northern California probably is characterized regionally by near-lithostatic fluid pressures at depth, but fracture zones with both low (*i.e.,* near-hydrostatic) and high (*i.e.,* near-lithostatic) fluid potentials probably exist at various depths from the surface. The Geysers dry-steam occurrence is envisioned as a fracture zone with low fluid potentials by virtue of a decrease in transmissive characteristics of a fracture system with depth, in a local region of high heat flow, possibly caused by the existence at depth of a magma chamber.

An abundance of direct fluid-pressure measurements within the Great Valley section of the Sacramento Valley demonstrates the existence of high fluid potentials. The only direct fluid-pressure measurement that has been made within the Great Valley section in the central or southern San Joaquin Valley indicates high fluid potentials. The regional chemistry of the lower Tertiary waters of the San Joaquin Valley (membrane effluent type) suggests that these waters have been extruded from a widely distributed series of mudstones and other rocks that are undergoing compaction. The presumed source for this widespread compacting sequence is the underlying Great Valley sediments with their postulated high fluid potentials.

It is concluded that the anomalous high fluid potentials of Tertiary rocks within folds on the west side of the San Joaquin Valley reflect indirectly the presence at depth of high fluid potentials in the underlying Great Valley section. The origin of the folds is attributed to dynamic tectonic compression caused by current deep-seated linear diapirism of Great Valley mudstones and related rocks that possess near-perfect plastic properties by virtue of their near-lithostatic fluid pressures. The closed gravity minimum over the south end of South Dome-Lost Hills anticline is postulated as being the result of a diapir of serpentine or similar material.

It is postulated that a fault zone, named herein the "West Side" fault, probably exists at depth along the west side of the Central Valley. This buried fault is envisioned as having an intermittent near-surface expression in the form of faults such as the Midland fault, or long linear folds such as the Kettleman folds. Diapirism along this fault is presumed to be responsible for these folds.

Subsidence along the West Side fault is postulated as having occurred contemporaneously with deposition of the Great Valley sequence and thus provided a local trough in which the thick (maximum 60,000 ft) Great Valley section was deposited. The depositional barrier between the Franciscan and Great Valley sequences is postulated as a zone of serpentinite-ultrabasic rocks that intruded intermittently to form a sediment trap on the continental slope throughout Jurassic-Cretaceous geosynclinal deposition.

The final conclusion reached is that an extensive geographic zone is present in which the pore-fluid pressures of the thick Franciscan and Great Valley geosynclinal sediments reach near-lithostatic values. This zone is 400-500 mi long and 25-80 mi wide; it is bounded on the west by the San Andreas fault and the granitic Salinas block, on the east by the buried West Side fault and the granitic Sierran-Klamath block, on the south by the granitic San Emigdio-Sierran block; the northern boundary is interpreted as being the northern termination of the San Andreas fault in the Cape Mendocino region. Structural deformation of this zone by diapirism and thrusting is facilitated by the lithic plasticity caused by high fluid pressures. Known diapirism and thrusting and possible diapiric folding suggest a late Cenozoic age for the development of the high fluid potentials.

The origin of the anomalous fluid pressures adjacent to the San Andreas fault is attributed to compression between the granitic Sierran-Klamath and Salinas blocks resulting from late Cenozoic extension of the central Great Basin in Nevada and Utah. The San Andreas is a transform fault which separates the independent stress field of the Pacific plate (Salinas block) that is moving northwestward relative to the North American plate (Sierran-Klamath block and the Great Valley-Franciscan sediments). The Sierran-Klamath block also is moving westward or southwestward by continued late Cenozoic central Great Basin extension; this westerly motion is terminated by compression of

[1]Manuscript received, January .30, 1969; revised and accepted, December 4, 1972.

[2]Department of Geology and Geophysics, University of California.

Thanks are due Petroleum Research Corporation of Denver, Colorado, Occidental Petroleum Corporation, Buttes Gas and Oil Company, and the University of California for financing parts of this research. Many oil companies and a state agency have contributed data without which this study would have been impossible. I thank all those groups, and particularly Standard Oil Company of California and Union Oil Company of California. I also thank A. A. Meyerhoff, E. M. Tidwell, Clyde Wahrhaftig, Donald L. Turner, O. Frank Huffman, Samuel H. Clarke, Jr., and Alexis W. Moisseeff for critical reading of this manuscript.

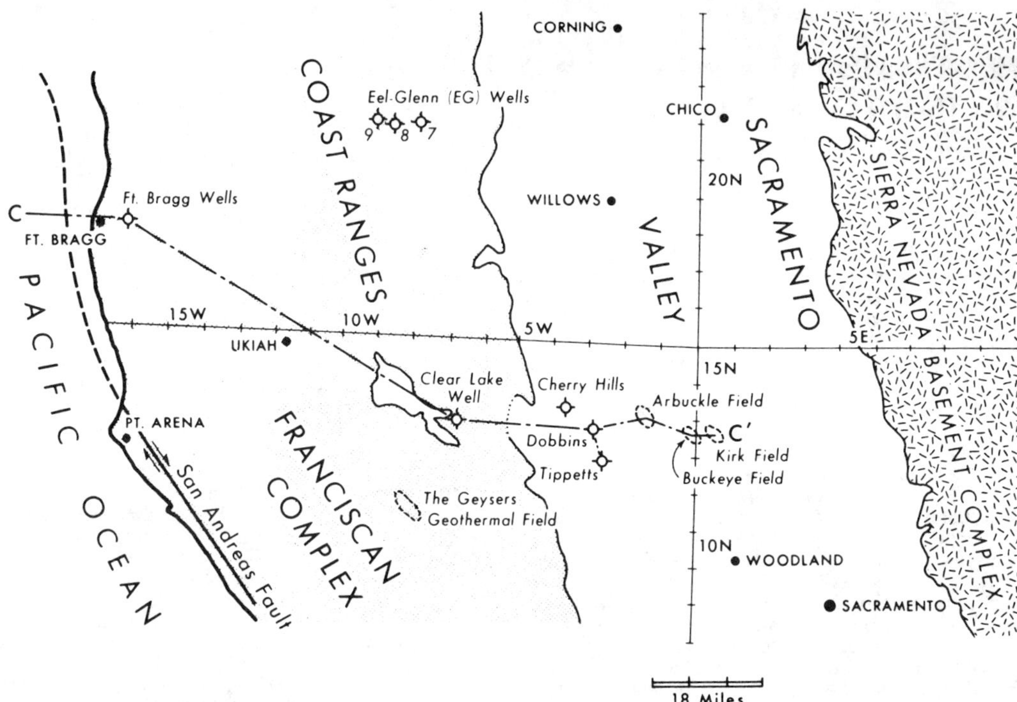

Fig. 1—Index map showing location of pertinent sites and cross section *C-C′*, Sacramento Valley and northern Coast Ranges, California.

the rocks on both sides of the San Andreas. This compression has the greatest effect within the Franciscan and Great Valley shale mass just east of the fault; the effect is greatly reduced within the granitic basement and overlying sediments of the Salinas block west of the fault but has been responsible for folding of the sedimentary veneer. The high fluid potentials are caused by the squeezing of this belt of highly compressible shales east of the San Andreas in a vise whose jaws are formed of relatively incompressible granite; these anomalous fluid potentials are envisioned as being late Cenozoic phenomena dynamically active today.

Diapirism and diapiric folding instead of thrusting have been the preferred modes of late Cenozoic structural deformation within this high fluid potential belt. The dominance of diapirism is attributed to the limited crustal shortening related to the development of this compressive field, as opposed to the dominance of the shearing stresses related to plate movements on both sides of the San Andreas fault. Diapirism and more limited thrust faulting related to the current generation of high fluid potentials may develop in the future.

Among the possible consequences of the existence of this postulated extensive zone of near-lithostatic fluid pressures are the shallow-focus earthquakes and extensive aftershocks along the San Andreas fault. The near-continuous fault creep along the San Andreas and related Calaveras and Hayward faults also may be a result of these postulated high pore-fluid pressures adjacent to these faults.

An important implication of this paper is the demonstration that fluid pressures within rocks can serve as extremely sensitive and unique strain gauges for the detection of local or regional structural movements.

Introduction

The central thesis of this paper is that near-lithostatic fluid pressures exist today within an extensive part of the California Coast Ranges, that their presence is due to dynamic tectonic forces, and that certain geologic phenomena are present and others may come into being as a result of the high fluid potentials.

Anomalous high fluid pressures are present within the pore space of sedimentary rocks in the Coast Range province of California (Fig. 1). The area where these high pressures exist is confined to the east side of the San Andreas fault north of the Garlock fault. The existence and distribution of these high fluid potentials can be documented best in the Sacramento Valley, where many petroleum exploratory and development wells have been drilled in the Great Valley sedimentary section. Most of these wells have penetrated only part of the Cretaceous section; few wells have been drilled into the Jurassic sediments. Many of these wells have found anomalously high fluid potentials, particularly on the west side of the Valley. The results of earlier studies on the origin and distribution of these high fluid potentials in the Sacramento Valley have been presented orally at technical meetings (Berry, 1965).

In this paper I attempt to describe the regional

distribution of these anomalous fluid pressures and their probable tectonic significance. Many areas and concepts are discussed. Extensive sections of the paper are concerned with a particular locality, phenomenon, or concept that I consider important. This paper contains an abundance of new evidence, as well as speculative interpretations for which I make no apologies. I hope I have succeeded in separating fact from opinion.

Data Sources

The fluid pressures have been determined principally from direct measurements made during drill-stem tests on individual wells. Copies of most of the actual drill-stem-test pressure charts were obtained and analyzed. Pressures from the shut-in pressure curves were extrapolated to infinite time by the method outlined originally by Horner (1951; also Dolan *et al.,* 1957; Perrine, 1956) in order to obtain the most accurate determination of the undisturbed static fluid pressure within the given sedimentary zone. Where the drill-stem test charts were not available, reported maximum values for the initial and final shut-in curves provided data as to the minimum static fluid pressure within the sedimentary section through which the drill-stem test was taken. In places valuable data on fluid pore pressure also can be obtained in scrutiny of the drilling history of a given well. Knowledge of the density of the mud column in a drilling well can provide data for determining the minimum, maximum, or actual fluid pore pressures if the well blows out or loses circulation, or the rock surrounding the well invades the well bore by plastic flow. Some valuable fluid pore-pressure information can be obtained when a well is shut in and a pressure gauge is installed at the wellhead. Accurate knowledge of the density of the fluid column occupying the well bore at the time of such a measurement is necessary for a reliable determination under such conditions. Several papers discuss various aspects of those alternative methods (Berry, 1959; Dickenson, 1953; Keep and Ward, 1934; Kok and Thomeer, 1955; MacGregor, 1965; Rubey and Hubbert, 1959; Tainsh, 1950; Thomeer, 1953, 1955; Thomeer and Bottema, 1961; Tkhostov, 1963).

All data from the Sacramento Valley were obtained from examination of the original pressure curves. Data from the San Joaquin Valley were obtained by analysis of pressure charts, ranges of mud weights under certain specific conditions, and surface shut-in pressures; data from the Franciscan rocks of northern California were obtained by mud weights and surface shut-in pressures. The types of data are not differentiated on the illustrations accompanying this paper.

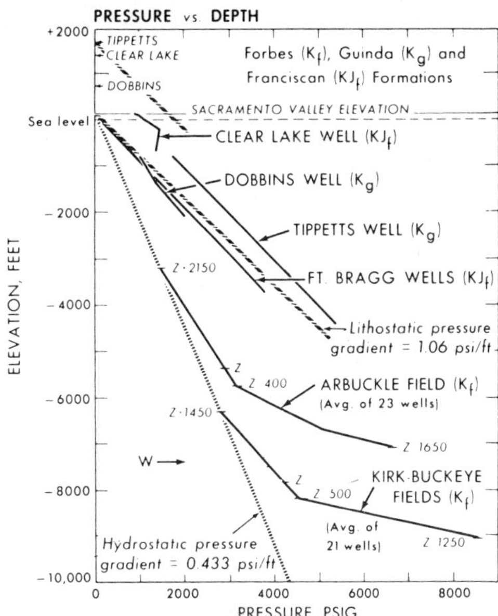

Fig. 2—Pressure versus depth plot in Sacramento Valley, California. Pressure-depth lines plotted for Arbuckle and Kirk-Buckeye gas fields are composite lines derived from series of pressure determinations in several wells. Formations from which fluid pressures were determined are indicated. "Z" is stratigraphic marker in Forbes Formation. Reference lithostatic and hydrostatic gradients are plotted from average surface elevation of Sacramento Valley (+50 ft). Surface elevations of Dobbins, Clear Lake, and Tippetts wells are indicated and reference lithostatic gradient is plotted from Tippetts surface elevation. Locations shown in Figure 1.

Fluid Potential Data—Northern California

A summary view of the pore fluid pressures from the Sacramento Valley is shown in Figure 2 by a plot of fluid pressures versus depth with respect to sea level. The solid lines show the distribution of pore fluid pressure versus depth at several different sites. The plots of the Kirk-Buckeye and Arbuckle gas fields are a composite of fluid-pressure determinations from many wells (Fig. 2). The plots for the Tippetts, Dobbins, and Clear Lake wells represent the fluid-pressure distribution at various depths in each of these wells.[3]

The stratigraphic unit from which the pore fluid pressure measurements were made is noted for each of the gas fields or wells. All pressure determinations in the Arbuckle and Kirk-Buckeye fields were from the Forbes Formation (Cretaceous); the position of the stratigraphic marker known as the "Z point" is noted, as well as the position of various points with respect to the Z Marker (*cf.* Z −400 [ft] or Z +1,250 [ft]). The pressure determinations from the Tippetts and Dobbins wells were made entirely within the

[3]Data for the Clear Lake well were supplied by Union Oil Company.

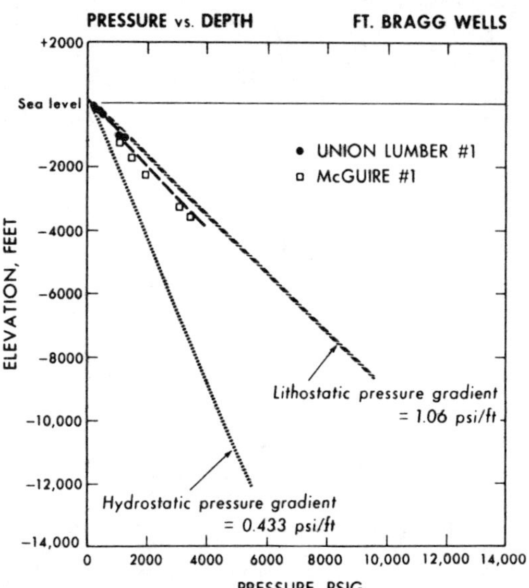

FIG. 3—Pressure versus depth plot of two wells near Fort Bragg, California. Reference hydrostatic and lithostatic gradients are plotted from surface elevation of these wells at approximately +300 ft.

Guinda Formation (Cretaceous); the Clear Lake well was drilled entirely within the Franciscan Formation (Cretaceous-Jurassic). Thus, all these wells have pore fluid pressure determinations from the Great Valley sedimentary sequence except for one from the Franciscan.

In a sense, the pressure versus depth plot shown in Figure 2 represents a cross section with respect to sea level of the fluid pore pressures at sites between the Kirk-Buckeye gas field on the east and the Clear Lake well on the west. The fluid pore pressures increase westward at any given datum elevation and increase downward to values progressively in excess of hydrostatic fluid-pressure values. These pore fluid pressures approach lithostatic pressure values at depth, but at progressively greater depths eastward. No fluid pressures measured equaled or were in excess of the lithostatic pressure for a given site and depth. The Tippetts, Dobbins, and Clear Lake wells show a pore fluid pressure distribution with depth that is essentially a lithostatic gradient (1.06 psi/ft) but at values of 500-1,000 psi below actual lithostatic values. Such a distribution may represent an equilibrium position in an environment where anomalous high pore fluid pressures are being generated. Should the pore fluid pressures exceed lithostatic pressure values plus the relatively small cohesive forces of the rocks, rupture (hydraulic fracturing) of the rocks would occur, with a consequent reduction of the pore fluid pressure to a value below the lithostatic pressure (Hubbert and Willis, 1957; Scott *et al.*, 1953).

The general distribution of the pore fluid potentials shown by Figure 2 is the same in all other sites investigated on the west side of the Sacramento Valley where wells have penetrated the Great Valley sedimentary section. Altogether, I have analyzed the pressure data from approximately 141 wells in 27 principal sites that have anomalously high fluid potentials. Thus, the increase of pore fluid pressures at depth and westward to near-lithostatic values is a regional and not a local occurrence. Local structure apparently does not control the presence of the high fluid potentials, although the development of fracture flow paths associated with local structural deformation does alter the pressure versus depth distribution at a given site by controlling the geometry of the intergranular and channel flow paths.

Three other wells drilled within the last several years extend the area of pore pressure observations west within the northern California Coast Ranges. The Cherry Hills well was a petroleum exploratory test in the Lower Cretaceous of the Great Valley sequence, approximately 5 mi east of the surface fault contact of the Franciscan-Great Valley section (Fig. 1). The fluid-pressure distribution in this well with depth was comparable to the Clear Lake, Dobbins, or Tippetts wells.

Two other wells, drilled in conjunction with geothermal steam exploration near Fort Bragg, are 9 mi east of the San Andreas fault (Fig. 1). The offshore location of the fault is taken from the work of Curray and Nason (1967) and Nason (1968). These wells penetrated the "Coastal Franciscan" of Bailey *et al.* (1964)—*i.e.*, the Cretaceous section containing K-feldspar-rich rocks of Franciscan-type facies.

The distribution of the pore fluid pressures versus depth (sea level datum) of the two Fort Bragg wells is shown on Figure 3 and is summarized on Figure 2. Both wells found very high fluid potentials. All the plots for the McGuire No. 1 well represent minimum pore fluid pressures. The values for the Union Lumber No. 1 contain one pair of maximum-minimum fluid pressure values at the same elevation; the other points represent minimum pore fluid pressures values. The pore fluid pressures in these wells are very nearly lithostatic, and most of the minimum values are only a few hundred psi below the lithostatic pressure. Pore fluid pressures approaching those of lithostatic values were within a few hundred feet (first determination at a depth of 405 ft) of the surface in the Union Lumber No. 1 well, which has near-lithostatic pressures at a shallower depth than in any other well examined. Though these wells were drilled for geothermal steam, no anomalously high temperatures were recorded during drilling.

Only a few deep wells have been drilled within the Franciscan and related Coastal Franciscan rocks of the California Coast Ranges because of unlikely prospects for petroleum. Thus, there are few sites on which to determine the regional distribution of the pore fluid potentials.

Three wells have been drilled by the California Department of Water Resources to depths of 1,350, 2,831, and 4,668 ft within Franciscan rocks along the line of the proposed Eel-Glenn River Diversion Tunnel, near the north end of the California Coast Ranges (Fig. 1). Data from these wells were supplied by Phil Lorens (personal commun.). Well EG-7 (SE corner Sec. 6, T21N, R8W, Glenn Co.; TD 4,668 ft; surface elevation 5,200 ft) had indications of high pore fluid potentials in the lower half of the hole; well EG-8 (C, Sec. 8, T21N, R9W, Glenn Co.; TD 2,821 ft; surface elevation 3,940 ft) had no indications of anomalous pore fluid potentials; well EG-9 (SW corner Sec. 1, T21N, R10W, Mendocino Co.; TD 1,350 ft; surface elevation 2,552 ft), drilled within the Black Butte fault zone in sheared Franciscan shale layers, had indications of high pore fluid potentials through much of the hole (Lorens, oral commun., October 3, 1967). Inflammable gas also was found in the last well.

Insufficient data are available to indicate the magnitude of the pore fluid potentials within these wells. The existence at depth of high potentials in the Franciscan rocks of this area, however, is confirmed by the drilling histories of the first and third wells, which record repeated instances of the holes being squeezed shut.

The Geysers (Fig. 1) geothermal field is the other area in northern California where deep wells have penetrated the Franciscan. Drilling has found dry steam at relatively low fluid pressures along a restricted zone that extends northwest-southeast parallel with the tectonic grain of the region, from depths as shallow as 500 ft to depths as great as 6,000 ft (Garrison, 1972). No high fluid pressures have been found. The Geysers is on the southwest flank of a large closed minimum Bouguer gravitational anomaly (California Division Mines and Geology, 1966). It is also on the flank of a roughly circular regional area, 10-20 mi in diameter, of probable high heat flow, with numerous Quaternary volcanic eruptive centers. An active magma chamber may underlie this area at depth. The origin of the dry steam has been the subject of speculation by several authors (Berry, 1967a, b, in press; Craig, 1963, 1966; McNitt, 1961, 1963, 1965; White *et al.*, 1971) but until the plumbing system at The Geysers is understood accurately, it will be impossible to describe with any precision the origin of the dry steam or to account for the lack of anomalously high pore fluid potentials which seem to characterize the Franciscan of northern California.

Fracture Flow—High Fluid Potential Areas

In the following paragraphs I will develop my views as to the problems related to fracture-type flow in a regional high fluid potential environment, such as I believe the Franciscan possesses. The Franciscan is extensively fractured and faulted. Many of these fractures and faults can be presumed to be sufficiently open to possess channel-type flow characteristics. If, indeed, the Franciscan of northern California is characterized by high pore fluid potentials, it is pertinent to understand how the fracture systems might affect the fluid potential distribution. These problems bear on the origin of the low fluid potentials and dry steam at The Geysers.

There is relatively little fluid-potential loss in flow along a given length of channel as opposed to an intergranular path. If a fracture channel extends from a zone of high fluid potential at depth to the air-water contact at the water table, the fluid potential distribution will depend on the transmissive characteristics of that channel. If interrelated series of fractures are present, the aggregate of the transmissive characteristics of the several fractures must be considered. Thus, if a near-vertical fracture channel increases in width with depth and penetrates a zone of high fluid potentials, the channel also would possess a high fluid potential (*i.e.*, in excess of the hydrostatic potential). The reverse is also true: if the total transmissive characteristics of a channel-type flow system decrease with increased penetration into zones of anomalous fluid potential, the fluid potential along the channel would be similar to the potential at the emergent end of the channel, and the fluid pressure gradient (not to be confused with the fluid pressure values) within the channel would be near-hydrostatic. Thus, high fluid potentials could be transmitted upward or low fluid potentials transmitted downward, depending upon the change in the transmissive characteristics of the channel with depth.

In the Franciscan, with its presumed increase of high fluid potentials with depth, fracture channels with a vertical component generally would have high fluid potentials (in excess of hydrostatic pressure values, but with a near-hydrostatic pressure gradient) if the total transmissivity of the channel-type flow system increased with depth; they would have fluid potentials similar to the surface elevations (*i.e.*, near-hydrostatic pressure values, also with a near-hydrostatic pressure gradient along the channel) if the channel-type transmissive characteristics decreased with depth.

Thus, it should be possible for wells drilled in the Franciscan to intersect fracture zones possessing very high fluid potentials and for other wells (or the same well at a different depth) to intersect other fracture zones with no abnormal fluid potentials. The Clear Lake well, for example (Fig. 2), began in Franciscan sandstones. At 1,630 ft, a fracture channel apparently was found. This channel was tested and the well flowed at an estimated rate of 30,000-35,000 bbl of hot water (210 degrees F) and 50,000-70,000 lb of steam per day. The water is in a liquid phase within the reservoir; steam was generated by flashing associated with production. The well is within the area of high heat flow in the northern California Coast Ranges. The well was shut in and an anomalously high shut-in pressure was recorded. The presumed fracture zone was cased off and the well was drilled approximately 400 ft deeper. Lower temperatures and fluid potentials were found at this greater depth than within the fracture zone. Such inversions are typical where channel flow occurs.

A possible example of the opposite type of condition is afforded by the California Department of Water Resources second deep well in the Eel-Glenn River Diversion project. This well (EG-8) was drilled to 2,830 ft in brittle silicified rock with no indication of high fluid potentials. The silicified rock has limited dimensions in map view (Jennings and Strand, 1960). The first of the deep Eel-Glenn River Diversion wells (EG-7) reached the base of this silicified zone at 1,200-1,300 ft, then penetrated 3,300 ft of low-grade metamorphosed shale (graphitic phyllite). The first well, like the second, found no high fluid potentials in the silicified rock, but did register very high pore fluid potentials in the metashale below the silicified rock. Presumably the silicified rocks can sustain fractures, and the total transmissibility of the fracture system extending to the surface through them decreases with depth. A decrease in transmissive properties almost certainly would occur in the Franciscan shales with the increase of their pore fluid pressures with depth to near-lithostatic values. The shale layers would become more plastic with increased depth—an environment which would tend to close fractures—whereas near-hydrostatic fluid pressure values exist in the fractures of the overlying silicified rock. Related phenomena may be responsible for the origin of the dry steam at The Geysers with its extremely low fluid potential. This possibility has been explored in another paper (Berry, in press). Thus, within the Franciscan, local areas or zones of near-hydrostatic fluid potentials within fractured rocks have been found, and others may be expected to exist. Presumably, abnormally high fluid potentials bound such zones both laterally and at depth, but such zones should be common within the extensively fractured Franciscan and related rocks.

Conclusions—Northern California

Evidence demonstrates that high fluid potentials are the normal regional occurrence in the west-side Great Valley section. Conflicting evidence in the Franciscan of northern California suggests that high fluid potentials also are normal at depth within those rocks, but that, in local fractured zones, near-hydrostatic fluid potentials extend to varied depths from the surface. The alternative interpretation is that the Franciscan rocks consist of a heterogeneous assemblage of abnormally high and near-hydrostatic fluid potential zones.

Figure 4 shows in east-west cross section along *C-C′* (Fig. 1) the regional distribution of the fluid potentials, from the middle of the Sacramento Valley to the San Andreas fault. These connected potentiometric surface lines are lines of equipotential; fluid flow is perpendicular to these lines. Flow is thus dominantly upward through the northern Coast Ranges. The depth where the pore fluid pressures are approximately equal to the lithostatic load pressure is plotted with a heavy line. In terms of the notation developed by Hubbert and Rubey (1959) and Rubey and Hubbert (1959), this is the depth where the ratio of the pore fluid pressures to the lithostatic pressures is approximately 1. These pore fluid pressures never are equal to the calculated lithostatic pressure, but approach with depth to within 500 psi of this calculated value, then develop a gradient equal to the lithostatic-pressure gradient. Their distribution with depth then becomes parallel with the lithostatic values.

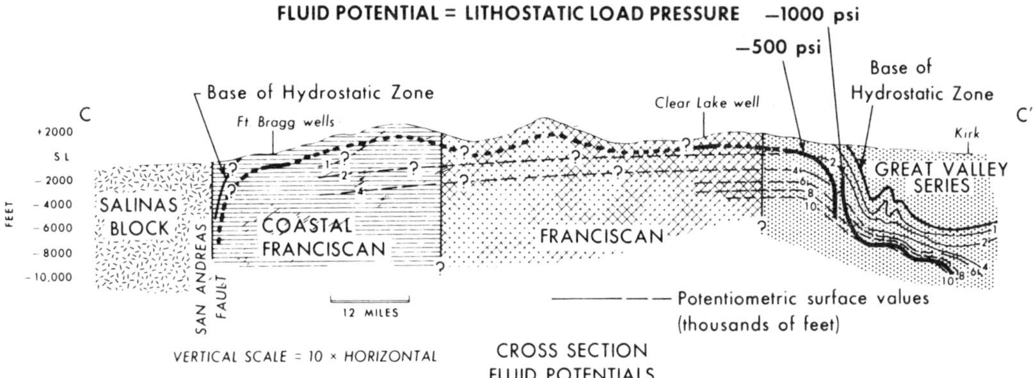

Fig. 4—Cross section *C-C'* showing distribution of fluid potentials across different geologic provinces. Values contoured represent potentiometric surfaces at particular sites. (See Hubbert, 1953, for difference between potentiometric and piezometric surfaces.) Good control exists for contours except in western 5 mi of Great Valley series; only sparse data are available for Franciscan. Degree of uncertainty is indicated by conventional symbols of solid and dashed lines and question marks. Labeled lines indicate boundary between zones with near-hydrostatic fluid pressures and those of higher values; hydrostatic zone parallels potentiometric surface contours.

The Coast Ranges east of the San Andreas fault are shown as having pore fluid pressures essentially equal to lithostatic values at relatively shallow depths. In accordance with my preferred interpretation, the zone of lithostatic pore fluid pressures is overlain and bounded on the east and west by a transitional zone where the pore fluid pressures are intermediate between hydrostatic and lithostatic values; this transitional zone is in turn overlain and bounded by a zone where the pore fluid pressures are hydrostatic. The distribution of the transition zone east of the San Andreas fault between the hydrostatic pore fluid pressure domain of the Salinas block and the lithostatic domain of the Franciscan-Great Valley sediments is entirely interpretative.

The distribution of the pore fluid pressure domains is well controlled within the Sacramento Valley as far west as the surface contact between the Franciscan and Great Valley sequences. Within the Franciscan and Coastal Franciscan sequences there are only two sites of control. It is quite probable that, because of extensive fracturing in these rocks, the shallow distribution of these fluid pressure domains is more complicated than is shown. Linear zones where the hydrostatic pore fluid pressure domain extends downward from the surface for hundreds to thousands of feet probably are common.

No pore fluid pressure data have been obtained along this line of cross section within the Salinas block, which is offshore at this site. Farther south, many deep wells have been drilled in the landward part of the Salinas block (in the Point Arena area, the Point Reyes region, the Santa Cruz Mountains, and the Salinas Valley-Carrizo Plains area). I have examined the drilling histories of most of these wells and have found no anomalously high pore fluid pressures. Near-hydrostatic conditions prevail within this province, which is floored by the crystalline rocks of the Santa Lucia Granite and the Sur Series metasediments. Several wells within the Salinas Valley have been drilled to depths below 10,000 ft in Tertiary rocks without finding anomalously high fluid potentials. Other wells have penetrated the crystalline basement—one well for nearly 1,000 ft—without finding anomalous high fluid potentials. There has been enough drilling to state with considerable confidence that the Salinas block is characterized by hydrostatic or near-hydrostatic pore fluid pressures.

Figure 5 presents the rock distribution and

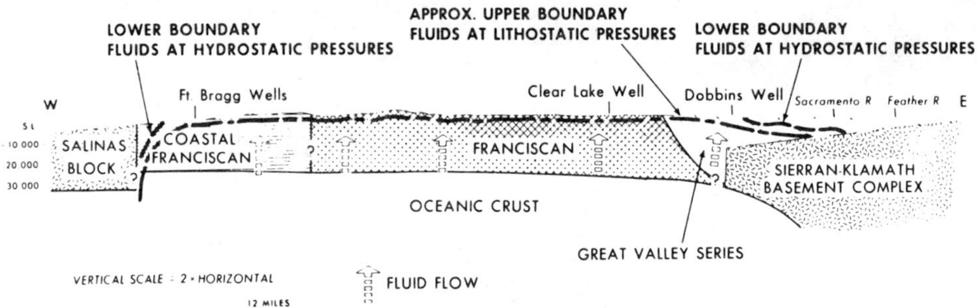

Fig. 5—Structural cross section *C-C′* showing Pacific Ocean to Sacramento Valley near-lithostatic and near-hydrostatic fluid pressure boundaries. Vertical contact between coastal Franciscan and Franciscan is arbitrary. Arrows indicate direction of fluid transport. Location of cross section shown Figure 1.

zone of near-lithostatic pore fluid pressures on a regional scale. The zone of near-lithostatic pore fluid pressures is found within both the miogeosynclinal (Great Valley sequence) and eugeosynclinal (Franciscan) type rocks. These dominantly mudstone sediments or similar rocks of low permeability lie east of the granitic Salinas block and west of the granitic Sierran-Klamath basement.

Fluid Potentials and Aqueous Chemistry—San Joaquin Valley

In the San Joaquin section of the Great Valley, the same Jurassic-Cretaceous sedimentary section is present as in the Sacramento Valley, but a Tertiary sequence commonly more than 10,000 ft thick overlies the Cretaceous section over much of the valley. The Tertiary sequence has many permeable sandstone beds interlayered with shale and siltstone. The sequence generally has high fluid transmissibilities and is characterized by near-hydrostatic pore fluid potentials. Only a few wells on the west side of the San Joaquin Valley have reached the Cretaceous section; thus, there is little direct knowledge of the fluid potentials within the Cretaceous section.

The most convincing evidence that the Cretaceous sedimentary section within the San Joaquin Valley has high fluid potentials stems from the chemical similarity of water from the Tertiary rocks to that of water from the Great Valley sequence in the Sacramento Valley. In general, the Great Valley sequence waters possess all the characteristics associated with membrane-effluent waters and probably are derived from tectonic compaction of the shales (Berry, 1959, 1966, 1967a, b, 1969; Buneev *et al.*, 1947; Engelhardt and Gaida, 1963; Hanshaw, 1962, 1964; Kharaka, 1971; Korzhinsky, 1947; Kryukov *et al.*, 1962; Lomtadze, 1954; McKelvey *et al.*, 1957; Siever *et al.*, 1965; Warner, 1964; White, 1965). Criteria (Berry, 1969; Kharaka, 1971; White, 1965) for an effluent origin are a low total concentration and a relatively high content of NH_4^+, B, I^-, and HCO_3^- with respect to either Na or Cl compared with seawater. The waters also have a relatively low Ca/Na ratio and a relatively high pH compared with most deep subsurface waters. An average analysis of 32 representative samples of subsurface water (Table 1) from the Sacramento-Great Valley sequence shows a dominance of NaCl waters, with the addition of other effluent-type chemical species. In general, the total concentration decreases and the relative content of effluent ingredients with respect to Na or Cl increases with depth.

The lower Tertiary waters of the San Joaquin Valley possess a striking similarity in chemical composition to water in the Sacramento-Great Valley sequence. The former generally have a low total concentration which decreases with depth; moreover, their chemical content is typical of membrane-effluent waters. These membrane-effluent waters are present regionally in the San Joaquin lower Tertiary rocks regardless of whether a local high fluid-potential anomaly exists within the rocks. The conclusion is inescapable that lower Tertiary waters of the San Joaquin are derived principally by tectonic compaction of the underlying Great Valley sequence; they are expelled pore waters which have been driven upward into the overlying relatively high-permeability lower Tertiary rocks, where, to some extent, they have been mixed with meteoric water.

Fluid-Potential Data—Kettleman Hills Folds

Most deep wells over the major anticlinal folds on the west side of the San Joaquin Valley have

Table 1. Average Analysis of 32 Subsurface Water Samples from Cretaceous Rocks of Sacramento Valley

Radicals	Average Mg/L	% of Total	Average Meq/L	% of Total	No. of wells
Na + K	6,606.8	36.20	287.31	46.62	26
NH_4	31.0	0.17	1.61	0.26	19
Ca	223.2	1.22	11.59	1.88	26
Mg	85.5	0.47	7.09	1.15	26
Ba	5.2	0.03	0.18	0.03	9
Fe	1.5	0.01	--	--	10
SO_4	133.0	0.73	2.92	0.47	26
Cl	10,452.1	57.28	294.75	47.82	26
CO_3	36.3	0.19	1.21	0.20	3
HCO_3	525.2	2.88	8.61	1.40	26
B_4O_7	68.5	0.38	0.88	0.14	20
I	28.2	0.16	0.20	0.03	16
SiO_2	50.9	0.28	--	--	12
Total solids	18,246.4		616.35		32

	Average value	No. of wells
Specific gravity at 60^oF	1.0156	20
Resistivity (Ohm-cm at 75^oF)	34.47	26
pH	7.60	26

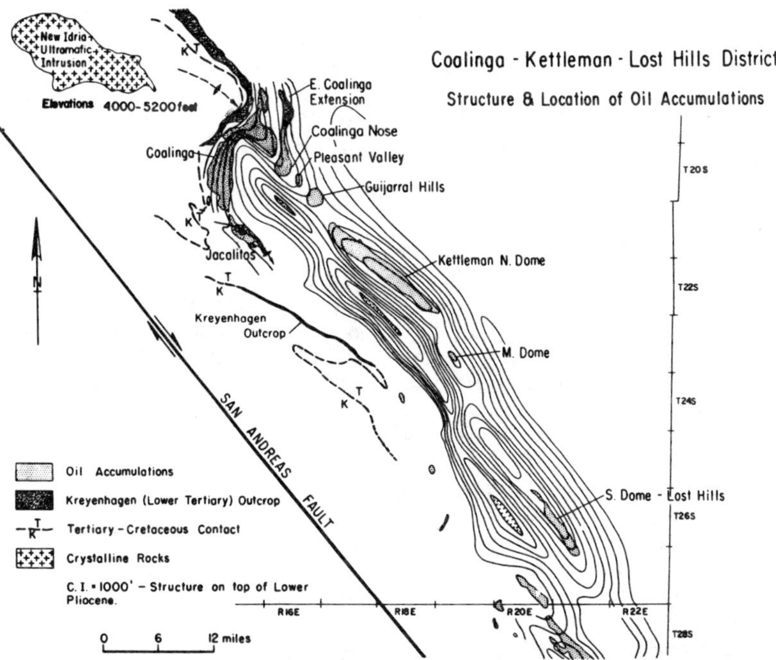

FIG. 6—Structural contour map of Coalinga-Kettleman Hills-Lost Hills area. Structural datum is top of lower Pliocene. Outcrop of Kreyenhagen Shale and Tertiary-Cretaceous contact adapted from Hoots *et al.* (1954).

anomalously high pore fluid potentials at various depths and stratigraphic positions within the Tertiary section. Fluid potential data are available on the three major folds of the greater Kettleman Hills system (Fig. 6). South Dome and Lost Hills appear to be separate folds on the surface, but subsurface data show that this feature is a single anticline at depth. Figure 7 represents a generalized potentiometric surface map for the middle Miocene sedimentary rocks (Temblor and equivalents), and Figure 8 is a generalized potentiometric surface map for the Eocene. These maps show linear closed equipotential features surrounding what appear to be linear high pore fluid pressure sources. The strong hydrodynamic forces indicated by the maps greatly influence petroleum accumulation on the Kettleman folds.

Figure 9 is a pore fluid pressure-depth plot for wells drilled essentially along the axis of South Dome-Lost Hills anticline, and Figure 10 is a plot for wells on the flanks and noses. Figure 11 shows the distribution of the pore fluid pressures with depth for wells on Middle Dome and one well on North Dome.

Figure 12 is a composite plot of the general pore fluid pressures at various depths for the three anticlines. Generally, the pore fluid datum pressure or pore fluid potential increases with depth and with stratigraphic penetration. The fluid potentials over South Dome-Lost Hills anticline are highest over the crestal part and decrease very abruptly in any given zone perpendicular to the axis. The fluid potentials decrease gradually from the crest along the axis toward the plunging noses. This decrease is particularly well documented off the southeastern end of Lost Hills-South Dome.

The average pore fluid pressure gradient plotted is approximately 0.7 psi/ft. The average slope, although less than that of the average lithostatic gradient (1.06 psi/ft), is extraordinarily high. Fluid pressures that exceed the average values for a given depth are present in three wells. One in the Cretaceous beds on North Dome anticline (at 13,500 ft) is the only valid fluid pressure measurement for the Cretaceous in the central or southern San Joaquin Valley. The water chemistry of these beds is similar to the Cretaceous waters of the Sacramento Valley. The fluid pressure measured is only about 100 psi below the lithostatic value for the given depth—a value similar to the maximum fluid pressures found at depth in the Sacramento Valley. Apparently, the Cretaceous fluid pressures have values that are near-lithostatic under North Dome anticline.

The two other most significant departures from the average pressure values are in well 12 at a datum depth of approximately 8,000 ft and in well 13 at approximately 11,000 ft (Fig. 8). Both these anomalies occur in the Leda Sandstone, a lens of limited lateral extent at the Oligocene-Miocene boundary. The pore waters with anomalously high fluid potentials also have anomalously high concentrations of dissolved solids. The origin of the high fluid pressures in the Leda Sandstone is not certain; one of two possibilities is indicated. The Leda is a subsurface lens, enclosed within siltstone and shale. Its waters are concentrated chemically (70,000 ppm) to values considerably in excess of those in the waters in stratigraphically higher and lower zones (20,000 and 8,000 ppm, respectively). The waters of the Leda have a chemical composition characteristic of membrane hyperfiltrated water (White, 1965). Thus, physical-chemical phenomena resulting from the semipermeable membrane properties of shales appear responsible for the chemical anomaly and may be a factor via chemical osmosis in the origin of the anomalous fluid potential. A back pressure, of osmotic origin, should have developed progressively within the Leda as a result of the buildup of its chemical concentration, but to what degree it contributes to the excess fluid pressures is unknown.

It is also probable that the Leda possesses fluid pressures higher than other Kettleman stratigraphic zones at comparable depths because of differences in transmissibility and the limited lateral low-transmissibility boundaries of the Leda lens compared with other sandstones of greater regional extent. It may be that the fluid pressures in the Leda represent the sum of the fluid pressure in the adjoining shales at the same datum depth plus the osmotic back pressure. Thus, a mechanical as well as a physical-chemical mechanism may be responsible for these excessive fluid potentials.

The significance and possible origin of high fluid potentials associated with the anticlinal folds are uncertain. Traditionally, high fluid pore pressures associated with anticlinal folds have been interpreted as having a local source (*i.e.*, tectonic compression of the sediments involved in the fold itself). Most such anticlines are relatively young, and several authors (*e.g.*, Suter, 1954; Tkhostov, 1963; Watts, 1948) have stated or implied that such postulated tectonic compression is a result of continuing compression of the fold form itself.

The extraordinarily high fluid pressures of the Kettleman anticlines cannot be explained by physical-chemical phenomena, but suggest the current operation of dynamic tectonic phenom-

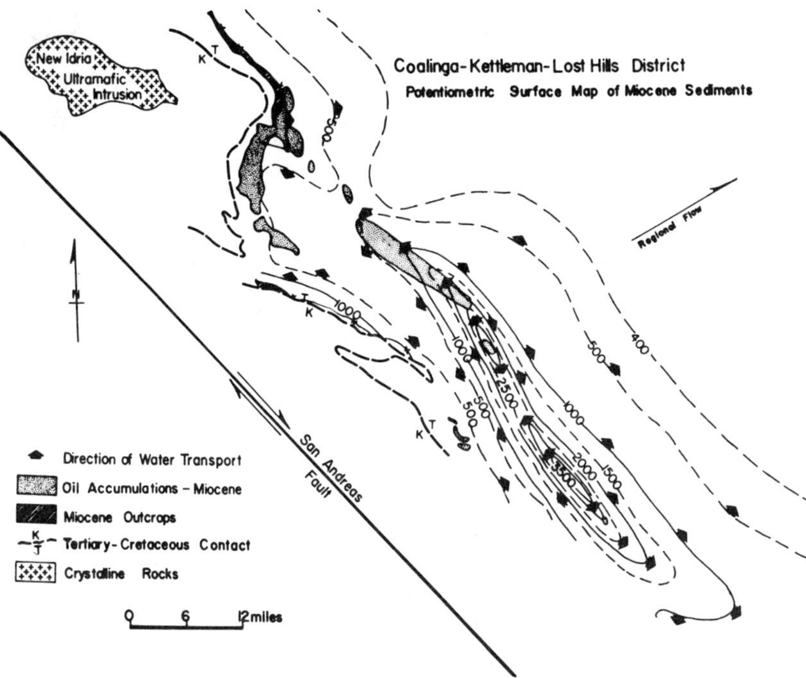

FIG. 7—Generalized middle Miocene (Temblor) potentiometric surface map of Coalinga-Kettleman-Lost Hills. Outcrop elevations of middle Miocene are same as values of potentiometric surfaces where they intersect outcrop. Regional hydrodynamic transport is northeast.

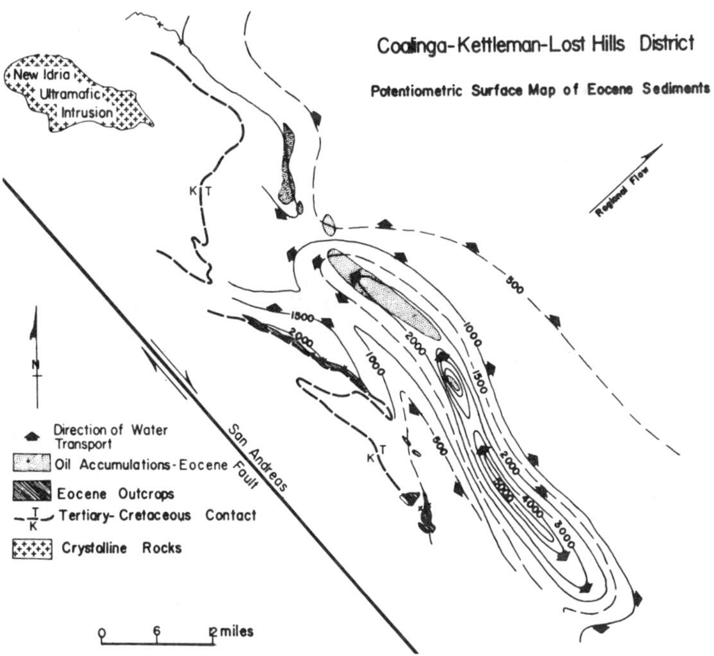

FIG. 8—Generalized Eocene (Gatchell, McAdams, Domengine) and Eocene-Oligocene (Kreyenhagen) potentiometric surface map of Coalinga-Kettleman-Lost Hills. Outcrop elevations of Eocene are indicated by potentiometric surface values. Regional hydrodynamic transport is northeast.

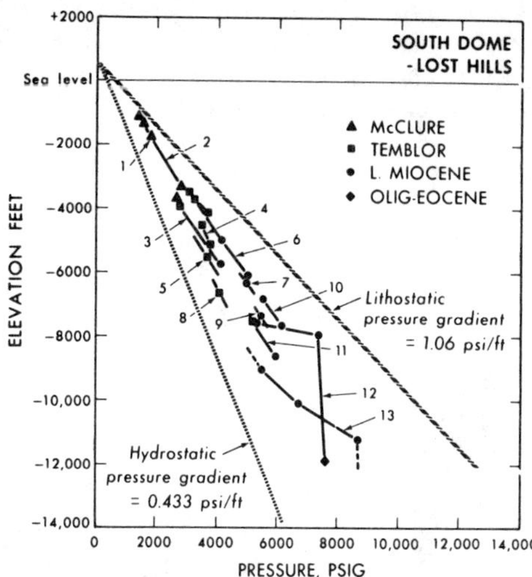

Fig. 9—Pressure versus depth plot of axis of South Dome-Lost Hills anticline, showing distribution of pore fluid pressure with respect to elevation below sea level for various wells. Hydrostatic and lithostatic gradients are plotted from surface elevations of about +400 ft. Slope of lines with only one point is estimated. Lines represent wells: (1) Texas 2 Overall; (2) Lincoln 1-A Theta; (3) Tidewater 1 Williamson; (4) Universal Consolidated 49; (5) Standard of California 58-4 Cahn; (6) General Petroleum 33-11; (7) Standard of California 16; (8) Ohio Oil 1 Smith; (9) Continental Oil 28-7 Gatchell; (10) Universal Consolidated 265; (11) Occidental Petroleum 131x-2 Hellman; (12) Standard of California 4-2; (13) Occidental Petroleum 27-27.

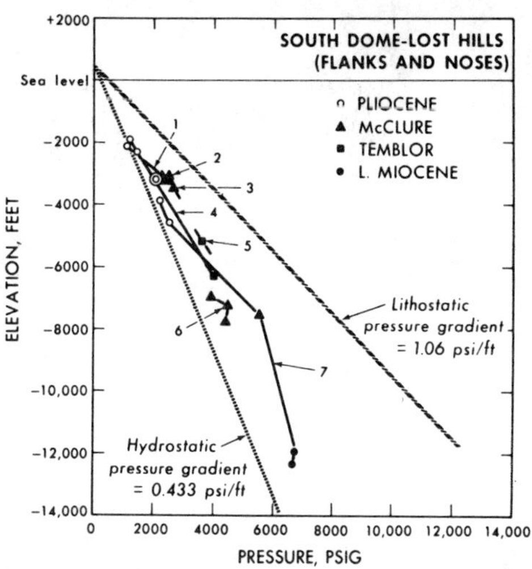

Fig. 10—Pore-fluid pressures are plotted with respect to elevation below sea level for wells drilled along flanks and at ends of noses of South Dome-Lost Hills anticline. Surface elevation is approximately +400 ft from which reference hydrostatic and lithostatic gradient lines are plotted. Numbered lines represent: (1) Standard of California 59 Cahn; (2) Standard of California 4-148 Cahn; (3) Texas Co. 1 Martin; (4) Buttes Gas and Oil 1 Vaughey-Occidental Lands; (5) California Lands 2 Occidental Lands; (6) Standard of California 3 Lost Hills Extension; (7) Standard of California 45 Van Sicklin.

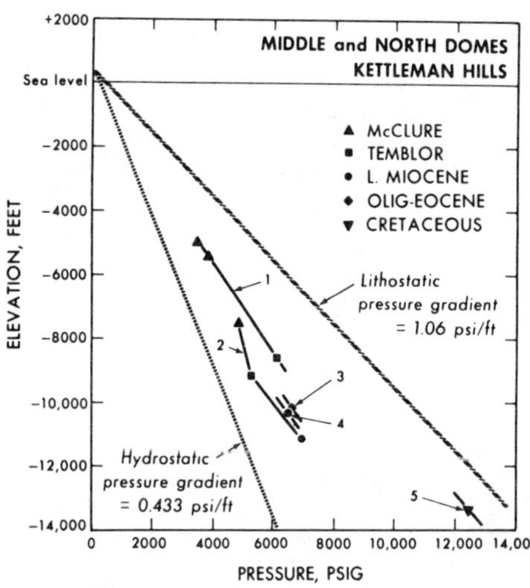

Fig. 11—Pore-fluid pressures plotted with respect to elevation below sea level for wells at Middle Dome and one well at North Dome. Surface elevation is approximately +500 ft from which reference hydrostatic and lithostatic gradient lines are plotted. Numbered lines represent, on Middle Dome: (1) Petroleum Securities 1; (2) Standard of California 613; (3) Middle Dome 38-19V; (4) Standard of California 68-4; on North Dome: (5) Kettleman North Dome 423.

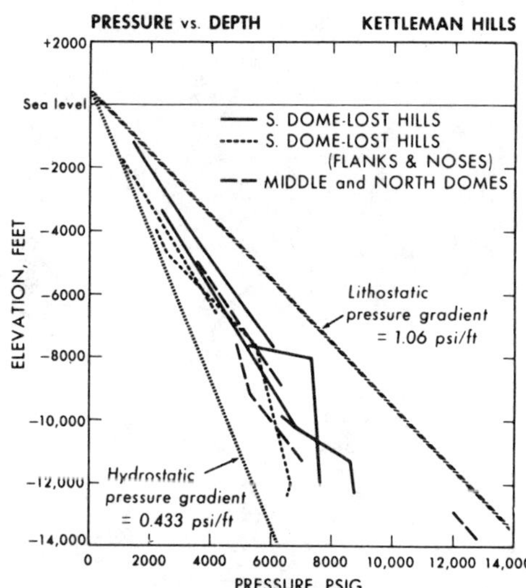

Fig. 12—Composite pressure versus depth plot at South Dome-Lost Hills anticline, Middle and North Domes. Reference hydrostatic and lithostatic gradient lines are plotted for average surface elevation of +400-500 ft.

ena. To explain the mechanism two alternatives are possible: (1) local compression of the sedimentary mass in the fold form, or (2) regional compression at depth of a much larger sedimentary mass, with fractures whose transmissive characteristics decrease upward, thereby transmitting the high fluid potentials of the underlying mass to higher and younger stratigraphic zones. Such fractures might be best developed within major anticlinal features.

Calculations give a probable order of magnitude of the rate of water transport through various stratigraphic zones radially away from the Kettleman anticlines. Accurate pressure measurements exist, drilling provides relatively good data on the thickness, distribution, and permeability of the stratigraphic units. If the source for the high fluid potentials is tectonic compaction of the rocks in the fold form, the total water potentially available for transport is the water occupying the pore space within the fold form. The sandstone is relatively incompressible, hence the shale must provide the water available for compaction.

I have tried to set up mathematical models to test the reasonableness of this postulated tectonic compaction of shale within the fold form as an adequate mechanism to provide water at the rates indicated by the separate calculations and have concluded that it is impossible to support the indicated flow rates solely by current tectonic compaction. The calculations indicate that the rate of water supply via tectonic compaction is too small by several orders of magnitude.

Kettleman Hills—Geologic History of Folds

The age of folding of the Kettleman anticlines can be determined rather accurately from well data. Lost Hills-South Dome began folding after deposition of the Temblor during McClure deposition (late Miocene). Folding continued into the Pliocene, and more than 80-90 percent of the folding was achieved by mid-Pliocene time. Folding has continued at a very subdued rate since then.

The folding on Middle Dome also began during the late Miocene and apparently continued through the Pliocene and Pleistocene, as judged from the surface exposures. Folding on North Dome is the latest; it was very moderate in the late Miocene, but became intense during the Pliocene and Pleistocene. Dips of 35 degrees in the Pleistocene testify to the recent date of the folding. Woodring *et al.* (1940) also deduced from surface work that Lost Hills and South Dome had an earlier history of folding than did Middle and North Dome. If the high fluid potentials of these folds were caused by tectonic compression of the rocks within the fold form, the youngest fold should have the highest fluid potentials; however, South Dome-Lost Hills has higher fluid potentials than North Dome when compared either by stratigraphic zone or simply by position within the fold form.

Description of Folds—Kettleman Hills and Great Valley

No method comes to mind for establishing definitively the origin of the high fluid pressures within these folds. A better grasp of the problem may be afforded by examining the possible origin of the folds. From studies of the Kettleman system and the other linear faults and folds that form an intermittent, nearly continuous zone along the west side of the Great Valley, I conclude that the folds are not the result of lateral compression. They are long, narrow, hairpinlike features. Kettleman South Dome-Lost Hills, for example, has more than 10,000 ft of structural relief perpendicular to the fold axis in a lateral distance of only 7 mi. The axial length is 34 mi.

On the west side of the Great Valley there is no series of parallel folds with a progressive decrease of fold magnitude as typically characterizes most fold belts. Rather, there is only a single line of folds or faults that can be traced discontinuously in the surface or subsurface north from Lost Hills-South Dome through the two northern Kettleman folds and Coalinga anticline. This line may be present in a fault across the Vallecitos syncline and is next picked up in the Tracy fault, south of the Stockton fault; it continues north of the Stockton fault along the Midland fault, which connects via the Dunnigan Hills fault to the faulted Rumsey Hills anticline and, farther north, with the Sites anticline. No clear trace of this line is found at the surface farther north. South of Lost Hills-South Dome, the axes of the folds begin to be offset to the west in a curved, semi-en echelon fashion, and here more than one line of folding appears. The fold axes feather out on the southeast and terminate against the White Wolf fault. This line of folds and faults along the west side of the Valley is distinctly separated, by a broad syncline or homoclinal belt, from the series of Coast Range uplifts adjacent to the San Andreas fault in the southern half of the Great Valley, such as Mount Diablo, the Diablo Range, and the Temblor Range.

The three Kettleman folds are not, in fact, truly en echelon, as described in the literature. The axes of the southern folds progressively rotate slightly clockwise with respect to more northerly folds. The folds are asymmetric, but not in a uniform direction. Kettleman North Dome and Middle Dome are asymmetric on the southwest, the steep flank being on the southwest side, whereas Lost Hills-South Dome has the steep flank on its northeast side. This reversal of asymmetry also suggests that the folds are not a result of simple lateral compression.

The two northern Kettleman folds are cut by many tensional faults on the surface, as shown on the geologic map of these structures prepared by Woodring *et al.* (1940). It is evident from extensive drilling that these tensional faults die out at very shallow depth. All surface faulting on North and Middle Domes dies out in the upper Miocene shales; the Temblor (middle Miocene) is essentially unfaulted.

Tensional faults are much less developed over the South Dome part of South Dome-Lost Hills, and none is visible on the surface in the Lost Hills area. The South Dome tensional faults also die out abruptly with depth. The Pliocene at South Dome-Lost Hills is essentially unfaulted. Some faults are present in the upper Miocene, but drilling is not sufficiently dense to reveal any pattern in deeper beds. Some high-angle reverse faults may be present at depth in middle Miocene or older beds under South Dome-Lost Hills and Middle Dome.

The fault pattern of the Kettleman folds suggests the fault model of Dallmus (1958), in which an upper zone (or cone) of tension over an anticline contains tensional faults that die out with depth. The zone of tension is separated from a deeper sone of compression by a region of no stress and no faulting—the "neutral stress surface" of Dallmus. High-angle reverse faults characterize the compressional zone. The zone of no stress should be at different depths with respect to the fold form in different folds. It is probably very near the surface over Lost Hills-South Dome and is at greater depth over Middle and North Domes. Similarly, the vertical extent of any given stress domain should differ from anticline to anticline.

Clark (1929) first suggested that a fault might lie at depth in the basement along the general surface trace of the three Kettleman folds. Gester and Galloway (1933) suggested the possible en echelon arrangement of the Kettleman folds. Woodring *et al.* (1940) also described the en echelon fold arrangement from detailed surface mapping. They suggested that the Kettleman-Coalinga fold was formed by north-south lateral compression, which also was responsible for movement on the San Andreas. They envisioned a zone of weakness at depth to which they attributed the apparent en echelon Kettleman fold arrangement. De Sitter (1956) treated the Kettleman anticlines as a classical example of the production of en echelon folds by lateral compression.

Moody and Hill (1956) suggested that the greater Coalinga anticline (and presumably its southern extension, the Kettleman anticlines) might be a result of secondary lateral compressive forces set up by right-lateral movement of the San Andreas fault. They reported a true en echelon arrangement with respect to the San Andreas, so that secondary compressional forces could produce this fold trend. However, the greater Coalinga-Kettleman fold in fact is oriented essentially parallel with the San Andreas. The Kettleman folds are not in a true en echelon arrangement with respect to one another, nor do they suggest development by lateral compressive forces.

The presumption of the en echelon arrangement has been made by most workers on the basis of surface exposure. However, the part of the fold form of South Dome-Lost Hills which is exposed does not give an accurate impression of the subsurface fold form. Detailed subsurface maps with datum on the top of the Temblor clearly show the clockwise rotation of the Kettleman folds from north to south.

West Side Fault

There is no evidence for right-lateral movement on a fault at depth underneath or alongside these folds; limited right-lateral motion could exist. However, I believe that a fault, or fault zone, with extensive vertical movement is highly probable. In fact, I postulate that a large fault beneath these folds marks the boundary between the Sierran-Klamath granitic continental basement on the east and the Franciscan province on the west along the entire west side of the Great Valley. Presumably oceanic crust underlies the Franciscan material. I designate this postulated buried feature as the West Side fault. Its location and nature are illustrated in Figure 13. In places there is a surface expression of this buried structure in the form of linear lines of folds and faults; in other places, it has no surface or shallow subsurface expression. South of North Dome, this linear feature is broken up, with the southern extension being offset toward the west. The offset increases southward until the line of folds terminates against the White Wolf fault, with an upthrown granitic basement on its south side.

I long suspected, from my own work, the probable existence of such a fault at depth under the Kettleman-Coalinga folds. Its probable regional nature and relation to the deposition of the Great

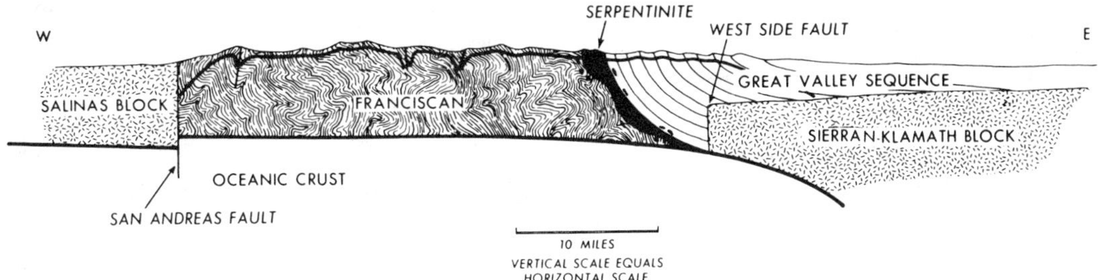

Fig. 13—Diagrammatic structural cross section of California Coast Ranges and Central Valley. Distribution of near-lithostatic fluid pressure domain is indicated by line.

Valley sequence were suggested to me by C. Anders Bengtson. Such a fault is necessary to accommodate the thick Great Valley section (maximum section 60,000 ft) unless major bowing—for which there is no evidence—occurs on the pre-Great Valley depositional surface. Wherever drilling has reached the Sierran-Klamath basement (a maximum depth of 15,000 ft), the surface is essentially homoclinal. Reflection seismographic studies have shown that the Sierran-Klamath basement continues its near-homoclinal westward dip as far as the line of the west-side folds and faults; west of that line, reflections are not recovered from the crystalline basement within the depths penetrated by the compressive waves (Barbat, 1971). I postulate that the depositional relation of the Great Valley sequence to the Sierran-Klamath block is as shown in Figure 13, and subsidence continued along this West Side fault zone during deposition of the Great Valley sequence.

The Midland and Tracy faults are interpreted as being relatively shallow expressions of this deep fault. The relation is particularly well documented for the Midland fault through the Eocene sedimentary sequence (American Association of Petroleum Geologists, 1951; Safonov, 1962).

Origin of Folds

The other near-surface manifestation of the postulated West Side fault consists of long, narrow, sharply-folded anticlines. Diapiric intrusions, principally of shale, along and above this fault zone are believed to have been the mechanism controlling the formation of the folds. Kingma (1958) suggested a somewhat similar origin for piercement structures in New Zealand. Diapiric intrusion of plastic materials might be favored along such a line of displacement. By my interpretation, this line would be bounded by granitic rocks on the east side and on the west dominantly by mudstones with pore fluids approximately at lithostatic pressures. Such rocks would behave as perfect or near-perfect plastic materials (Handin *et al.*, 1963). Such a geologic environment would be ideal for the formation of extensive diapirs.

It would be perfectly plausible for serpentine to be injected diapirically in this position. As shown in Figure 13, the West Side fault is presumed to extend downward to a rind of serpentine which separates the Great Valley from the Franciscan sequence.

Serpentine diapirs are common along the San Andreas and other faults of the Coast Ranges, and I postulate that a serpentine diapir is present at depth under the Lost Hills part of Kettleman South Dome-Lost Hills anticline. Figure 14 shows a residual gravity (Bouguer anomaly) and generalized structure-contour map for the three Kettleman anticlines. Kettleman North Dome, Middle Dome, and the northern end (South Dome part) of South Dome-Lost Hills anticline show the expected gravity maxima over the crests of the anticlines. However, there is a definite closed gravity minimum over the south end (the Lost Hills part) of the structure. Barton (1938) first described this anticlinal gravity minimum, which Barton (1938, 1944) and Boyd (1946) interpreted as a localized concentration of low-density diatomaceous shale of late Miocene age. Since Barton and Boyd advanced this hypothesis many deep wells have been drilled and logged by modern electrical and other devices, and they show that there is no local thickening in the upper Miocene shale sufficient to explain the gravity minimum. Moreover, the amount of diatomaceous material in the upper Miocene section is substantially less than in other Monterey Shale sections farther west in the Coast Ranges. McCulloh (1967) inferred that the Lost Hills gravity minimum might be related to the shallow Pliocene oil accumulation, but I doubt this inference. I propose, rather, that the gravity minimum is attributable to a serpentine or possibly even a

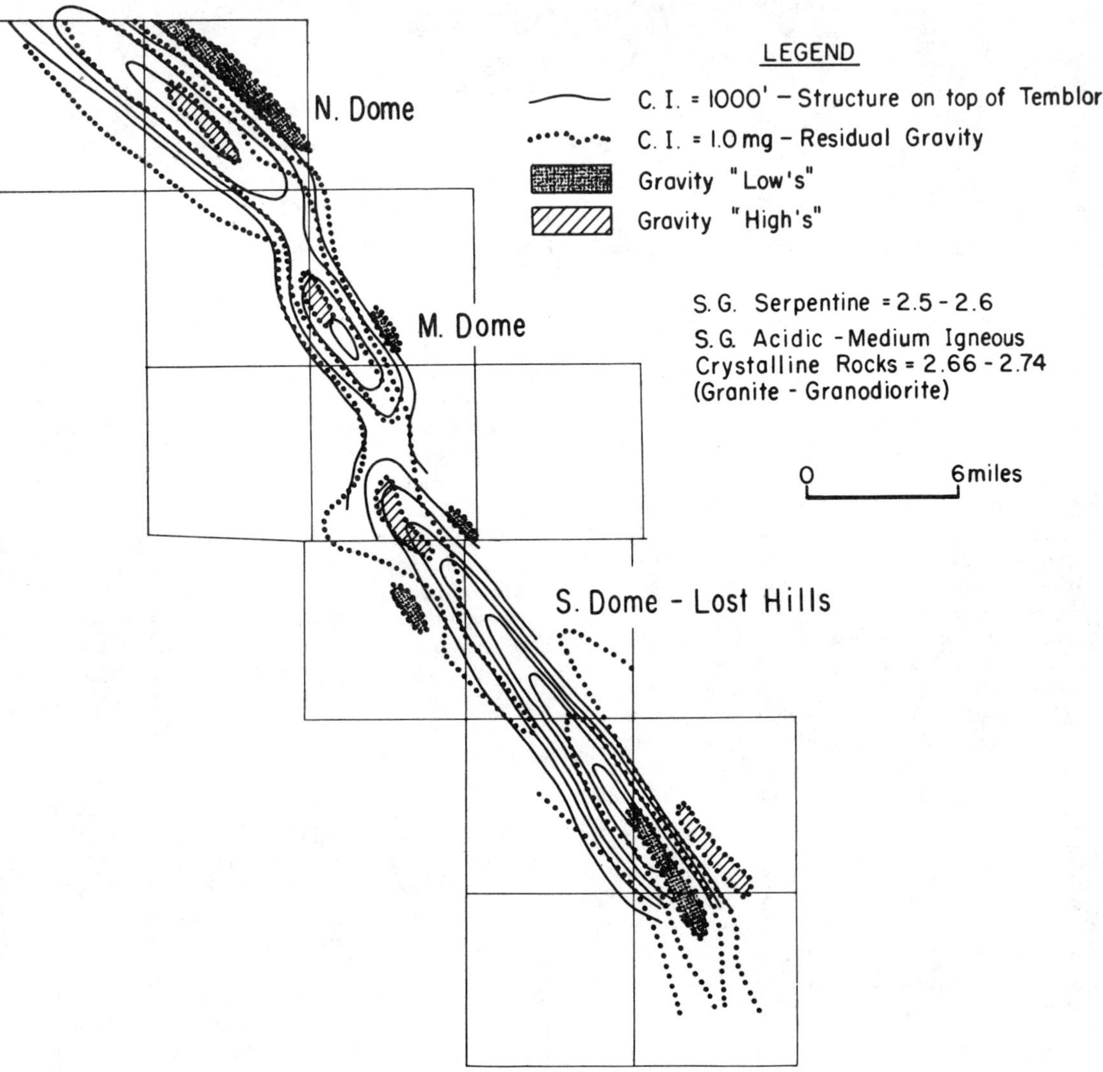

Fig. 14—Gravity map of South Dome-Lost Hills anticline, Middle and North domes (from Boyd, 1946), showing residual gravity (Bouguer anomaly) and generalized structure contours for three anticlines of greater Kettleman Hills district. Structural datum is top of middle Miocene Temblor. CI = 1,000 ft.

shale diapir at depth. The density of serpentine (specific gravity 2.5-2.6) is sufficiently low with respect to granitic rocks (specific gravity 2.65-2.75) and dense sedimentary rocks (specific gravity of mineral content 2.65-2.7) to cause such a gravity minimum. In fact, gravity minima are associated with known serpentine diapiric intrusions in the Coast Ranges (Byerly, 1966). A magnetic study of South Dome-Lost Hills should prove or disprove the postulated presence of serpentine at depth; serpentine bodies give positive magnetic anomalies.

Other data confirm the hypothesis on the postulated West Side fault, the origin of the near-surface west-side folds and faults, and the distribution at depth of the Franciscan, Great Valley, and Sierran-Klamath lithologic groups. Mercury is produced with oil from Cymric oil field located on a west-side fold southwest of South Dome-Lost Hills (Bailey *et al.*, 1961; Stockman, 1947). In the Coast Ranges, mercury is only known to occur in association with Franciscan rock and serpentine; thus, it is a good presumption that Franciscan-type sediments are present at depth under or adjacent to the Cymric field. Seismic investigations have detected an anomalous mushroom-shaped core at depth under a prominent anticline on the west side of the San Joaquin Valley. At least one deep well has found at great depth a rock which resembles the lithology of the Franciscan or Knoxville.

Origin of High Fluid Potentials in Folds

Active diapirism at depth under the west-side folds would mean that the sediments overlying the diapir would be compacted tectonically. I am inclined to believe that tectonic compression related to this mechanism probably is adequate to account for the major amount of the anomalous high fluid pressures found in the Tertiary rocks over these anticlines.

Fracturing may be a minor factor in the overall three-dimensional distribution of the fluid potentials, but fractures do not remain open at depth in this environment. The pore fluid pressures almost certainly increase with depth under these anticlines until they approximate lithostatic pressure values; this means that the rocks become increasingly more plastic and less capable of sustaining open fractures with depth.

I suspect that the presence at depth of high-angle reverse faults and the pattern of their distribution may be important in determining the degree to which high fluid potentials are generated in these folds. Such faults would provide very low transmissibility boundaries for stratigraphic zones with relatively high transmissibility. Folds that possess such faults at depth would tend to develop much higher fluid potentials when subjected to tectonic compaction than folds where such faults are absent. The relative vertical distribution of the high fluid potentials in any fold form undergoing tectonic compaction should be related directly to the vertical extent and frequency of such faults. Certain folds in other geologic provinces which have anomalously high fluid potentials also have high-angle reverse faults. (See discussion concerning Ventura Avenue anticline in following section.) The presence of such faults, serving as transmissibility barriers at depth at South Dome-Lost Hills and Middle Dome, probably explains why my mathematical models to test the reasonableness of tectonic compaction as the cause of these high fluid potentials indicated that the rate of water supply was too small via tectonic compaction and Darcy flow through reservoir rocks with no lateral transmissibility boundaries within, or adjacent to, the fold form.

The question remains as to why the anticlines with the earlier folding histories now have the highest fluid potentials. The answer simply may be that South Dome-Lost Hills and Middle Dome have high-angle reverse faults at depth through their middle and lower Tertiary sedimentary sections, whereas such faults are absent at North Dome. It also may be that, if diapirism and folding are still active to some degree in all of the folds, the fold whose deformation began first would tend to have the higher fluid potentials. If so, a current dynamic state of continuing deformation is necessary. The magnitude of deformation required to maintain high fluid potentials via continuing tectonic compaction would be significantly less than that necessary to produce such high values initially. The high fluid potentials of Lost Hills-South Dome, for example, should have developed during the time of major folding (*i.e.*, from late Miocene through early Pliocene time). If the subsequent rate of folding is sufficient to generate fluid pressures in excess of their decay by Darcy flow, then the fluid potentials at depth would remain at high values.

The size of the fold apparently is not a critical factor in determining the magnitude of the fluid potentials. North Dome has the lowest fluid potentials of the three Kettleman anticlines, but it is larger than Middle Dome and smaller than South Dome-Lost Hills.

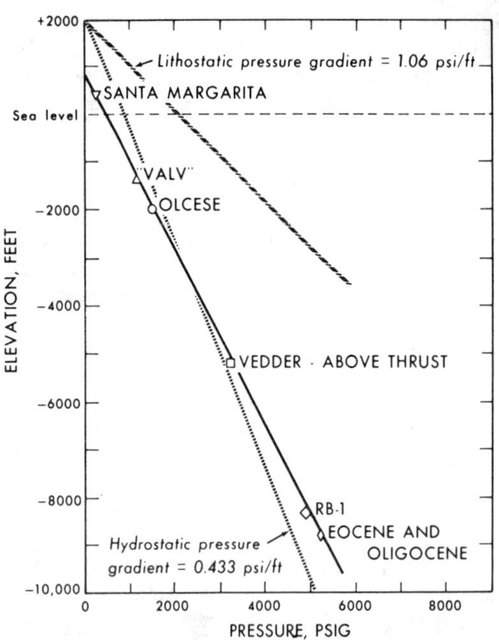

Fig. 15—Pressure versus depth plot of Wheeler Ridge anticline showing distribution of fluid pressure with datum elevation from various stratigraphic zones at Wheeler Ridge anticline (location on Fig. 16). Reference lithostatic and hydrostatic gradients are plotted from surface elevation at approximately +2,000 ft.

Other High Fluid Potential Areas

Several other areas of anomalously high fluid potentials exist in the California Coast Ranges. One is Wheeler Ridge anticline just west of the Tejon embayment at the southern end of the San Joaquin Valley, south of the White Wolf fault and north of the Garlock fault (Fig. 17). The Franciscan and Great Valley sequences are absent, and Eocene rocks lie on Mesozoic granitic rocks. Figure 15 shows a plot of the fluid pressures on Wheeler Ridge anticline. The fluid pressures in the Eocene and Oligocene rocks are only slightly above hydrostatic values with potentials of +3,000 ft; the fluid potentials decrease upward and at shallow depths are below hydrostatic values with respect to the surface elevation of approximately 2,000 ft. In papers presented orally (Berry, 1960; Berry and Hanshaw, 1960), I have postulated that the slightly high fluid potentials of the Wheeler Ridge area are caused simply by gravitational flow through fractures from the granitic outcrop at elevations of 5,000 ft or more, 10 mi south in the San Emigdio Mountains. Such an origin is possible, but I now question it. The development of abnormal fluid potentials is a function both of the magnitude and duration of the applied stress and the transmissibility of the rocks. The rocks of the Wheeler Ridge area have relatively high transmissibility. If they were undergoing active tectonic compaction today, they could develop only moderately elevated fluid potentials and those only at considerable depth. I suspect that tectonic compaction is in fact operative today in this Wheeler Ridge area.

Anomalously high fluid pressures that reach nearly lithostatic values at depth are found within the Ventura Avenue oil field in the Ventura basin (Levorsen and Berry, 1967, p. 409-410; Watts, 1948). Extensive drilling through the rest of the Ventura basin and in parts of the Transverse Ranges has not shown any other localities with anomalously high fluid potentials. Unlike the Coast Ranges north of the White Wolf and east of the San Andreas faults, the Transverse Ranges do not possess any regional development of anomalously high fluid pressures. I concur with Watts (1948) who attributed the anomalously high fluid pressures of Ventura Avenue anticline to local tectonic forces (*i.e.*, continued folding of the anticline with resulting compaction). The Ventura Avenue anticline is in the thick Pliocene trough of the Santa Clara Valley between two opposing thrusts—the San Cayetano thrust on the north and the Oak Ridge thrust on the south (*see* Bailey and Jahns, 1954, cross section, p. 95-96; Bailey, 1954). Both thrusts presumably are still active. More than 16,500 ft of Pliocene beds are known to be involved in the Ventura Avenue fold (California Division of Oil and Gas, 1961; Watts, 1948). It seems probable that the opposing thrusts have caught part of the Pliocene rocks in the sedimentary trough in a compressive vise; the result probably is a discontinuous anticlinal fold that dies out at depth. The anomalous high fluid pressures thus may reflect only the continued local tectonic compaction of the sediments involved in this fold form. A well-developed and mapped system of high-angle reverse faults is present at depth within the Ventura Avenue anticline. The anomalously high fluid potentials are developed within this reverse fault system. The faults probably serve as lateral transmissibility barriers; their presence probably facilitates greatly the development of the high fluid potentials.

Recent drilling along the north edge of the Los Angeles basin has revealed a series of en echelon anticlines along Wilshire Boulevard from downtown Los Angeles to Santa Monica; this trend is essentially parallel with the south edge of the Santa Monica Mountains. The pore fluid pressures at depth on some of these structures have been reported to be slightly in excess of hydrostatic values. The magnitude of these excessive pressures apparently is small, probably in the range of the anomalous fluid pressures at Wheeler Ridge. The slightly anomalous potentials could originate either from small continuing tectonic movements of the Santa Monica Mountains and related features, or from a subsurface gravitational flow system from the Santa Monica Mountains; I favor a tectonic origin.

Regional Description

My interpretation of the geographic distribution of the several fluid potential domains is shown in Figure 16. Figure 17 is a generalized map of the geographic distribution of the Franciscan and granitic provinces in, and adjacent to, the California Coast Ranges. Their distribution in the Transverse Ranges is uncertain. The map reflects my interpretation of the probable distribution of these rock types within and offshore from the Los Angeles basin (*cf.* Schoellhamer and Woodford, 1951). Following the suggestion of many workers, I have interpreted the Catalina Schists of the Los Angeles basin to be Franciscan equivalents. Figures 16 and 17 show that there is only one regional area in California of anomalously high fluid pressures that approach lithostatic values. This area is underlain by a thick sedimentary mass in a particular structural setting. The anomalous fluid pressures are confined to the mudstone and siltstone of the Franciscan and Great Valley sequences, except under some anticlines associated with the West Side fault. The structural setting is the area bounded on the west by the San Andreas fault, on the east by the postulated West Side fault, and on the south by the Garlock fault; the northern boundary is undetermined, but it is presumed to be the northern termination of the San Andreas fault in the Cape Mendocino area. The western and southern boundaries are interpreted as being sharply terminated by the bounding faults; the eastern boundary, however, extends within the Great Valley sequence a short distance east of the buried West Side fault (Fig. 5).

Evidence for the hydrostatic pressure domains shown in Figure 16 is based in part on direct fluid pressure measurements of near-hydrostatic fluid pressures in the Los Angeles and Ventura basins, the Salinas Valley, and parts of the Sacramento and San Joaquin Valleys. I have not studied the hydrogeology of the area southwest of the Nacimiento fault and north of the Transverse Ranges (*i.e.*, the Santa Maria province) in similar detail. Careful reconnaissance investigations, however, reveal no anomalous fluid potentials or water chemistry attributable to abnormally high fluid pressures.

No pressure data are available from the Mojave province. The rocks of that province are dominantly granitic or volcanic or arkosic; the development of any measurable fluid pressure anomalies within such rocks generally would be very limited, even if they were undergoing active compaction. Formation-water chemistry data would not be diagnostic within such high transmissibility rocks.

In addition to the criteria indicated, a phenomenon that I believe accompanies the regional presence of anomalous high fluid potentials is the continuing release from solution in the ascending waters of a dominantly methane-rich free gas phase. The origin of the dry gas province of the Sacramento Valley is related, in my opinion, to this phenomenon. The absence of dry gas areas and active gas seeps over the provinces that have dominantly near-hydrostatic fluid pressures suggests further confirmation of their fluid-pressure environment. However, the absence of a dry gas province on the west side of the San Joaquin Valley cannot be explained at present.

In summary, the near-lithostatic pore fluid pressures are not related to any particular lithologic type or metamorphic grade. Near-lithostatic fluid pressures are present in the Franciscan and Great Valley sequences only in a specific structural setting. In the Santa Maria, Ventura, and Los Angeles basins, both Franciscan and/or Great Valley sequence rocks are present with near-hydrostatic fluid pressures.

Alternative Interpretations of High Fluid Potentials

Only two interpretations can explain the regional presence of these near-lithostatic fluid pressures—tectonic compaction or temperature-driven dehydration reactions. It is my opinion that the origin is related to currently active tectonic processes. Gravitational flow, physical-chemical phenomena, and fossil pressures cannot cause these particular anomalies. Gravitational flow is impossible, as there is no site in the California Coast Ranges with sufficient elevation to account for the highest fluid potentials found at depth. Physical-chemical phenomena also are impossible, as there is no indication of sufficient potential differences within the chemical, electrical, or thermal environments to explain the extraordinary anomalies.

A first generation of anomalously high fluid potentials probably developed during the deposition of the Franciscan and Great Valley sequences by continued subsidence, deposition, and gravitational compaction. High fluid potentials of similar origin are present today at depth in Gulf Coast rocks (Cannon and Craze, 1938;

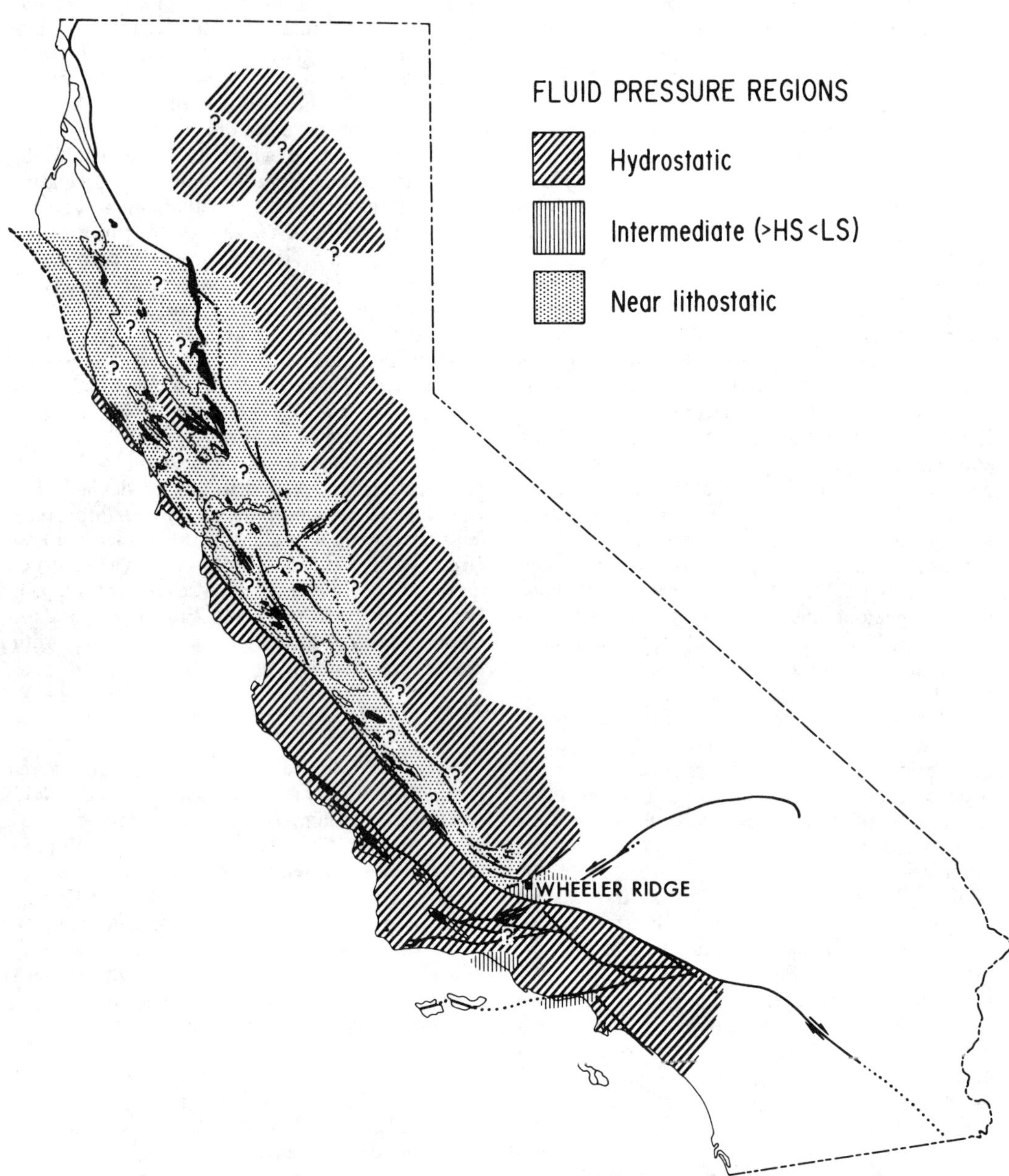

Fig. 16—Geographic distribution of fluid-potential zones in California. Principal faults are indicated.

Dickenson, 1953; Dickey *et al.*, 1968; Hubbert and Rubey, 1959; Hottman, 1967; Jones, 1967; MacGregor, 1965; Rubey and Hubbert, 1959; Wood, 1967). The explanation of the present anomalies of the California Coast Ranges as fossil remnants of an early generation is rejected on several grounds.

1. The observed distribution of anomalously high fluid potentials is not in accord with the pattern to be expected if fossil pressures were a consideration. The higher potentials should match the areas with the greatest amount of sedimentation, but they do not. Their distribution is clearly related to the present Coast Ranges.

2. Of greatest importance in any consideration of fossil fluid pressures is time. How long can an unbalanced fluid pressure be maintained? The limiting factor is the permeability of the shales. Laboratory experiments and calculations of in-place permeability have been made based on various assumptions. The decay rates of fossil fluid pressures are critically dependent on the assumed transmissive nature of the shales. The results indicate that within the ranges of shale permeability (10^{-3} to 10^{-9} md) commonly assumed to be representative, fossil fluid pressures cannot be sustained for long periods of geologic time. Granted that most laboratory measurements probably give higher permeabilities than are representative, it still is unlikely that significant anomalous fluid pressures would be preserved in a disequilibrium environment for periods longer than tens or hundreds of thousands of years. Bredehoeft and Hanshaw (1968) and Hanshaw and Bredehoeft (1968) have discussed some of the problems of fluid pressure maintenance and decay in low permeability sediments. It is manifestly impossible for the present anomalously high fluid pressures of the Sacramento Valley-Coast Ranges province to have originated in and persisted since Cretaceous-Eocene time—a span of approximately 60 m.y.

There are three principal arguments against temperature-driven dehydration reactions as an explanation for the anomalous fluid potentials. If current thermal dehydration were responsible, the fluid equipotential surfaces should parallel isothermal surfaces. Figures 3 and 4 show that the fluid equipotential surfaces and the near-lithostatic fluid pressure zone are not parallel with the ground surface but rise toward the west. My studies show that the isothermal surfaces are essentially parallel with the ground surface and, locally, rise sharply to form small closed features whose origin I attribute to forced thermal convection by channel flow of water upward along faults. Thus, the regional isothermal surfaces are not parallel with the fluid equipotential surfaces. Moreover, there are several different regions in the California Coast Ranges that have Franciscan and/or Great Valley rocks at metamorphic grades similar to those found within the belt of anomalously high fluid potentials but where near-hydrostatic conditions prevail. The pronounced discordance of the fluid potentials in rocks of the same metamorphic grade in different structural settings argues against any ancient metamorphic contribution to the fluid potential anomalies.

Regional Interpretation—Structural Relations

The cross section in Figure 13 places the structural relations pertaining to the belt of near-lithostatic fluid pressures in the California Coast Ranges in perspective. This section is drawn along the general line *C-C′* indicated in Figure 1, but a section similar in its principal features could be drawn anywhere through the central California Coast Ranges. The cross section is diagrammatic and interpretative, but an attempt was made to draw the Great Valley-West Side fault area to scale. I have no accurate knowledge of the thicknesses of the Franciscan or of the depths to the oceanic crust across this region. Thompson and Talwani (1964) placed this depth at approximately 12 km and the depth to the mantle has been estimated at 20-25 km (Healy, 1963; Eaton, 1966). I have constructed the section so that the continental crust is slightly thicker west of the fault in the opinion that, if the Salinas block represents a segment of granitic continental crust displaced to the northwest along the San Andreas fault, it should be thicker than the adjoining Franciscan sequence which presumably lies on oceanic crust. The oldest part of the Great Valley sequence also should have been deposited on oceanic crust; such a relation now has been found in the field by Bailey *et al.* (1970).

The buried West Side fault and the faulted Franciscan-Great Valley contact are shown on the cross section. Serpentinites are present along the fault, but their origin is uncertain. It is important to the geology of the California Coast Ranges to understand the origin of the serpentinites and the probable geographic distribution of the Franciscan and Great Valley sequences at the time of deposition.

Dietz (1963) advanced the hypothesis that uncoupling of the oceanic from the continental crust would occur during the onset of the tectonic phase of the orogenic cycle after the geosynclinal sediments had accumulated as continental rises against the continent margins. He further specu-

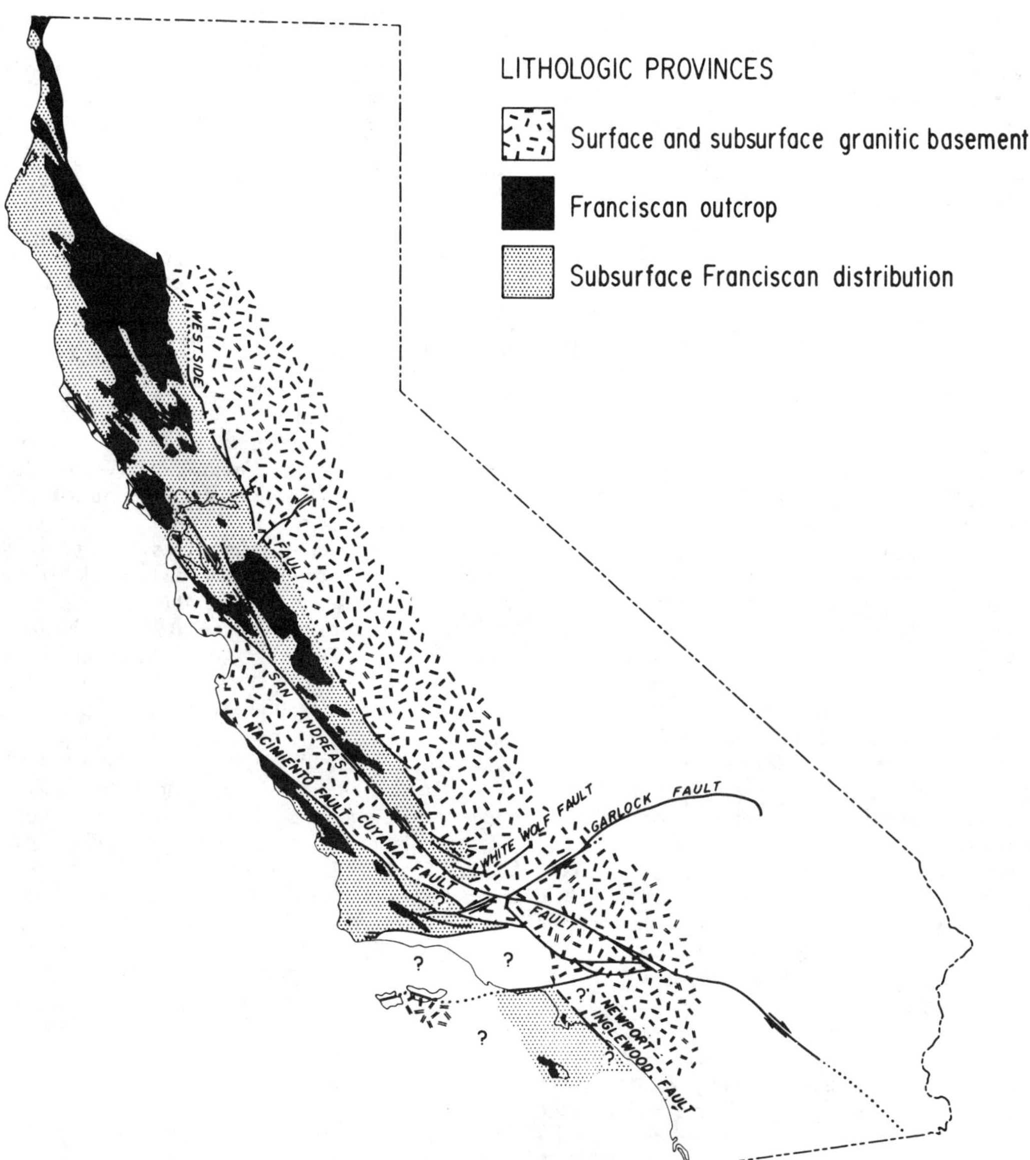

Fig. 17—Map of regional surface and subsurface geographic distribution of granitic and Franciscan provinces and of major faults in California.

lated that the oceanic crust and its underlying mantle would be shoved under the continental crust by oceanic crustal spreading; in this process, according to Dietz, blobs of ultramafic material from the upper surface of the mantle would be scraped off, intruded into the geosynclinal rocks, and later altered to serpentinite. Hess (1965) suggested that the oceanic crust is composed principally of serpentinite with a 1-2-km veneer of basalt; Coleman (1971) rejected this suggestion.

Some aspects of Dietz's hypothesis are appealing. It seems plausible that uncoupling, in the sense that Dietz used that term, of the oceanic and continental masses, accompanied by sliding of oceanic material under the continental margin, might have preceded deposition of the near-contemporaneous eugeosynclinal Franciscan and miogeosynclinal Great Valley sequences and controlled the site of their deposition at the continental margin. The critical point, however, is whether or not near-continuous blobs of ultrabasics-serpentinite would be available, either by being scraped off or by being intruded into the overlying material. If so, it would be possible for a near-continuous zone of serpentinite, or ultrabasic intrusions altered to serpentine as advocated by Coleman (1971), to form on the continental slope parallel with the continent margin. If intrusion and diapiric rise of the serpentine-ultrabasic material accompanied continued subsidence and deposition along the continental margin, it would be possible for this serpentinite zone to have been present as a sediment trap dividing the nearshore miogeosynclinal (Great Valley sequence) depositional environment from the seaward eugeosynclinal region. A somewhat similar proposal was made by Bailey and Blake (1969). The recurrence of intermittent sequences of oceanic basalt flows (spilites) throughout the Franciscan lends credence to the concept of intermittent but recurrent rise of ultrabasics-serpentinite. Similar sediment traps, bounded presumably by fault blocks of granitic material, have been demonstrated by seismic profiling to be present off the California Coast (Curray, 1966; Curray and Nason, 1967).

The postulated existence of such a serpentinite barrier at the time of deposition of the Franciscan and Great Valley sequences would have several consequences. The first is that it would dispose of two of the critical problems in California geology—the age and origin of the serpentinite belt that now separates Franciscan from Great Valley sediments, and the spatial problem as to how and where these two geosynclinal sequences were deposited contemporaneously with their distinctly separate facies. The simplicity of this proposed solution is appealing.

According to this model, the serpentine belt would be in fault contact with both the Franciscan and Great Valley sequences—which it is. Various degrees of thrusting along the serpentinite belt, subsequent to deposition of the Great Valley sequence, would rotate the western edge of the Great Valley sequence into its present structural position. One consequence of this postulate is that detrital serpentinite should exist at various sites and in various stratigraphic zones in Great Valley, as well as in the Franciscan sequences. At one such site near Wilbur Springs (T14N, R5W, Colusa Co.), an abundance of detrital serpentinite has been mapped in the Knoxville of the Great Valley sequence (Averitt, 1945; Lawton, 1956; Lockwood, 1971; Moisseeff, 1966; Owens, 1941; Taliaferro, 1943). West of Lake Berryessa, detrital serpentinite appears to be present at the Knoxville-Lower Cretaceous contact (A. W. Moisseeff, written commun., April 1968). Farther south, detrital serpentinite may be present at depth in the unexposed lower intervals of the Great Valley section. Lockwood (1971) suggested that some elongate serpentinite bodies in the Franciscan, now regarded as sills, may be sedimentary beds of ultramafic detritus. Detrital serpentinite has been found in sediments of various ages adjacent to the serpentinite core of the New Idria diapir (Eckel and Myers, 1946). This postulate concerning the original spatial distribution of the Franciscan and Great Valley sequences and the intervening serpentinite belt should be amenable to confirmation or rebuttal by field and subsurface investigations.

Direct analogues of the West Side fault are present, in my opinion, elsewhere in California. The Nacimiento fault separating the Franciscan Santa Maria from the granitic Salinas province is one of the proposed analogues to this West Side fault. Page (1970) postulated that the Sur-Nacimiento fault zone originated at the former margin of the continent and represents the boundary between the granitic Salinas block and the Franciscan trench deposits. South of the La Panza granitic mass, the Cuyama and Ozena faults may become analogues to the Midland fault of the Rio Vista region, which is proposed as the near-surface expression of the deep West Side fault. The analogy can be carried further—namely, the thick Eocene section described by Vedder *et al.* (1967) between the Cuyama-Ozena and Nacimiento faults would be the direct analog of the thick Eocene section of the Rio Vista area. These analogues are not unlikely if segments of the Nacimiento and Cuyama-Ozena faults were formed during Mesozoic time west of the continental crust and south of the present Great Valley, and subsequently were shoved northwest by right-lateral movement on the San Andreas fault. I postulate also that the Newport-Inglewood fault zone in the Los Angeles basin and its probable southerly offshore extension as far as southern Baja California is the southern analogue to the proposed West Side fault (Fig. 17).

Time of Development—High Fluid Potentials

The presence of near-perfect plastic properties in this extensive belt of rocks shows that they have been, and still could be, capable of structural deformation by diapirism and low-angle thrust faulting. Coleman and Ho (1967), working with shallow sediments of the Gulf Coast, showed that clayey sediments with relatively high pore fluid pressures do form diapirs. Diapirs already have formed to some extent within the Coast Ranges. Several authors have suggested that the New Idria (Eckel and Myers, 1946; Oakeshott, 1965, 1968; Page, 1966) and Mount Diablo (Colburn, 1961; Oakeshott, 1965, 1968; Page, 1966; Pampeyan, 1963; Taff, 1935) areas are diapirs of Franciscan material injected into the Great Valley and younger sequences. Along the buried segments of the West Side fault, diapirism may be the mode of structural deformation that is principally responsible for the elongate, sharp line of folds on the west side of the Great Valley. Large-scale diapirism is involved, together with an earlier thrust, in the regional uplift of the Diablo Range (Page, 1966), which has Franciscan rocks exposed in its core, and may be involved in large uplifts such as the Temblor Range. I also suspect that diapirism may be responsible for the erratic and limited distribution of some of the enigmatic Franciscan localities where rocks of high-grade metamorphic facies are found. Nearly all the eclogite and glaucophane-schist areas, for example, are very localized and in fault contact with other Franciscan rock assemblages.

The time of development of these known and suspected diapirs is of interest. Diapirism within shales and other low-permeable rocks probably is a direct indication of the presence at depth of near-lithostatic pore fluid pressures. Thus, the times of diapir movement must be a direct record of the time of development of such abnormally high fluid potentials. If folding on the anticlines on the west side of the Great Valley is a result of diapirism at depth, as I suspect, then their folding histories also would reflect the times when such high fluid potentials are developed at depth.

The presence of sedimentary detritus suggests that the core of the New Idria diapir was sufficiently active tectonically to be uplifted and exposed to surface erosion during Late Cretaceous, middle Miocene, and Pliocene times (Eckel and Myers, 1946). Unconformities indicate pre-Paleocene, middle Eocene, and middle Pliocene folding of the Mount Diablo diapiric anticline; the Franciscan core of the diapir was not exposed at the surface until late Pliocene or early Pleistocene. Dickinson's work at Table Mountain (1966) suggested the presence of an anticline with a diapiric core that was active tectonically during the Pliocene-Pleistocene. The general history of the folds on the west side of the Great Valley is that they began in the Miocene to Pliocene; the Pliocene-Pleistocene represents the time of greatest folding intensity. Folding, which is still active around certain structures, may have begun on a given fold at any time from early Miocene to middle Pliocene. Folds with Miocene development generally had limited folding in the early Miocene and only modest deformation in the middle Miocene; folding became much more marked in the late Miocene and generally increased in the Pliocene.

Thus, known diapirism and folding on the west side anticlines suggest the development of two episodes of abnormally high fluid potentials at depth—an earlier episode in the pre-Paleocene and Eocene, and a later episode beginning gradually in the middle Miocene and increasing steadily in intensity into the Pliocene-Pleistocene. It is difficult to judge the relative intensity of any post-Pleistocene folding other than to say that it exists at some sites.

There also is evidence for limited late Cenozoic thrusting in the Coast Ranges. Page (1966) reviewed the data for a post-Miocene thrust and a separate post-Pliocene-Pleistocene thrust east of the San Andreas fault, and suggested that more careful mapping probably would reveal more; he interpreted the limited data on these thrusts as being compatible with the general northeast-southwest shortening responsible for the Coast Range folding during Pliocene-Pleistocene time.

Recent workers (Bailey *et al.*, 1964; Berkland, 1971; Blake *et al.*, 1967; Brown, 1964a, b; Brown and Rich, 1961; Dickinson, 1965; Ghent, 1964; Irwin, 1960, 1964, 1966; Swe and Dickinson, 1970) have suggested that the Great Valley sequence has been thrust at low angles westward over the Franciscan sequence in several localities. The data on the age of the postulated thrusting are limited, but it probably occurred in the early Cenozoic. Swe and Dickinson (1970) argued for an Oligocene age.

It is probable that the current structural belt of near-lithostatic fluid pressures was not in existence during Late Cretaceous-early Tertiary time and hence could not have been responsible for the development of the low-angle thrusts. Two separate episodes of high fluid-pressure genesis within the geosynclinal Franciscan and Great Valley sedimentary sequences are suggested by the history of diapirism, folding, and thrusting. The first episode should have occurred contemporaneously with deposition of these sediments from Late Jurassic through Cretaceous times. Rapid subsidence and continued deposition in the geosynclinal environments of low-transmissibility mudstone and siltstone, whose maximum thickness may range from 50,000 to 60,000 ft, imply that continuing gravitational compaction would have generated anomalously high fluid potentials at depth. The situation would be analogous to the occurrence of anomalously high fluid potentials today at depth within the shaly section of the thick Gulf Coast sedimentary wedge. With the cessation of subsidence and termination of deposition in Late Cretaceous-late Eocene time in the geosyncline of the California Coast Ranges, the dynamic period of generation of anomalous high fluid potentials by gravitational compaction should have ceased. Decay of the fossil anomalous fluid potentials then should have occurred—probably quite rapidly. Early Cenozoic tectonic compression also may have contributed to the development and maintenance of these high fluid potentials. The first generation of these anomalous potentials would have been responsible for the plastic properties of the rocks in any diapirism or extensive low-angle thrusting during early Tertiary time. The second cycle of anomalous fluid potentials developed principally in the late Cenozoic, within those Franciscan and Great Valley rocks caught and squeezed between the Salinas and the Sierran-Klamath block. The documented diapiric folding and thrusting during the late Cenozoic (Miocene and particularly Pliocene-Pleistocene) are probably direct registrations of the development of anomalously high fluid potentials at these times by this proposed mechanism.

Regional Interpretation—Origin of High Fluid Potentials

The structural model in Figure 13 indicates that the sedimentary mass with the near-lithostatic fluid pressures is caught between eastern and western granitic blocks. If this part of the continental crust is in compression, the two relatively incompressible granitic blocks would act like a vise closing on a blob of wet clay. If the clays were confined in three dimensions, compaction caused by tectonic processes would increase the fluid potential of the waters occupying the interstitial pores. If an increase in tectonic pressure under isothermal conditions can initiate metamorphic-type dehydration reactions, such reactions would contribute to the origin of the anomalous fluid potentials.

The sediments above the Sierran-Klamath granitic block possess anomalously high fluid potentials for a distance east of the buried West Side fault (Fig. 13). This probably is due to the fact that the granitic block rises to the surface over a long distance along a gently sloping homocline. Compression and crustal shortening between the Sierran-Klamath and Salinas granitic blocks would involve the rocks on top of the homoclinal-sloping Sierran block, as well as those caught in the main part of the vise west of the West Side fault. A sharp vertical boundary to the zone of near-lithostatic fluid pressures probably is present on the west side of the sedimentary mass adjacent to the near-vertical San Andreas fault and the granitic Salinas block. Thus, compressive forces within the Salinas block would be transmitted through the relatively rigid granitic basement.

Geodetic studies (Burford, 1967) of the Coast Ranges in the vicinity of the San Andreas fault north of the Garlock fault and south of Hollister demonstrate maximum northeast-southwest contraction and some northwest-southeast contraction or extension. Contraction probably exerts compressive stress in all horizontal directions. Compton's work (1966) in the Santa Lucia Range within the Salinas block indicated that the surface Pliocene-Pleistocene folds require a northeast-southwest basement shortening by compression of approximately 12 percent. His petrofabric analysis indicated that the Pliocene—Pleistocene lineations in the near-surface rocks should be produced by a deep-rooted northeast to southwest flowage in the lower crust or upper mantle.

The most attractive concept to me concerning the origin of the compressive forces that could produce this regional zone of anomalously high fluid potentials is east-west spreading, during late Cenozoic time, of the Basin-Range province in Nevada and Utah opposite the belt of high fluid potentials in the California Coast Ranges. The Great Basin as a whole has been an area of active east-west crustal extension during most of the middle and late Cenozoic. The entire Great Basin can be divided into a southern area of large extension with dominantly mid-Cenozoic movement; a central region with large extension of principally late Cenozoic movement; and a northern region with moderate extension, where movement also occurred in the late Cenozoic.

The central Great Basin (Nevada-Utah) is well known for its late Cenozoic extension. Estimates of the amount of east-west extension range from 30 to 60 mi (Hamilton and Myers, 1966; Stewart, 1971) to 100 mi at the latitude of Yerrington, Nevada (Proffett, 1971, 1972), to approximately 180 mi (Thompson, 1959). Proffett's estimate is based on detailed mapping and appears to me to be the most accurate. Stewart (1971) believed that this extension commenced at a low rate approximately 17 m.y. ago; he stated that the rate increased considerably about 11 m.y. ago, and increased still more during the last 7 m.y. Proffett (1971, 1972), however, believed that most of the extension occurred between 12 and 4 m.y. ago, and has been at a greatly reduced rate for the last 4 m.y. He stated that extension is greater on the east and west sides of the Great Basin, and is least in the middle.

Opposing northeast-southwest forces must be present to cause a zone of compression in the California Coast Ranges resulting in the high fluid potential belt east of the San Andreas. If the Sierran-Klamath block has been moved—and still is moving—westward, or more probably southwestward, by continued extension of the central Great Basin, then the Salinas block west of the San Andreas must have been an independent stress field to provide forces that resist the westerly movement of the Sierran-Klamath Block. The San Andreas is an active right-lateral strike-slip fault along which there has been movement for a long time. It now appears virtually certain that the present San Andreas is a transform fault, in terms of the hypothesis of plate tectonics, and represents the present boundary between the North American and Pacific plates (Atwater, 1970; McKenzie and Parker, 1967; Morgan, 1968; Vine and Wilson, 1965). This concept is extremely important. There is an entirely different stress field in the Pacific plate that is moving northwestward relative to the North American plate and is separated by the San Andreas fault. In addition, the Sierran-Klamath block is moving westward or southwestward by continued central Great Basin extension; this westerly motion is terminated by compression of the rocks on both sides of the San Andreas. These compressive effects are more limited within and overlying the granite basement of the Pacific plate (Salinas block) than within the Franciscan and Great Valley shale mass directly east of the fault on the western edge of the North American plate. As a result, the belt of highly compressible shales just east of the San Andreas fault is caught in a closing vise whose jaws are formed of relatively incompressible granite. I conclude that the observed zone of near-lithostatic fluid pressures in central and northern California is dominantly a late Cenozoic phenomenon resulting from extension of the central Great Basin, and that it is still active today.

There are no data concerning the fluid potentials adjacent to the San Andreas in the Mojave province of southern California. No low transmissibility rocks are adjacent to the San Andreas in that region, and extension in the southern Great Basin on the east occurred principally in the mid-Cenozoic and is essentially nonexistent today. Farther east, however, the Rio Grande graben area has undergone late Cenozoic extension and is still active. Greatly reduced compression as a result of this extension might cause the development of abnormally high pore fluid pressures in a very limited zone at considerable depth adjacent to the San Andreas in the Mojave region.

The moderately high fluid potentials of the Wheeler Ridge area most probably are related to the compression derived from local northward thrusting along the Pleito, Wheeler Ridge, and perhaps other thrust faults in the well-known vee formed by the intersection of the active right-lateral San Andreas fault with the active left-lateral Garlock fault.

Geologic Phenomena Related to High Fluid Potential Zone

One of the most important implications of this paper is the demonstration that fluid pressures within rocks can serve as extremely sensitive strain gauges recording the distribution of local or regional structural movements. Bodvarsson (1970) concluded that pore-fluid pressures serve as sufficiently sensitive strain gauges to permit the detection of minor dilational effects, such as are created by tides and seismic waves. The magnitude of any abnormally high pressure anomaly is dependent on the permeability of the rocks, as well as the magnitude and duration of the applied stress and the compressive properties of the rock. In practical terms, the presence of an extensive low-permeability section of mudstone or shale permits the development of large anomalies that can be detected and mapped most easily by direct pressure measurements or, indirectly and with less precision, by the presence of extensive amounts of formation water interbedded with and/or adjacent to large shale masses whose chemistry is typical of a relatively low-salinity, membrane-effluent type water. Though not a subject of this paper, large-scale structural dilation probably is responsible for the development of extensive areas of abnormally low fluid potentials within low-permeability rocks such as are present within the Cretaceous sediments of the San Juan basin of northwestern New Mexico (Berry, 1959); again the fluid potentials provide a unique strain gauge for determining such movements.

Various phenomena may be caused by the existence of this belt of sediments with near-lithostatic fluid pressures. Most of these phenomena are related to rheological properties of rocks with such fluid pressures. Such rocks will behave as near-perfect plastic materials (Handin *et al.*, 1963; Hubbert and Rubey, 1959; Rubey and Hubbert, 1959).

Certain areas along the San Andreas fault are unique in that they are characterized by nearly continuous fault creep and a high frequency of low-magnitude earthquakes. The principal area of fault creep is in central California between the Pinnacles National Monument and the Hollister-Watsonville area; movement also characterizes parts of the Hayward and Calaveras faults which join the San Andreas in the Hollister region (Bolt *et al.*, 1968; Burford, 1971). Nason (1971) has made a definitive review of this fault creep movement. The greatest amount of creep is centered near Hollister. Byerlee and Brace (1970, 1972) suggested that the presence of high pore-fluid pressures may account for stable creeping motion exhibited by some faults, such as the San Andreas-Hayward—Calaveras system in the locality described above. My data do not permit any definitive assessment of the actual pore-fluid pressures in this particular area; my regional interpretation, however, suggests that the Franciscan and Great Valley sections in this vicinity should have near-lithostatic pore-fluid pressures at depth. The area of principal creep motion is located where the currently active Calaveras-Hayward system departs from the also active San Andreas system. The stresses induced by opposing motions in this region may generate particularly high pore-fluid pressures at particularly shallow depths, and thus may facilitate creep-type motion. More precise delineation of the pore-fluid pressures with depth is needed adjacent to the San Andreas system to attempt any explanation as to why certain segments of this system deform by relatively stable fault-creep motion, whereas others do not.

Another phenomenon that characterizes San Andreas earthquakes is that they are essentially shallow-focus shocks (5-10 km). Only a small percentage (McEvilly *et al.*, 1967, cited 25 percent) of the total energy related to a particular earthquake along the San Andreas is released in the initial shock; the rest of the energy is released in a long series of aftershocks (McEvilly, 1966; McEvilly *et al.*, 1967). The energy in each of these aftershocks decreases steadily. Some workers have suggested that the shallow focus means that the San Andreas may be a shear zone extending only to shallow depth; current thinking in terms of plate tectonics, which considers the San Andreas as a transform fault (*cf.* Atwater, 1970; Dickinson *et al.*, 1972; McKenzie and Parker, 1967; Morgan, 1968; Vine and Wilson, 1965), clearly requires shear through the crust to depths far below the depth of focus of the earthquakes. I suggest that the origin of these seismic phenomena is related to the presence of the abnormally high pore-fluid pressures adjacent to the fault at depth. Benioff (1962) suggested that shallow-focus earthquakes might be produced if a plastic substance were present on at least one side of the fault. Such criteria should exist along the San Andreas fault in central and northern California, and might be present on a much more limited basis in southern California. The depth of focus of the earthquakes may be the depth at which the zone of near-lithostatic fluid pressures intersects the fault.

Instead of low-angle thrust faulting, diapirism and diapiric folding have been the favored style of late Cenozoic deformation within this high fluid potential belt, even though more detailed surface and subsurface information, particularly in northern California, might reveal that thrusting is more important than it now appears. The probable reason for this dominance of diapirism must be the fact that the compression that generates these abnormal fluid potentials adjacent to the San Andreas, as a result of central Great Basin extension, simply is not large enough to cause sufficient crustal shortening for large-scale development of thrust faults. This compression must be secondary in magnitude to the shearing stresses related to movement of the Pacific and North American plates that cause the San Andreas transform fault. Diapirism must be the favored mode of deformation when rocks have become plastic from the development of high pore-fluid pressures but where crustal shortening is not the dominant condition; low-angle thrusting, however, must be the preferred deformation style where crustal shortening dominates and rocks with plastic properties are present or develop. The stage is still set for further extensive structural deformation. In a belt 400-500 mi long and 25-80 mi wide (narrower on the south and wider on the north) most of the rocks have near-perfect plastic properties. East-west compressive forces probably are responsible for tectonic compaction. The setting appears ideal for future diapirism and the possible development of low-angle thrust

faulting. The coast of California promises to continue to be a lively and dynamic region.

Selected References

American Association of Petroleum Geologists, 1951, Cenozoic correlation section from northside Mt. Diablo to eastside Sacramento Valley through Rio Vista-Thornton-Lodi gas fields, California: Am. Assoc. Petroleum Geologists, Pacific Sec., Correlation Sec. 1.

Atwater, T., 1970, Implications of plate tectonics for the Cenozoic tectonic evolution of western North America: Geol. Soc. America Bull., v. 81, p. 3513-3536.

Averitt, P., 1945, Quicksilver deposits of the Knoxville district, Napa, Yolo, and Lake Counties, California: California Jour. Mines and Geology, v. 41, no. 2, p. 65-89.

Bailey, E. H., and M. C. Blake, Jr., 1969, Tektonicheskoye razvitiye zapadnoy Kalifornii v pozdnem mezozoye (Tectonic development of western California in the late Mesozoic): Geotektonika, no. 3, p. 17-30.

——— ——— and D. L. Jones, 1970, On-land Mesozoic oceanic crust in California Coast Ranges, *in* Geological Survey Research, 1970: U.S. Geol. Survey Prof. Paper 700-C, p. C70-C81.

——— W. P. Irwin, and D. L. Jones, 1964, Franciscan and related rocks, and their significance in the geology of western California: California Div. Mines and Geology Bull. 183, 177 p.

——— P. D. Snavely, and D. E. White, 1961, Chemical analyses of brines and crude oil, Cymric field, Kern County, California: U.S. Geol. Survey Prof. Paper 424-D, p. D306-D309.

Bailey, T. L., 1954, Geology of the western Ventura basin, Santa Barbara, Ventura, and Los Angeles Counties, Map Sheet 4, *in* Geology of Southern California: California Div. Mines Bull. 170, scale 1 in. to 6 mi.

——— and R. H. Jahns, 1954, Geology of the Transverse Range province, Southern California, *in* Geology of Southern California: California Div. Mines Bull. 170, p. 83-106.

Barbat, W. F., 1971, Megatectonics of the Coast Ranges, California: Geol. Soc. America Bull., v. 82, p. 1541-1562.

Barton, D. C., 1938, Gravitational methods of prospecting: Science of Petroleum, v. 1, p. 366-381.

——— 1944, Lost Hills, California—an anticlinal minimum, *in* D. C. Barton, Case histories and quantitative calculations in gravimetric prospecting: Am. Inst. Mining and Metall. Engineers Tech. Pub. 1760; Petroleum Technology, v. 7, no. 6, p. 22-49.

Benioff, H., 1962, Movements on major transcurrent faults, Chap. 4, *in* Continental drift: New York, Academic Press (Internat. Geophysics Ser., v. 3), p. 103-133.

Berkland, J. O., 1971, New occurrence of Cretaceous and Paleocene strata within Franciscan terrane of the northern Coast Ranges, California (abs.): Geol. Soc. America Abs. with Programs, v. 3, no. 2, p. 81-82.

Berry, F. A. F. 1959, Hydrodynamics and geochemistry of the Jurassic and Cretaceous Systems in the San Juan basin, northwestern New Mexico and southwestern Colorado: Ph.D. dissert., Stanford Univ., 192 p.

——— 1960, Geologic field evidence suggesting membrane properties of shales (abs.): Am. Assoc. Petroleum Geologists Bull., v. 44, p. 953-954.

——— 1965, Origin and tectonic significance of high fluid pressures, Central Valley and Coast Range, California (abs.): Am. Assoc. Petroleum Geologists Bull., v. 49, no. 3, p. 335.

——— 1966, Proposed origin of subsurface thermal brines, Imperial Valley, California (abs.): Am. Assoc. Petroleum Geologists Bull., v. 50, no. 3, p. 644-645.

——— 1967a, Geothermal brines, *in* Natural gas, coal, ground water—exploring new methods and techniques in resources research—8th Western Resources Conf., Colorado School Mines, 1966: Western Resources Papers, v. 8, Colorado Univ. Press., p. 155-169.

——— 1967b, Role of membrane hyperfiltration on origin of thermal brines, Imperial Valley, California (abs.): Am. Assoc. Petroleum Geologists Bull., v. 51, p. 454-455.

——— 1969, Relative factors influencing membrane filtration effects in geologic environments, *in* Geochemistry of subsurface brines: Chem. Geology, v. 4, p. 295-301.

——— (in press), Geothermal resources.

——— and B. B. Hanshaw, 1960, Geologic evidence suggesting membrane properties of shales (abs.): 21st Internat. Geol. Cong., Copenhagen, p. 1-209.

Blake, M. C., Jr., W. P. Irwin, and R. G. Coleman, 1967, Upside-down metamorphic zonation, blueschist facies, along a regional thrust in California and Oregon: U.S. Geol. Survey Prof. Paper 575-C, p. C1-C9.

Bodvarsson, G., 1970, Confined fluids as strain meters: Jour. Geophys. Research, v. 75, p. 2711-2718.

Bolt, B. A., C. Lomnitz, and T. V. McEvilly, 1968, Seismological evidence on the tectonics of central and northern California and the Mendocino escarpment: Seismol. Soc. America Bull., v. 58, p. 1725-1767.

Boyd, L. H., 1946, Gravity-meter survey of the Kettleman Hills-Lost Hills trend, California: Geophysics, v. 11, p. 121-127.

Bredehoeft, J. D., and B. B. Hanshaw, 1968, On the maintenance of anomalous fluid pressures: I., Thick sedimentary sequences: Geol. Soc. America Bull., v. 79, p. 1097-1106.

Brown, R. D., Jr., 1964a, Thrust-fault relations in the northern Coast Ranges, California: U.S. Geol. Survey Prof. Paper 475-D, p. D7-D13.

——— 1964b, Thrust-fault relations in the northern Coast Ranges, California: U.S. Geol. Survey Map GP-481.

——— and E. I. Rich, 1961, Geologic map of the Lodoga quadrangle, Glenn and Colusa Counties, California: U.S. Geol. Survey Oil and Gas Inv. Map OM-210, scale 1:48,000.

Buneev, A. N., P. A. Kryukov, and E. V. Rengarten, 1947, Experiment in squeezing out of solutions from sedimentary rocks: Akad. Nauk SSSR Doklady, v. 57, no. 7, p. 707-709 (in Russian).

Burford, R. O., 1967, Strain analysis across the San Andreas fault and Coast Ranges of California: Ph.D. dissert., Stanford Univ., 74 p.

——— 1971, Fault creep and related seismicity along the San Andreas fault system between Pinnacles National Monument and San Juan Bautista (abs.): Geol. Soc. America Abs. with Programs, v. 2, no. 3, p. 90-91.

Byerlee, J. D., and W. F. Brace, 1970, Modification of sliding characteristics by fluid injection and its significance for earthquake prevention (abs.): Am. Geophys. Union Trans., v. 51, no. 4, p. 423.

——— and ——— 1972, Fault stability and pore pressure: Science, v. 62, p. 657-660.

Byerly, P. E., 1966, Interpretations of gravity data from the central Coast Ranges and San Joaquin Valley, California: Geol. Soc. America Bull., v. 77, no. 1, p. 83-94.

California Division Mines and Geology, 1966, Gravity map of Geysers area: Mineral Inf. Service, v. 19, no. 9, p. 148-149.

California Division of Oil and Gas, 1961, California oil and gas fields maps and data sheets—Pt. 2, Los Angeles-Ventura basins and central coastal regions, p. 729-793.

Cannon, G. E., and R. C. Craze, 1938, Excessive pressures and pressure variations with depth of petroleum reservoirs in the Gulf Coast region of Texas and Louisiana: AIME Trans., Petroleum Development and Technology, v. 127, p. 31-38.

Clark, B. L., 1929, Tectonics of the Valle Grande of California: Am. Assoc. Petroleum Geologists Bull., v. 13, p. 199-238.

Colburn, I. P., 1961, The tectonic history of Mount Diablo, California: Ph.D. dissert., Stanford Univ., 276 p.

Coleman, J. M., and C. Ho, 1967, Early diagenesis and compaction in clays, *in* 1st Symposium on abnormal subsurface pressure, Proc.: Louisiana State Univ., p. 23-50.

Coleman, R. G., 1971, Petrologic and geophysical nature of serpentinite: Geol. Soc. America Bull., v. 82, p. 897-917.
Compton, R. R., 1966, Analyses of Pliocene-Pleistocene deformation and stresses in northern Santa Lucia Range, California: Geol. Soc. America Bull., v. 77, p. 1361-1380.
Craig, H., 1963, The isotopic geochemistry of water and carbon in geothermal areas, *in* Nuclear geology on geothermal areas (Spoleto, 1963): Pisa, Italy, Consiglio Nazionale delle Ricerche, Laboratorio di Geologia Nuclearee, p. 17-53.
——— 1966, Superheated steam and mineral-water interactions in geothermal areas (abs.): Am. Geophys. Union Trans., v. 47, no. 1, p. 204.
Curray, J. R., 1966, Geologic structure on the continental margin, from subbottom profiles, northern and central California, *in* Geology of northern California: California Div. Mines and Geology Bull. 190, p. 337-342.
——— and R. D. Nason, 1967, San Andreas fault north of Point Arena, California: Geol. Soc. America Bull., v. 78, p. 413-418.
Dallmus, K. F., 1958, Mechanics of basin evolution and its relation to the habitat of oil in the basin, *in* L. G. Weeks, ed., Habitat of oil: Am. Assoc. Petroleum Geologists, p. 883-931.
De Sitter, L. U., 1956, Structural geology: New York, McGraw-Hill, 552 p.
Dickenson, G., 1953, Geological aspects of abnormal reservoir pressures in Gulf Coast Louisiana: Am. Assoc. Petroleum Geologists Bull., v. 37, p. 410-432.
Dickey, P. A., C. R. Shriram, and W. R. Paine, 1968, Abnormal pressures in deep wells of southwestern Louisiana: Science, v. 160, p. 609-615.
Dickinson, W. R., 1965, Folded thrust contact between Franciscan rocks and Panoche Group in the Diablo Range of central California (abs.): Geol. Soc. America Spec. Paper 82, p. 248-249.
——— 1966, Table Mountain serpentinite extrusion in California Coast Ranges: Geol. Soc. America Bull., v. 77, p. 451-472.
——— D. S. Cowan, and R. A. Schweickert, 1972, Test of new global tectonics: discussion: Am. Assoc. Petroleum Geologists Bull., v. 56, p. 375-384.
Dietz, R. S., 1963, Collapsing continental rises: an actualistic concept of geosynclines and mountain building: Jour. Geology, v. 71, no. 3, p. 314-333.
Dolan, J. P., C. A. Einarsen, and G. A. Hill, 1957, Special applications of drill-stem test pressure data: AIME Petroleum Trans., v. 210, p. 318-321.
Eaton, J. P., 1966, Crustal structure in northern and central California from seismic evidence, *in* Geology of northern California: California Div. Mines and Geology Bull. 190, p. 419-426.
Eckel, E. B., and W. B. Myers, 1946, Quicksilver deposits of the New Idria district, San Benito and Fresno Counties, California: California Jour. Mines and Geology, v. 42, no. 2, p. 81-124.
Engelhardt, W. V., and K. H. Gaida, 1963, Concentration changes of pore solutions during compaction of clay sediments: Jour. Sed. Petrology, v. 33, no. 4, p. 919-930.
Garrison, L. E., 1972, Geothermal steam in The Geysers-Clear Lake region, California: Geol. Soc. America Bull., v. 83, p. 1449-1468.
Gester, G. C., and J. Galloway, 1933, Geology of Kettleman Hills oil field, California: Am. Assoc. Petroleum Geologists Bull., v. 17, p. 1161-1193.
Ghent, E. D., 1964, Petrology and structure of the Black Butte area, Hull Mountain and Anthony Peak quadrangles, northern Coast Ranges, California: Ph.D. dissert., Univ. California, Berkeley, 229 p.
Hamilton, W., and W. B. Myers, 1966, Cenozoic tectonics of the western United States: Rev. Geophysics, v. 4, p. 509-549.
Handin, J., R. V. Hager, Jr., M. Friedman, and J. N. Feather, 1963, Experimental deformation of sedimentary rocks under confining pressure—pore pressure tests: Am. Assoc. Petroleum Geologists Bull., v. 47, p. 717-755.
Hanshaw, B. B., 1962, Membrane properties of compacted clays: Ph.D. dissert., Harvard Univ., 113 p.
——— 1964, Cation-exchange constants for clays from electrochemical measurements, *in* Clays and clay minerals, 12th Natl. Conf., Atlanta, Georgia: New York, Macmillan Co., p. 397-421.
——— and J. D. Bredehoeft, 1968, On the maintenance of anomalous fluid pressures, Pt. 2, Source layer at depth: Geol. Soc. America Bull., v. 79, p. 1107-1122.
Healy, J. H., 1963, Crustal structure along the coast of California from seismic-refraction measurements: Jour. Geophys. Research, v. 68, p. 5777-5787.
Hess, H. H., 1965, Mid-oceanic ridges and tectonics of the seafloor, *in* Submarine geology and geophysics: London, Butterworth, Colston Papers 17, p. 327-332.
Hoots, H. W., T. L. Bear, and W. D. Kleinpell, 1954, Geological summary of the San Joaquin Valley, California, Pt. 8 *in* Chap. 2 *of* R. H. Jahns, ed., Geology of southern California: California Div. of Mines Bull. 170, p. 113-129.
Horner, D. R., 1951, Pressure build-up in wells: 3d World Petroleum Cong. Proc., Leiden, Netherlands, E. J. Brill, Sec. 11, p. 503-521.
Hottman, C. E., 1967, Occurrence and characteristics of abnormal subsurface pressures in the northern Gulf basin, *in* 1st Symposium on abnormal subsurface pressure, Proc.: Louisiana State Univ., p. 1-6.
Hubbert, M. K., 1953, Entrapment of petroleum under hydrodynamic conditions: Am. Assoc. Petroleum Geologists Bull., v. 37, p. 1954-2026.
——— and W. W. Rubey, 1959, Mechanics of fluid filled porous solids and its application to overthrust faulting, pt. 1 *of* Role of fluid pressure in mechanics of overthrust faulting: Geol. Soc. America Bull., v. 70, p. 115-166.
——— and D. G. Willis, 1957, Mechanics of hydraulic fracturing: Jour. Petroleum Technology, v. 9, p. 153-168; Discussion and reply: AIME Trans., Tech. Pub. 4597, June.
Irwin, W. P., 1960, Geologic reconnaissance of the northern Coast Ranges and Klamath Mountains, California, with a summary of the natural resources: California Div. Mines and Geology Bull. 179, 80 p.
——— 1964, Late Mesozoic orogenies in the ultramafic belts of northwestern California and southwestern Oregon: U.S. Geol. Survey Prof. Paper 501-C, p. C1-C9.
——— 1966, Geology of the Klamath Mountains province, *in* Geology of northern California: California Div. Mines and Geology Bull. 190, p. 19-38.
Jennings, C. W., and R. G. Strand, 1960, Geologic map of California Ukiah sheet: California Div. Mines, scale 1:250,000.
Jones, P. H., 1967, Hydrology of Neogene deposits in the northern Gulf of Mexico, *in* 1st Symposium on abnormal subsurface pressure, Proc.: Louisiana State Univ., p. 91-207.
Keep, C. L., and H. L. Ward, 1934, Drilling against high rock pressures with particular reference to operations conducted in the Khaur field, Punjab: Inst. Petroleum Jour.
Kharaka, Y. K., 1971, Simultaneous flow of water and solutes through geological membranes: Ph.D. dissert., Univ. California, Berkeley, 263 p.
Kingma, J. T., 1958, Possible origin of piercement structures, local unconformities, and secondary basins in the eastern geosyncline, New Zealand: New Zealand Jour. Geology and Geophysics, v. 1, p. 269-274.
Kok, P. C., and J. H. M. A. Thomeer, 1955, Abnormal pressures in oil and gas reservoirs: Geologie en Mijnbouw, August.
Korzhinsky, D. S., 1947, Filtration effect in solutions and its role in geology: Acad. Sci. URSS Bull., Ser. Geol., no. 2, p. 35-48.
Kryukov, P. A., A. A. Zhuchkova, and E. V. Rengarten, 1962,

Changes in the composition of solutions pressed from clays and ion exchange resins: Akad. Nauk SSSR Doklady, v. 144, no. 6, p. 167-169.

Lawton, J. E., 1956, Geology of the north half of the Morgan Valley quadrangle and the south half of the Wilbur Springs quadrangle: Ph.D. Dissert., Stanford Univ., 223 p.

Levorsen, A. I., and F. A. F. Berry, 1967, Geology of petroleum, 2d ed.: San Francisco, W. H. Freeman, 724 p.

Lockwood, J. P., 1971, Sedimentary and gravity-slide emplacement of serpentinite: Geol. Soc. America Bull., v. 82, p. 919-936.

Lomtadze, V. D., 1954, About role of compaction processes of clayey deposits in the formation of underground waters: Akad. Nauk SSSR Doklady, v. 98, no. 3, p. 451-454.

MacGregor, J. R., 1965, Quantitative determination of reservoir pressures from conductivity log: Am. Assoc. Petroleum Geologists Bull., v. 49, p. 1502-1511.

McCulloh, T. H., 1967, Mass properties of sedimentary rocks and gravimetric effects of petroleum and natural-gas reservoirs: U.S. Geol. Survey Prof. Paper 528-A, p. A1-A50.

McEvilly, T. V., 1966, The earthquake sequence of November, 1964, near Corralitos, California: Seismol. Soc. America Bull., v. 56, p. 755-773.

——— W. H. Bakun, and K. B. Casaday, 1967, The Parkfield, California, earthquakes of 1966: Seismol. Soc. America Bull., v. 57, p. 1221-1244.

McKelvey, J. G., K. S. Spiegler, and M. R. J. Wyllie, 1957, Salt filtering by ion-exchange grains and membranes: Jour. Phys. Chemistry, v. 61, p. 174-178.

McKenzie, D. P., and R. L. Parker, 1967, The north Pacific: an example of tectonics on a sphere: Nature, v. 216, p. 1276-1280.

McNitt, J. R., 1961, Geology of The Geysers thermal area, California: Rome, U.N. Conf. on New Sources of Energy, paper G/31, 18 p.

——— 1963, Exploration and development of geothermal power in California: California Div. Mines and Geology Spec. Rept. 75, 45 p.

——— 1965, Review of geothermal resources, *in* Terrestrial heat flow: Am. Geophys. Union Mon. Ser. 8, p. 240-266.

Moisseeff, A. W., 1966, The geology and the geochemistry of the Wilbur Springs quicksilver district, Colusa and Lake Counties, California: Ph.D. dissert., Stanford Univ., 214 p.

Moody, J. D., and M. J. Hill, 1956, Wrench-fault tectonics: Geol. Soc. America Bull., v. 67, p. 1207-1246.

Morgan, W. J., 1968, Rises, trenches, great faults, and crustal blocks: Jour. Geophys. Research, v. 73, p. 1959-1982.

Nason, R. D., 1968, San Andreas fault at Cape Mendocino, *in* Conference on geologic problems of San Andreas fault system, Stanford, California, 1967, Proc.: Stanford Univ. Pubs. Geol. Sci., v. 11, p. 231-240.

——— 1971, Investigation of fault creep slippage in northern and central California: Ph.D. dissert., Univ. California, San Diego, 231 p.

Oakeshott, G. B., 1965, Diapiric structures in the Diablo Range, California (abs.): Am. Assoc. Petroleum Geologists Bull., v. 49, p. 354.

——— 1968, Diapiric structures in Diablo Range, California, *in* Diapirism and diapirs: Am. Assoc. Petroleum Geologists Mem. 8, p. 228-243.

Owens, J. S., 1941, The geology of parts of Colusa and Lake Counties, California: M.S. thesis, Univ. California, Berkeley.

Page, B. M., 1966, Geology of the Coast Ranges of California, *in* Geology of northern California: California Div. Mines and Geology Bull. 190, p. 255-276.

——— 1970, Sur-Nacimiento fault zone of California: continental margin tectonics: Geol. Soc. America Bull., v. 81, p. 667-689.

Pampeyan, E. H., 1963, Geology and mineral deposits of Mount Diablo, Contra Costa County, California: California Div. Mines and Geology Spec. Rept. 80, 31 p.

Perrine, R. L., 1956, Analysis of pressure-building curves, *in* Drilling and production practice: Am. Petroleum Inst., p. 482-509.

Proffett, J. M., Jr., 1971, Late Cenozoic structure in the Yerington district, Nevada and the origin of the Great Basin (abs.): Geol. Soc. America Abs. with Programs, v. 3, no. 2, p. 181.

——— 1972, Nature, age, and origin of Cenozoic faulting and volcanism in the Basin and Range province (with special reference to the Yerrington district, Nevada): Ph.D. dissert., Univ. California, Berkeley, 297 p.

Rubey, W. W., and M. K. Hubbert, 1959, Overthrust belt in geosynclinal area of western Wyoming in light of fluid pressure hypothesis, pt. 2 *of* Role of fluid pressure in mechanics of overthrust faulting: Geol. Soc. America Bull., v. 70, p. 167-205.

Safonov, A., 1962, The challenge of the Sacramento Valley, California, *in* Geologic guide to the gas and oil fields of northern California: California Div. Mines and Geology Bull. 181, p. 77-97.

Schoellhamer, J. E., and A. O. Woodford, 1951, The floor of the Los Angeles basin, Los Angeles, Orange, and San Bernardino Counties, California: U.S. Geol. Survey Oil and Gas Inv. Map OM 117, scale about 1 in. to 2 mi.

Scott, P. O., W. G. Bearden, and G. C. Howard, 1953, Rock rupture as affected by fluid properties: AIME Trans., Tech. Pub. 3540.

Siever, R., K. C. Beck, and R. A. Berner, 1965, Composition of interstitial waters of modern sediments: Jour. Geology, v. 73, p. 39-73.

Stewart, J. H., 1971, Basin and Range structure: a system of horsts and grabens produced by deep-seated extension: Geol. Soc. America Bull., v. 82, p. 1019-1043.

Stockman, L. D., 1947, Mercury in three wells at Cymric: Petroleum World, February, p. 37.

Suter, H. H., 1954, The general and economic geology of Trinidad, British West Indies—Colonial geology and mineral resources: London, Colonial Geol. Surveys Mineral Res. Div., 134 p.

Swe, W., and W. R. Dickinson, 1970, Sedimentation and thrusting of late Mesozoic rocks in the Coast Ranges near Clear Lake, California: Geol. Soc. America Bull., v. 81, p. 165-187.

Taff, J. A., 1935, Geology of Mount Diablo and vicinity: Geol. Soc. America Bull., v. 46, p. 1079-1100.

Tainsh, H. R., 1950, Tertiary geology and principal oil fields of Burmah: Am. Assoc. Petroleum Geologists Bull., v. 34, p. 840-852.

Taliaferro, N. L., 1943, Franciscan-Knoxville problem: Am. Assoc. Petroleum Geologists Bull., v. 27, p. 109-219.

Thomeer, J. H. M. A., 1953, Problems arising from shale unstability in oil well drilling: Geologie en Mijnbouw, May.

——— 1955, The unstable behavior of shale formations in bore holes and its control by properly adjusted mud-flush quality: 4th World Petroleum Cong. Proc., Rome.

——— and J. A. Bottema, 1961, Increasing occurrence of abnormally high reservoir pressures in boreholes, and drilling problems resulting therefrom: Am. Assoc. Petroleum Geologists Bull., v. 45, p. 1721-1730.

Thompson, G. A., 1959, Gravity measurements between Hazen and Austin, Nevada—a study of Basin-Range structure: Jour. Geophys. Research, v. 64, p. 217-229.

——— and M. Talwani, 1964, Crustal structure from Pacific basin to central Nevada: Jour. Geophys. Research, v. 69, no. 22, p. 4813-4837.

Tkhostov, B. A., 1963, Initial rock pressures in oil and gas deposits (translation from Russian by R. A. Ledward): New York, Macmillan, 118 p.

Vedder, J. G., H. D. Gower, H. E. Clifton, and D. L. Durham, 1967, Reconnaissance geologic map of the central San Rafael Mountains and vicinity, Santa Barbara County, California:

U.S. Geol. Survey Misc. Geol. Inv. Map 1-487, scale 1:48,000.

Vine, F. J., and J. T. Wilson, 1965, Magnetic anomalies over a young oceanic ridge off Vancouver Island: Science, v. 150, p. 485-489.

Warner, D. L., 1964, An analysis of the influence of the physical-chemical factors upon the consolidation of fine-grained plastic sediments: Ph.D. dissert., Univ. California, Berkeley, 136 p.

Watts, E. V., 1948, Some aspects of high pressure in the D7 zone of the Ventura Avenue field: AIME Trans., v. 174, p. 191-200.

White, D. E., 1965, Saline waters of sedimentary rocks, *in* Fluids in subsurface environments: Am. Assoc. Petroleum Geologists Mem. 4, p. 342-366.

——— L. J. P. Muffler, and A. H. Truesdell, 1971, Vapor-dominated hydrothermal systems compared with hot-water systems: Econ. Geology, v. 66, p. 75-97.

Wood, J. J., 1967, Detection methods for abnormal fluid pressures, *in* 1st Symposium on abnormal subsurface pressure, Proc.: Louisiana State Univ., p. 51-80.

Woodring, W. P., R. B. Stewart, and R. W. Richards, 1940, Geology of the Kettleman Hills oil field, California, stratigraphy, paleontology, and structure: U.S. Geol. Survey Prof. Paper 195, 170 p.